PROGRESS IN BIOMASS CONVERSION

VOLUME 2

Academic Press Rapid Manuscript Reproduction

PROGRESS IN BIOMASS CONVERSION

VOLUME 2

Edited by

KYOSTI V. SARKANEN

College of Forest Resources
University of Washington
Seattle, Washington

DAVID A. TILLMAN

Envirosphere Company
Bellevue, Washington

ACADEMIC PRESS 1980

A Subsidiary of Harcourt Brace Jovanovich, Publishers

NEW YORK LONDON TORONTO SYDNEY SAN FRANCISCO

ACADEMIC PRESS, INC.
111 Fifth Avenue, New York, New York 10003

United Kingdom Edition published by
ACADEMIC PRESS, INC. (LONDON) LTD.
24/28 Oval Road, London NW1 7DX

ISSN 0192-6551

ISBN 0-12-5355902-0

PRINTED IN THE UNITED STATES OF AMERICA

80 81 82 83 9 8 7 6 5 4 3 2 1

Contents

Contributors

Numbers in parentheses indicate the pages on which authors' contributions begin.

John B. Grantham (1), U. S. Forest Service, Seattle, Washington 98155

Jack O. Howard (1), Pacific Northwest Forest and Range Experiment Station, Portland, Oregon

William D. Kitto (145), Envirosphere Company, Bellevue, Washington 98004

S. Kresovich (89), Battelle Columbus Laboratories, Columbus, Ohio 43201

E. S. Lipinsky (89), Battelle Columbus Laboratories, Columbus, Ohio 43201

K. V. Sarkanen (127), College of Forest Resources, University of Washington 98195

G. F. Schreuder (59), College of Forest Resources, University of Washington 98195

W. Ramsey Smith (181), College of Forest Resources, University of Washington 98195

D. A. Tillman (59), Envirosphere Company, Bellevue, Washington 98004

Bruce Zobel (37), School of Forest Resources, North Carolina State University, Raleigh, North Carolina 27650

Preface

Belated recognition of the contribution and potential of biomass fuels has come, bringing with it a myriad of research programs and problems. Consequently, conferences and conflicts abound. Efforts exist now where but five years ago ennui prevailed. The stimulus, of course, is the price of over $30/ bbl oil. At the heart of the biomass fuels and chemicals consideration is the need to bring more systems, and new systems, to the marketplace. Then the impact of biomass on national energy problems can be increased.

The current cornerstone of biomass fuel utilization is wood. This form of biomass is storable "on the stump," and it can be harvested without any particular regard to season. Further, it is the basic raw material for the vast, essential forest products industry that produces lumber, plywood, pulp and paper, particleboard, and numerous other products. An ample supply of mill residues is, therefore, generated; and those residues supply fuel for the boilers and kilns of many plants.

Because the mill residues are virtually totally consumed now, as they are produced, systems must be developed to increase the efficiency of their utilization. Simultaneously forest residuals must play an increasing role in energy supply. Wood must be grown at a more rapid rate. Concommitantly, environmental issues must be addressed rather than ignored.

Beyond wood, crops such as sugar cane show promise, particularly in systems that can meet needs of food and fuel in a flexible manner. With the development of new separation schemes, a wider variety of crops can be converted into liquid fuels, chemicals, and energy. These issues of supply and use are addressed by the chapters in this volume. Some answers are given, and more questions are posed.

The biomass fuels and chemicals field is ever-changing and at a rapid rate. Thus concepts that appeared novel half a decade ago are now entering commercial practice. And more innovative approaches are emerging. Preparation of this volume required considerable cooperation from many people—the contributors, idea generators, and helpful critics. We wish to acknowledge the help of such friends and colleagues here. As we move forward with more efforts in this series, we welcome such continued assistance.

LOGGING RESIDUE AS AN ENERGY SOURCE

John B. Grantham
Jack O. Howard

U. S. Forest Service
Seattle, Washington

Progress in Biomass Conversion

ISBN 0-12-535902-0

I. INTRODUCTION

A. *Summary*

1. Nearly 50 million tons of wood and bark residue, in pieces four inches and larger in diameter, remain on the ground annually after logging operations in the United States. The energy potential of this residue is roughly equivalent to 100 million barrels of oil, but the residue material is widely scattered and highly variable in size and shape. The difficulty and cost of collecting logging varies with its size and location to the extent that only part of the total may be recovered regardless of the cost of other fuels. Logging residue in pieces less than four inches in diameter is estimated to amount to somewhat more than the tonnage of larger material but is even more expensive to collect in most situations.

2. Whole-tree harvesting and chipping operations are expected to increase as a means of more efficiently recovering more wood and bark volume from each acre logged. The practice of whole-tree chipping, however, has two disadvantages, namely: (1) It reduces the opportunities to sort wood for maximum value use, *i.e.*, poles, sawlogs, veneer bolts, bark-free pulp chips, etc. (2) It increases the risk of removing too much fine material that is essential to maintain the timber growing capacity of the land.

3. As the practice of sorting wood for maximum value use is extended to currently unused material, the fraction remaining for fuel will be most readily available to those doing the sorting, namely forest product manufacturers.

4. Forest products manufacturing firms, particularly pulp and paper manufacturers, are the most likely users of the fuel fraction of former logging residue for other reasons, including such considerations as: (1) Forest industry plants are, in general, located close to timber supplies. (2) The in-plant energy requirements of forest industry firms make it possible to produce energy efficiently in small increments, *e.g.*, 2 megawatt capacities.

(3) The experience of forest industry firms in purchasing, handling, and storing wood in many forms makes them less concerned with the problems of maintaining adequate wood fuel supplies.

5. Wood supplies available as logging residue can be supplemented by wood available in low-value hardwood stands that have been uneconomical to harvest; and in trees now lost through cumulative annual mortality or through catastrophic losses to insect epidemics, windstorms, major fires, or disease. It is expected that such material, like logging residue, will be used selectively to obtain maximum value, but that substantial tonnages will be available as fuel.

6. The anticipated practice of sorting currently unused wood for a variety of products may appear to limit the quantities available for fuel. On the other hand, selective use to enhance value may permit the recovery of wood and bark that is not economically available for use as fuel only.

B. *Basic Considerations*

Logging residue left on the ground after timber harvesting operations is of increasing interest as a source of energy. The estimated total weight of wood and bark left following harvesting operations in the United States during 1976 amounted to nearly 50 million oven-dry tons, with an energy equivalent of roughly 0.85×10^{15} Btu (quads).

To assess the energy potential of logging residue, however, it is necessary to consider such important factors as: (1) quantities, characteristics and locations of the material; (2) anticipated changes in the management policies and practices of forest land owners or managers; (3) developments in harvesting equipment and systems; (4) changing demands for all wood products, but especially for pulp, paper, and composite wood products; and (5) the delivered cost of logging residue as compared with other raw materials and fuels.

II. ESTIMATED QUANTITIES OF LOGGING RESIDUE

Although estimates of the current quantities of logging residue are based on limited sampling, the estimates may be sufficiently accurate to assess their overall energy potential. National or regional inventories of logging residue are useful primarily to indicate overall potential. Any planned use of residue wood and bark will depend on the quantity, characteristics, and cost of the residue available to a specific location. Further, it is likely that estimates of logging residue available to a specific site will be combined with estimates of other fuels such as wood manufacturing residues, urban wood waste, agricultural residues, and fossil fuels.

A major problem in assessing the energy potential of logging residue is definitional. In the broadest sense, logging residue is all material remaining on the ground following a harvesting operation. Included in the definition are logs, stumps, tops, limbs, and foliage. Some reports also include trees, usually smaller ones, that are killed but not cut. The final definition used in any given study should reflect the objective of the assessment being made. This discussion focuses only on the wood and bark above a one-foot high stump, and in pieces which are four inches or larger in diameter. It is this portion of the residue that provides the greatest opportunity for recovery as an energy source.

The estimated quantity of logging residue (four inches and larger) in the United States is shown by region in Table I. (See Figure 1 for regional boundaries). These data represent an aggregation of published estimates by experts in each region. Consequently, the level of accuracy of the final figures is impossible to determine.

Omission of pieces less than four inches in diameter does not mean that this material has no value for energy. This smaller material is excluded because of the difficulty in recovering and handling smaller material with today's harvesting equipment. Thus a greater cost is associated with handling these small pieces. One noncash cost is associated with the need to maintain nutrient supplies on any forest site. A large proportion of the nutrients are concentrated in the tops, limbs, and foliage of trees.

There are other sources of material associated with timber harvesting having potential for energy production. In the eastern part of the United States, partial cutting is the predominant harvesting practice. In many areas, small trees are left standing, some of which may not be desirable components of the future stand. Also left standing may be larger trees not commonly used for conventional wood products. These trees, large and small, could be removed as part of a residue recovery effort. In addition, many dead trees, deemed to not have sufficient product value to warrant removal, would be available for energy purposes.

Where clearcut harvesting is practiced, the trees described above would normally be felled and would, therefore, be included in the logging residue figure. As can be seen, residue figures for the East do not include all of the trees included in the estimates for the West, where clearcutting is widely practiced. A recent report by Carpenter indicates that these "standing residual trees" may account for about

TABLE I. Quantity of Logging Residue (4+ inches and larger in diameter) in the United States[a]

	Volume[b] (thousand cubic feet)	Weight[b,c] (oven dry tons)
North		
Northeast	528,500	8,400,000
North Central	244,100	3,900,000
South		
Southeast	679,200	10,360,000
South Central	578,400	8,820,000
Intermountain	248,100	3,400,000
Pacific Coast[d]	1,038,100	14,430,000
Total U.S.	3,324,500	49,310,000

[a]See Fig. 1 for region boundaries.

[b]Includes bark.

[c]Using average density factors (#/cu. ft) of: North (31.8); South (30.5); Intermountain (27.4); and Pacific Coast (27.8); these factors reflect relative proportion of hardwoods and softwoods in each region.

[d]Includes Alaska and Hawaii.

15 million tons in the East. This quantity, reported as excess nongrowing stock, amounts to nearly 50% of the volume of logging residue for that area.

Other sources of wood, such as from thinning operations, might be available for energy production if a local market existed. Whether included with the logging residue or not, the total quantities involved certainly need further consideration as a potential source of raw material for energy.

While this chapter only deals with logging residue, in reality, the entire resource must be considered in decisions of allocation to the various competing uses. The trend in forestry is toward greater utilization of the soil resource

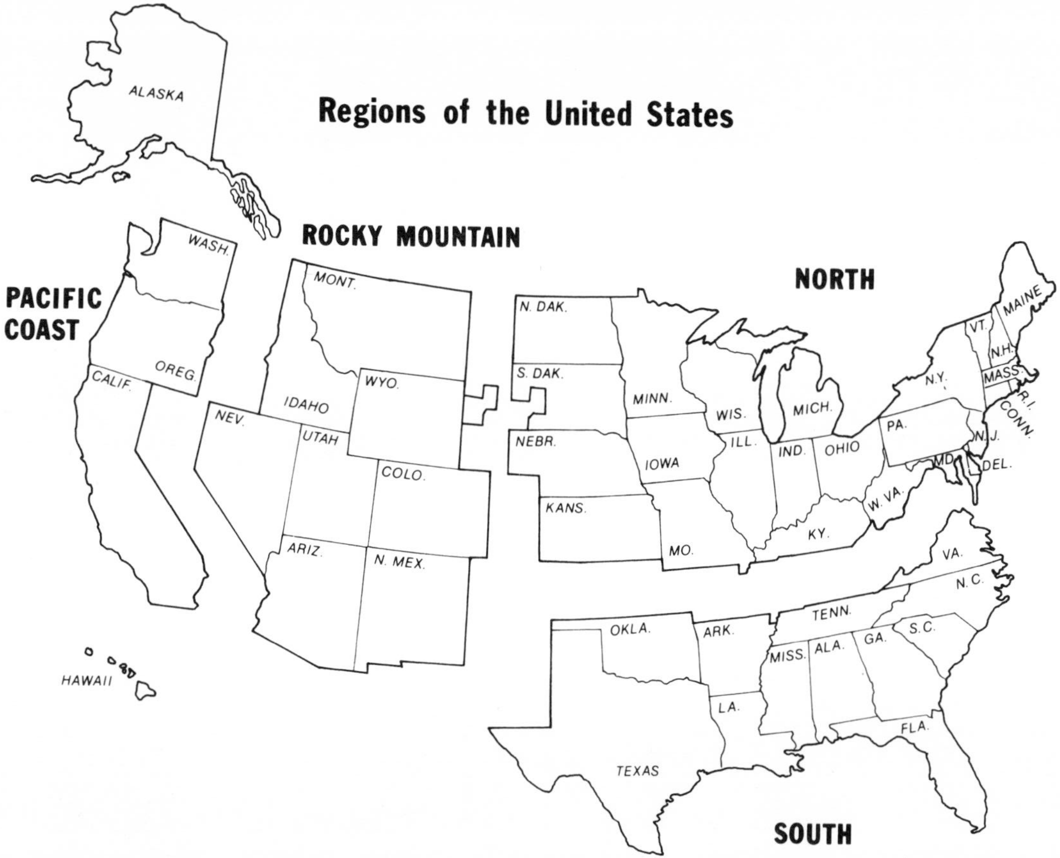

FIGURE 1. Regions of the United States.

and each tree grown on that soil. Greater utilization means less material left on the ground and more material at the processing centers. A portion of this wood will no doubt be used as fuel, by the processor or by independent energy producers. The final allocation of the forest resource is one made in light of a myriad of economic and production criteria. Fuel is one of many wood products that must ultimately compete with all products for a renewable, but not unlimited, resource.

In light of these facts, it may be useful to refer to some recent estimates of the above-ground annual growth of wood and bark as summarized in Table II, and to the most recent estimate of annual timber removals and mortality as shown in Table III.

III. REGIONAL DIFFERENCES IN LOGGING RESIDUES

Regional differences in the quantities and characteristics of logging residue (Table I) are attributed to such factors as: (1) predominant species, (2) size and condition of the timber stands, (3) steepness and roughness of the terrain, and (4) the regional definition of what constitutes logging residue.

In general, hardwoods make up a large proportion of logging residue in eastern United States, while softwoods dominate in the west, as shown in Table IV.

With the exception of old-growth stands in western United States, hardwood stands tend to produce more residue than softwood stands. This is attributable largely to the branching characteristics of hardwoods. The main stem of most hardwoods, such as oaks, forks in multiple branches at a lower point on the tree than in softwoods, such as pine or fir. The result of this forking process is less recovery of straight logs suitable for product manufacture and a greater proportion of material ending up as residue.

Softwoods such as pines grown in plantations on short rotation, yield low volumes of residue, and this, too, contributes to differences in regional residue totals. Clark (1978) reports, in some detail, the differences in wood recovery for pine, oak, and yellow poplar in southern United States when logging for sawlogs only, for sawlogs and pulpwood, or when producing whole-tree chips.

The largest concentration of logging residue occurs in the Pacific Coast states, associated with harvesting old-growth timber. Residue quantities in excess of 100 oven-dry tons per acre have been experienced. Average quantities are lower, around 30-40 O.D. tons per acre. This average, however, is two to three times the average in eastern United States. Residue quantities per acre in the Intermountain area of the

TABLE II. Recent Estimates of Net and Gross Annual Growth of Wood and Bark on Commercial Forest Lands of the United States

	Spurr & Vaux[a]	Ellis & Wahlgren[b]	RPA Assessment[c]
	(millions tons, oven dry)		
Net growth			
Stemwood[d]			
Softwood (@ 2.7% of inventory)		146	169
Hardwood (@ 3.7% of inventory		130	156
Totals	295-455[e]	276	325
Annual mortality loss			
Softwood stemwood		38	32
Hardwood stemwood		29	28
Gross growth of stemwood		343	385
Branch and top wood, including mortality loss		73	
Cull and rotten trees (gross growth)		27	
Bark on stems of all trees		45	
Bark on branches and tops		11	
Totals		499	552
Foliage on groww growth		35	

[a]Spurr and Vaux, 1976. [b]Ellis and Wahlgren, 1978.
[c]USDA Forest Service, 1979. "Forest Statistics of the U.S., 1977."
[d]Main stem of all live, sound trees 5 inches and over in dbh.
[e]Economic potential--biological potential, respectively.

TABLE III. Estimated Growth, Removal, and Mortality of Wood and Bark in Growing Stock[a] on Commercial Forest Lands in the United States, 1976 (millions of tons, oven dry)

	Gross Growth	Mortality	Removals as Products[c]	Removals as Residue[d]	Removals as Other[e]
Softwoods	201	32	123	11	5
Hardwoods	184	28	47	10	13
Totals	385	60	170	21	18
Plus bark at 15% of wood	49	9	25	3	3

[a]Annual growth on all live, sound trees 5 inches and larger in diameter, breast high, not including stump and root wood estimated as an additional 20%.

[b]Annual losses due to competition, insects, disease, wind, and fire.

[c]Annual removals from growing stock for all products, including fuel wood. Rough, rotten, and salvable dead trees provided another 17 million tons of wood to products.

[d]Logging residue from growing stock--estimated to be about one-half of the total wood and bark left after harvesting.

[e]Primarily wood removed in clearing forest land for other uses.

Source: USDA Forest Service (1979), Forest Statistics of the United States, 1977.

TABLE IV. Sources of Logging Residue by Region (in wt %)

Region	Source: Hardwoods	Softwoods
North	81	19
South	58	42
Intermountain	–[a]	100
Pacific	8	92

[a]Less than one-half of one percent.

West are somewhat lower than those in the Pacific Coast states on average; however, the residue in the Intermountain area tends to be more uniform in size than that on the Pacific Coast.

A contributing factor to lower residue quantities in eastern United States is the generally more gentle terrain. Steep, broken terrain is common in the western United States; and it results in more breakage as trees are felled. This creates more residue. Additionally, the flatter terrain in the East allows for easier access by harvesting equipment, thereby reducing cost and allowing for greater utilization of each tree. Whole-tree logging, for example, will continue to be more widely practiced in the East, where the terrain restricts the choice of equipment less than it does in the West.

Ownership patterns differ significantly between the regions. The proportion of publicly owned land is much higher in the West than in the East, where private ownership accounts for approximately 86% of all commercial forest lands (U.S. Dept. of Agric., 1978). There are differences in management objectives between and within these two major ownership groups. While few studies have been made indicating the impact of owner objectives on residue quantities, it is likely that private land owners would have different strategies and standards for harvesting timber. It could be expected, therefore, that residue quantities would differ on that basis alone.

A study conducted in Arkansas by Porterfield did show that residue amounts on lands with "absentee ownership" were higher than on other privately owned lands. While owner intent may affect residue quantity, a greater concern may relate to the availability of that residue. Such concerns will be discussed in the balance of this chapter.

A gradual shift from old-growth to young-growth forests will reduce the per-acre concentration of logging residue. This change is of significance in the Pacific Coast states, where dead and cull trees now contribute heavily to residue quantities. Additionally, the residue from young-growth forests of the West will have a heavier proportion of tops and limbs and a smaller proportion of broken stemwood than do residues from old-growth forests.

IV. CHANGES ANTICIPATED IN FOREST MANAGEMENT POLICIES AND PRACTICES--AND THEIR EFFECT ON RESIDUE TYPE AND AVAILABILITY

Management policies and practices, in the long run, reflect market values for timber. Higher values promote better management, including greater attention to prompt regeneration, superior species selection, improved protective measures, stand improvement cuttings, and more complete utilization of the final harvest. As the demand grows for wood to supply construction materials, paper products, and energy, the resource value will increase and management practices will intensify on both public and private lands.

More intensive management, in turn, will lead to more complete timber utilization to obtain maximum returns and to aid future resource management. Closer utilization not only will provide more wood per acre; but will simplify regeneration and protection of the next timber crop by more complete removal of residue. The total wood supply will be increased further by thinnings and other stand improvement measures undertaken as part of more intense management.

Markets for paper, particleboard, fuel, and other products for which trees of small diameter can be used will be especially beneficial. Hardwood timber stands in nonindustrial private ownership, in particular, should benefit by a market for wood fuel that will accept all species and trees of less desirable form. Returns from this market will enable and encourage timberland owners to upgrade their stands and manage them for production of an array of timber products. This is but one example of how a wood fuel market can complement log or pulpwood production as well as compete for available wood supplies.

Management policies and practices on publicly owned timber lands may respond less directly to improved markets for lower quality timber because of the relatively higher proportion of attention that must be given to other forest values including recreation, esthetics, fish and wildlife habitat, the production of forage, and the protection of water supplies.

Management activities subsequent to harvesting may encourage the use of logging residue for energy production. Where clear-cutting is practiced, preparation of the area for the next crop of trees may require the removal or breakdown of remaining vegetation and residue. Such activity could be the first step in converting logging residue to fuel once this was a management objective.

An increasing amount of logging residue is being used for firewood. Currently, the use of logging residue for domestic heating is greater in the western United States, particularly near urban areas. This is brought about in part by the piling of residue, a requirement in many sales of public timber. Piling concentrates the material, thus providing easy access to large volumes of wood. Although the use of wood for home heating may be more common in eastern United States than in the West, generally other sources of wood (such as dead or undesirable standing trees) provide most of the firewood. Logging residue, which is more scattered in the East than in the West because of more extensive partial cutting, is less attractive as a domestic fuelwood source.

Unless logging residue is utilized in some manner shortly after its creation, site preparation activities (particularly burning) may destroy or modify the material. Likewise, access to some areas for recovery of residue may be prohibited once young trees are established following harvesting operations. Finally, certain areas are environmentally sensitive, and disturbances must be minimized. Conditions such as unstable soils require special equipment and handling during logging and may constrain the removal of logging residue.

Improved markets for fiber products and for energy will, however, permit both public and private timberland managers to achieve a higher degree of utilization on the lands that are cut over. Managers of national forests, in particular, will be able to specify more complete removal of available wood when a market for such material exists.

The harvest of young, vigorous timber stands under intensive management produces less defective and broken material than does the harvest of older, unmanaged stands. On the whole, however, a greater supply of timber is made available for several reasons, including: (1) more permanent road systems will provide easier access for improvement cuttings or salvage of damaged trees; (2) frequent thinnings or other intermediate cuttings will reduce natural mortality losses; (3) improved stand vigor through thinnings should reduce losses to insect epidemics; and (4) greater attention to hardwood stand management and utilization will increase the availability of hardwood residue.

As growing markets for whole-tree pulp chips, wood fuel, or other material of small dimensions permit land managers to require closer utilization, such requirements will produce changes in timber harvesting practices. Anticipated changes and their effect on the wood supply are discussed below.

V. MODIFIED HARVESTING SYSTEMS AND THE EFFECT ON WOOD RESIDUE AVAILABILITY

Harvesting systems are likely to be modified substantially during the next decade, particularly in the West. Modifications are expected in response to two major pressures. First, there is a need to harvest timber with less impact on the environment, especially on steep slopes or sensitive soils in the West--and to accomplish this economically. Second, there is a need to meet expanding demands for wood for solid products, fiber products, and energy wherever these demands occur.

Such modifications present significant challenges because they require handling material ranging from overmature, often defective, trees of tremendous size through small stems (produced in thinnings, stagnated stands, or insect-killed timber) to short pieces or chunks produced in logging. Furthermore, the job is complicated by the fact that the trees may occur on rough, steep slopes, sensitive soils, or even on ground that is wet during much of the year.

Despite the difficulties posed in changing harvesting equipment and methods, changes are occurring that will affect the availability of forest residue. One of the most obvious changes is the use of mobile in-woods chippers that recover wood from tree tops and limbs that otherwise would be left as residue. This development occurred in answer to a need for more pulpwood and to an acceptance of whole-tree chips that included bark. In-woods chipping has, in turn, encouraged whole-tree yarding because the whole tree could be fed to a chipper without the need for log bucking, limbing, or debarking; or the bucking could be done mechanically at roadside under more careful supervision. Furthermore, use of a chipper avoids a heavy accumulation of tops and limbs at the road, that can occur with whole-tree yarding.

Tops, limbs, and bark, which are not included in forest inventories, add to the available wood supply. This is a particularly significant source of wood in many hardwood forests, where whole-tree chipping may produce twice the quantity of wood per acre that was obtained by conventional logging.

Related developments--At least three pieces of related equipment are under development to recover even more harvesting residue in chip or shredded form. These related pieces of equipment include: (1) A mobile chipper combined with felling bar to chip small standing trees and logging debris after conventional harvest. The machine is designed for rock-free terrain with less than 30% slope (Koch and Nicholson, 1978). (2) A mobile shredder developed for site preparation, but to be combined with recovery of the shredded residue (O'Dair, 1978). (3) A mobile topwood shear to reduce large hardwood tops for grapple skidding and chipping at roadside (Mattson *et al.*, 1978).

Reduction of forest residue volume through more complete harvesting systems, in most cases, requires handling of many pieces of small diameter. Trees of small diameter or short length contain small volume and must be handled in multiples to maintain an economical production rate.

The problem of bunching small trees for skidding has been solved on ground with slopes up to 30% by use of feller-bunchers with accumulator shears (Davidison, 1978). Some of these heads will accumulate as many as 10-12 stems six inches in diameter, but it is more common to accumulate 2-5 stems of varying diameter, and then lay two or three such bunches together to make an appropriate load for a grapple skidder.

On slopes over about 30%, it is difficult to operate a feller-buncher, and some other means is needed to bunch small stems for yarding by cable or tractor. It has been suggested that once a high-production cable yarding machine is available for small timber, bunching hooks can serve the need to economically accumulate payloads with the yarder. Such a system will serve to capture much of what was formerly residue.

High-production yarders that can thin timber of small diameter on steep slopes with little damage will permit earlier and more frequent access to material that may be otherwise lost to natural mortality, especially in the West. Such equipment has been developed and should soon be in wide use.

Expected advances in logging technology will reduce the per-acre quantities of residue. In general, a new generation of more versatile and flexible harvesting equipment will lead to more wood removal from a given area. As will be discussed later, this more complete harvesting may not reduce the amount of wood available for energy production in all cases.

VI. ANTICIPATED COMPETITION FOR LOGGING RESIDUE

The very existence of logging residue does not ensure that the material is available for energy production. A determination of whether or not adequate raw material is available for energy production can only be made after a specific site has been chosen.

If the total demand for logs, pulpwood, and wood fuel doubles by the year 2000, as projected by comparing Tables V and VI, competition for all classes of wood will heighten. Such a development will not only absorb most of the surplus growth of softwood and hardwood timber, but will enable timber harvesters to remove much of the material currently left as residue. Increasing prices for wood will overcome the economic restraints on removal of logging residue that prevailed during the 1970s.

Because the major increase in wood demand is expected to be provided by the pulp and paper industry--for both raw material and fuel--wood of nearly any species and form will be acceptable. Acceptance of logging residue, improvement cuttings and disaster-killed timber will grow in response to market pressures and will, thereby, help timberland managers to achieve more complete utilization of the wood resource.

As previously discussed, close utilization favors improved management by removing physical impediments and providing greater financial incentives. Logging residue, standing dead timber, and many hardwood stands that have been considered unmerchantable contain a considerable range in tree quality. It is expected, therefore, that such materials will be subject to some sorting or beneficiation to recover the highest values under increasing demands. The opportunity to use formerly unmerchantable material for paper, particleboard or other products as well as for fuel increases the competition for logging residue and like materials. On the other hand, such diversified end uses may make wood fuels more competitive with coal or other fossil fuels, providing that a greater share of wood production costs are borne by the raw material for paper and board than by the wood fuel.

The usefulness of residue materials, however, will be influenced by the form in which they are delivered, e.g., the larger the piece size the greater its usefulness. The advantages of delivering residue in maximum size pieces should influence the way in which harvesting operations are conducted. For example, a formerly unmerchantable stem section may be left attached to a merchantable portion for more efficient handling. The separation of merchantable and unmerchantable material can then be made at a processing or sorting center. Again, if the unmerchantable portion of the stem contains some merchantable material, it may pay to separate this by sawing or peeling before the remainder of the previously unmerchantable stem section is chipped for fiber products, hogged for energy production, or converted to a miscellaneous product such as posts.

Added value recovered from merchantable portions of former residue material will help offset the cost of delivering the remaining portions for fuel. Use of any portion of forest residue materials for energy production improves the opportunity for more complete harvest with associated benefits to land management (fire hazard reduction, improved regeneration, easier access for future protection, etc.).

Dual or multiple use of residue, thus, may stimulate the removal of more material in harvesting operations and thereby increase the wood available for products, as well as that available for energy.

Manufacturing residue from sawmills and other wood processing plants has long been an important source of energy to forest industry plants because the cost of producing and delivering that residue for fuel was borne largely by the lumber or other primary product. Expansion of this common industry practice will improve the chances of delivering more fuel from former logging residue. Demands on former residue wood to supply small logs, pulp chips, particleboard furnish, and fuel may be complementary rather than competitive.

But, under any circumstances, an increasing value for wood fuel will serve to encourage and stimulate the recovery of more forest residue--beginning with that which is most easily reached economically, and extending to more remote sources as demand for wood products and wood fuel expands.

Long-term fuel supplies are of concern to any energy producer. Capital investments with 30 year lives are common for central power generating plants, and long-term commitments on fuel supplies are desirable, if not mandatory. The long-term availability of logging residue for energy is of particular concern.

Competition for logging residue is, however, so dependent on the market for all wood products that any long-term commitment of logging residue for specific use is virtually impossible. This is further complicated, in the case of public timber, by the fact that timber purchasers have the right to exercise their judgment on the disposal of all wood in the sale area. Previous attempts to specify the disposal of material unsuitable for products at the time of sale have been unsuccessful. The purchaser may be happy, at the time of sale, to have a fuel outlet for unmerchantable material. On the other hand, the purchaser, because of dynamic shifts in markets, reserves the right to dispose of unmerchantable material as he sees fit whenever the timber is cut.

Piling of unmerchantable material is a common practice on federal timber sales in the West. Although the practice adds substantially to the logging cost, it does reduce the risk of uncontrolled fire and improves the site for future management. Hopefully, this practice also encourages utilization of the unmerchantable material for pulpwood, firewood or other products. Attempts to carry this piling practice another step by mandatory removal of the residue wood to a concentration

TABLE V. Wood Requirements, Product Output, and the Disposition of Wood Manufacturing in the United States--1970 (in millions of short tons, oven dry)

Industry	Roundwood required[a]	Manufacturing residue output	Purchased residue to products[b]	Product output	Manufacturing residue to fuel	Unused residue
Lumber	98.0	65.0	(-21.0)	33.0	18.0	26.0
Plywood	17.4	9.7	(-7.4)	7.7	1.4	0.9
Pulp and paper	63.5	46.0	23.0	40.0[c]	46.5[d]	0.0
Misc. products	5.7	2.5	3.4[e]	6.6		2.5
Export logs	6.0			6.0[f]		
Export chips			2.0	2.0		
Domestic fuel	7.5			7.5		
Totals	198.1	(123.2)	(28.4)	102.8	65.9	29.4

[a] Includes an estimated 20 million tons of bark.

[b] A substantial proportion of the manufacturing residue generated in the lumber and plywood industries is sold to others, primarily the pulp and particleboard industries.

[c] Approximately 50% of total wood input (63.5-6.0 + 23.0 million tons).

[d] Dissolved wood solids in spent pulping liquor (40.5 million tons) plus 6 million tons of bark.

[e] Residue purchased for particleboard and related board products.

[f] Export logs and chips are considered as products since no processing is done in the United States.

TABLE VI. Projected Roundwood Requirements, Product Output, and the Use of Wood Manufacturing Residue in the United States--2000 (in millions of short tons, oven dry)

Industry	Roundwood required[a]	Manufacturing residue output	Purchased residue to products	Product output	Roundwood & manufacturing residue to fuel
Lumber[b]	128.0	77.0	(-39.0)	51.0	38.0
Plywood[c]	29.2	16.2	(-9.0)	13.0	7.2
Pulp and paper[d]	129.0 70.0	86.0	+37.0	80.0	86.0 70.0
Misc. products, incl. board[e]	6.0 1.5	2.5	0.0 +10.0	3.5 11.5	2.5
Export logs[f]	3.0			3.0	
Export chips[f]			+ 1.0	1.0	
Domestic fuel[g]	25.0				25.0
Totals	391.7	(181.7)	(48.0)	163.0	228.7

[a] Source: Phelps, 1979.

[b] Input is expected to be largely merchantable saw logs, but will include some utility logs, pole-size and dead timber. About 1/2 of the manufacturing residue generated is expected to be used for energy--mostly in-plant. The remaining residue will be used to produce pulp, particleboard, etc.

[c]Input is expected to be largely merchantable veneer logs but will include some utility logs. Nearly 1/2 of the residues generated are expected to be used for in-plant energy; the balance will be sold for particleboard furnish, pulp chips, etc.

[d]Input will include pulp logs and pulpwood, manufacturing residue in chip form, whole tree chips, urban wood waste, etc. Estimated breakdown is 199 million tons of wood from the forest for both fiber and fuel, plus 37 million tons of manufacturing residue for pulp production. Of the total, about 1/3 is expected in product output; 2/3 used in energy production.

[e]It is expected that 80% of the roundwood will be used for the production of poles, piling, mine timbers, etc. The balance plus the manufacturing residue purchased will be used for particleboard, hardboard, etc.

[f]Export logs and chips are considered as products since no domestic processing is anticipated.

[g]Fuelwood is expected to be largely from standing or down trees that are not inventoried because they are rough, rotten, dead or of undesirable species.

yard or point of use have so far been unsuccessful, and the timber sale purchaser retains the right of first use.

An alternative to commiting logging residue to a specific energy producer is to foster greater use of logging residue for energy production within the forest industry. The increasing cost of fossil fuels is already inducing increasing self-sufficiency within the forest industries, particularly in the pulp industry. Incentives to accelerate this trend would commit more logging residue to energy production without long-term commitments of wood fuel from one industry to another.

A further alternative is a joint venture between a utility and a forest industry--the utility providing the experience in electric power generation and distribution; the forest industry providing the wood fuel from its operations.

VII. DELIVERED COST OF LOGGING RESIDUE

In the absence of firm information on the cost of collecting and delivering logging residue, the best guide is the current cost of delivering merchantable logs or pulpwood in that locality. The reported cost of delivering logs or pulpwood should be as specific as possible in regard to locality and timber type, but if specific cost information is lacking, regional or area averages may be used and adjusted on the basis of local experience and judgment. Further adjustments necessary to convert merchantable wood costs to the cost of delivering logging residue are discussed on the following pages.

A. *Cost Breakdown*

To insure that all cost factors and all adjustments are considered in estimating the cost of logging residue, it will be useful to break down the costs into such elements as:

Stumpage--cost of wood on the stump or ground
Engineering (may be included with next element)
Construction of spur roads and landings
Equipment move-in, setup, move-out
Felling and bucking, or
Felling and bunching mechanically
Yarding or skidding--stump to landing at roadside
Loading, or
Chipping and loading

Miscellaneous, such as road maintenance, fire protection, supervision and overhead
Transportation

B. Estimating Cost Elements for Logging Residue

1. Stumpage. A major element in the cost of merchantable logs or pulpwood is the cost of stumpage which may range from \$0.25 per cubic foot (\$20 per cord; \$15-20 per O.D. ton) to \$2.00 per cubic foot (\$300 per thousand board feet, log scale). Stumpage cost may be disregarded so far as logging residue is concerned since any stumpage charge is likely to be nominal (less than \$0.01 per cubic foot; \$0.50 per cord). Thus all costs of timber growing and access for harvesting are borne by the merchantable wood that is produced. It also indicates the cost advantage of associating any harvest of logging residue for energy with the recovery of wood for other use.

2. Engineering, Spur Roads, Landings, and Equipment Moves. The costs of spur roads, landings, and equipment moves may be carried by the merchantable wood produced. Main access road construction, which may be a major cost against merchantable timber in the West, is never charged to logging residue because the road is constructed to access merchantable timber and is in place before residue is created. Road maintenance may be disregarded except in the West where it can be an appreciable charge.

3. Felling and Bucking. Because most logging residue is on the ground, there may be little or no cost for felling and bucking. This is in contrast to a salvage operation in standing dead timber or of residual standing trees following logging. Felling and bucking costs are generally small for down residues, but mechanical felling and bunching to reduce the cost of tree-length yarding in salvaging small timber may be a significant cost.

4. Yarding. Yarding is always a substantial cost in collecting logging residue. The cost per cord or per ton will depend on the harvesting equipment used and the crew size. These, in turn, are determined by the timber type and size and the terrain. Large timber and steep terrain will require large machines and cable systems, while flat ground, or moderate slopes, and smaller timber will permit the use of crawler tractors, rubber-tired skidders, or light cable systems. Crew size is more variable depending in part on the equipment selected and in part on the production rate that is required.

In any case, local experience in harvesting merchantable wood will provide a guide to the yarding equipment and crew needed. In fact, the same equipment and crew used to yard merchantable wood may be used to yard the logging residue.

This will avoid the need for a second yarding operation, although production efficiency will drop as residue of small diameter or short length is yarded. (See Fig. 2, adapted from Mifflin and Lysons, 1978.)

Crew experience and attitude, as well as crew size, can affect yarding cost. Where the same crew yards merchantable logs and residue material, their experience in yarding residue and their attitude toward such material may lower production and raise costs significantly. A minimum crew size is dictated by the equipment used, but there is latitude in how many men are used with a particular combination of equipment--depending upon the production required. Again, the crew's attitude or willingness of each member to do more than one job can affect the required crew size.

Once the equipment and crew size are established, the hourly cost of yarding can be determined by estimating from experience, the hourly ownership cost and operating cost of each machine. Mifflin and Lysons (1978) have provided a useful guide to estimate such machine costs. There is a further need to estimate the relationship between productive machine hours (PMH) and scheduled machine hours (SMH) since no machine and crew can work productively 100% of the time. Scheduled maintenance, break time for the crew, unscheduled delays for repairs, etc., may account for as much as one-third of the time. For example, Canadian cost studies of feller-bunching machines (Folkema, 1977) use a 60% utilization factor for unfavorable conditions and an 80% utilization factor for favorable conditions.

Considering all of the above factors, it could be assumed that the cost of yarding logging residue will be double that of yarding merchantable logs. Such an estimate, however, should be tempered by local experience of the estimator and by any confirming tests or studies that provide helpful information. Fortunately, there is a growing awareness of the need for confirming tests on the cost of recovering logging residue.

5. *Loading Cost*. Loading cost can be a significant cost in delivering logging residue. Because logging residue is generally smaller in size and less uniform than merchantable logs, the costs of loading logging residue are expected to be higher; perhaps by as much as 50%.

6. *In-Woods Chipping and Loading*. Chipping in the woods with a mobile chipper which blows the chips directly into a chip van is an alternative to loading individual pieces for size reduction at a point of use. A major advantage of in-woods chipping (if a market for whole-tree chips, including bark, is available) is that the saving in loading cost may more than offset the increased cost of chipping with a mobile chipper. Another advantage of in-woods chipping, particularly with wide-branching hardwoods or whole-tree yarding of

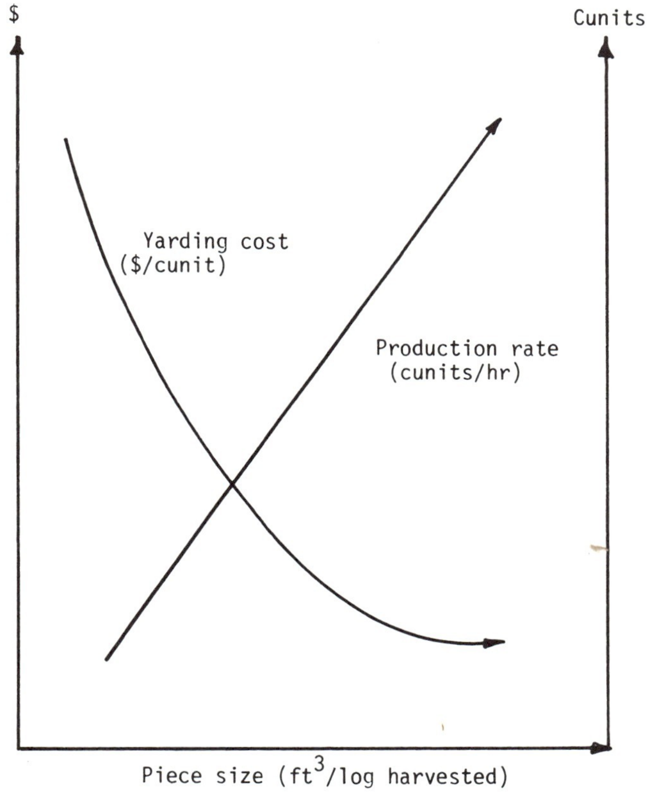

FIGURE 2. The influences of piece size on production rates and costs. Decreases in piece size reduce production rates and increase the cost/cunit (100 ft^3) of residues brought to the landing.

softwoods, is that more material is recovered from each acre worked. A disadvantage of in-woods chipping is the commitment made to whole-tree chips as the end product. Sorting opportunities are more restricted in the woods than at a plant.

7. *Miscellaneous Costs.* Miscellaneous costs that are chargeable to logging residue may include road maintenance, fire protection, and general expense or overhead. Depreciation not associated with a machine involved in the harvesting operation could be included here.

Road maintenance generally is charged on the same basis as is merchantable wood, namely, so much per ton or unit volume of wood hauled. If much of the haul is over forest roads rather than county or state highways, the charge can amount to one or two dollars per ton (oven-dry basis).

Fire protection costs are relatively small but real if a tanker truck or other fire equipment is required and a watchman is needed after working hours. General expense or overhead may cover the other costs that cannot be charged to a given machine or crew member. These supervisory or other general expenses may be estimated as a percent of direct labor costs or as an average cost per unit of production. They are, in most cases, a significant part of the total cost of moving residue from stump to truck.

8. *Transportation Costs.* Transportation costs are the most easily and accurately estimated of all costs considered here. More often than not, the transportation of logs and pulpwood is contracted at a specified cost per load or per ton, depending upon the haul distance. Rates generally will exceed average costs per ton/mile for highway haul because of slower travel possible on forest roads. Also, loading time is taken into account in determining the rate or may be added as a flat charge per load.

Basically, truck transportation costs are determined by what each truck and driver must earn per working day to pay ownership and operating costs for the truck, driver's wages, and a reasonable profit. Thus, load capacity, hauling time, and loading and unloading time are all important. The transportation cost, sometimes expressed as cost per ton per loaded mile will be higher for short hauls and poor roads than for long highway hauls.

Because trucks and trailers are normally loaded to the highway weight limitation, regardless of whether logs or chips are being hauled, any difference in the hauling cost per ton for the two classes of wood is apt to be due to differences in required loading time or vehicle capacity. Estimated transportation costs for whole-tree chips (Table VII) for Maine, Michigan, and Oregon are remarkably similar despite a substantial difference in drivers' wages. On a per ton, oven-dry, per loaded mile basis, the costs range from $0.1236 to

TABLE VII. Estimated Costs of Whole Tree Chip Production in Maine--1979[a]

Operation	Estimated Cost $/ton, green	Estimated Cost $/bone dry ton
Felling and bunching	1.91	3.82
Skidding	1.72	3.44
Chipping	1.88	3.76
Misc.	.98	1.96
Road building and maintenance	1.00	2.00
Transportation/45 miles	3.42	6.84
Totals	10.91	21.82

Base: 2 medium feller-bunchers @ 55% capacity
2 medium grapple skidders @ 69% capacity
1 large whole tree chipper @ 69% capacity
Labor @ $5.50/hr × 1.34 = $7.37/hr
Maint. and Repair: 20% investment cost for chipper
15% investment cost for other equipment
10% investment cost on intermittently-used equipment
5% investment cost on pickups
Fuel: 1.5 gal/hr feller-bunchers and skidders; 4 gal/hr chipper
Oil and Lube: 15% of fuel costs/hr
Hauling: $1.67/loaded mile for 22 ton load
1 BDU: 2.4 tons, green, or 1.2 tons, oven dry

[a]Source: Rust Engineering Co, 1979.

$0.1368. The hauling cost per ton, as hauled, would be half the costs shown because each estimate assumed the loads to be half water and half wood by weight--a common assumption for freshly-cut or green wood.

9. Credits for Recovering Logging Residue. The benefits of recovering logging residue may include savings in the cost of slash disposal that is normally charged to the primary logging operation. Slash disposal costs are those incurred in piling, burning, crushing or otherwise treating limbs, foliage, broken stem sections, etc., that interfere with regeneration of the timber stand or with the future protection and management of that new stand. Utilizing logging residue for energy production or other use will save a substantial amount for each acre covered. The credit earned per ton of residue removed, however, may be low as compared with the total cost of residue removal.

10. Summarizing Residue Recovery Costs. The costs of delivering logging residue to a point of use may be summarized by cost element. This practice can facilitate a comparison between the costs of harvesting merchantable wood and residue, and may simplify the task of applying log harvesting cost experience to estimates of residue recovery costs.

C. Examples of Cost Studies or Estimates

Cost studies of actual residue recovery are extremely limited when the scope of tree species, condition, locality, terrain, and other variables are considered. As of now, cost estimates based on harvesting merchantable logs, pulpwood, dead timber, or thinnings may provide the best source of information. Until such time as more studies of actual residue recovery are available, a few examples of timber appraisal, engineering cost estimates, and limited field studies may be helpful in estimating the delivered cost of residue. For that reason, some of this information is summarized on the following pages.

The costs of collecting logging residue and loading it for transport have been estimated in the course of estimating wood costs to a structural flakeboard plant in several parts of the United States (USDA, 1978). Although the range in the costs estimated by different authorities for the northeast, southwest, interior west, and Pacific coast is great, the approaches taken in the several estimates may provide some guidance. Estimates of 1976 delivered costs varied from $14.68 in Vermont (based on pulpwood costs) to $32.25 per ton of whole-tree chips (oven-dry basis) from residue in California. Other estimates varied from $21.29 per ton, O.D. in western Washington to $32.23 per ton in eastern Oregon.

One of the more recent and comprehensive estimates of logging residue, delivered as whole-tree chips, is that made by Rust Engineering Company (1979) for delivery to a plant in Westbrook, Maine. The estimated 1979 cost was $14.98 per ton, O.D., plus $6.84 per ton for transportation 50 miles. Some details of that estimate are shown in Table VII. 1975 costs reported by Biltonen *et al.* (1975) for whole-tree chips from an experimental hardwood thinning operation in Michigan as updated are $15.10 per ton, O.D., plus $6.40 per ton for transportation 50 miles (Table VIII). The close agreement between these costs shown in Tables VII and VIII is coincidental because cost elements differ for the two studies, as shown in Table IX. As a further consideration of cost elements that may be included, particularly in the West, Table IX includes timber appraisal estimates of the cost of producing whole-tree chips from small-diameter timber (such as dead lodgepole pine) in eastern Oregon.

VIII. ALTERNATIVE SYSTEMS FOR RESIDUE RECOVERY AND THEIR INFLUENCE ON WOOD COST

It has already been mentioned that any logging residue recovery effort, particularly in the West, is closely related to the primary logging effort which created that residue.

Access to the logging residue and the cost of recovery will be influenced strongly by how the recovery is associated with the primary logging. If the reduction in residue is accomplished by close initial logging and sorting of residue from merchantable wood at the roadside, an intermediate sorting yard, or at a forest utilization center, the cost of residue recovery is likely to be minimum. In any case, a portion of the residue recovery cost is borne by the primary logging as discussed previously.

Johnson *et al.* (1977) compare the costs incurred in delivering whole-tree chips a distance of 50 miles, assuming that the chipped residue is produced as: (1) a separate relogging (secondary operation), (2) an integral part of a sawlog operation, and (3) a mechanized tree length logging operation. Their estimated delivered costs were respectively: $27.50, $18.40, and $21.40 per ton, oven dry, for the three assumed approaches.

In 1970, more than 75% of the 123 million tons of residue created in processing logs to lumber, plywood, pulp, and miscellaneous products were used in the forest products industries for the production of pulp and board products (23%) or for fuel (53%) as shown in Table V. The 24% of manufacturing residue unused in 1970 is being rapidly committed to use as a

TABLE VIII. Mechanized Thinning, Upper Michigan, Nov. 1974 (Red maple, sugar maple, yellow birch)

Operation	1975 Cost[a]	Estimated 1979 Cost[b]	
	$/green ton	$/green ton	$/ton o.d.[f]
Felling and bunching[c]	1.81	2.22	4.44
Skidding[d]	1.62	1.98	3.96
Chipping[e]	1.97	2.41	4.82
Miscellaneous	.77	.94	1.88
Transporting	2.61	3.20	6.40
Totals	8.78	10.75	21.50

[a]Labor @ $4.50/hr × 1.5 = $6.75 × 1.225 = 8.27 vs. 14.13 in PNW.
@ $5.50/hr × 1.5 = $8.25 × 1.225 = 10.11 vs. 14.92 in PNW.
Labor 39% of total cost.

[b]1975 × 1.225 based on increase in producer price index.

[c]Felling and bunching 135 stems/PMH 85/SMH
productive machine hr (PMH) scheduled machine hr (SMH)

[d]10.5 stems/turn × 12.75 turns/PMH 6.83 turns/SMH = 15.57 tons, green/SMH

[e]35 gross tons/PMH 17 gross tons/SMH.

[f]1 ton, o.d. = 2.0 tons, green
av. 55 pcf

56.2% red maple @ 52.0 pcf.
23.2% yellow birch @ 59.4 pcf.
20.6% sugar maple @ 58.7 pcf.

Source: Biltonen, et al., 1975.

TABLE IX. Comparative Estimated and Experimental Costs of Whole-Tree Chip Production ($/ton, o.d.)

Cost Element	*Maine*[a]	*Michigan*[b]	*R-6*[c] *Appraisal 1979*
Fell and buck or bunch	*3.82*	*4.44*	*6.22*
Yard or grapple skid	*3.44*	*3.96*	*7.38*
Chip and load	*3.76*	*4.82*	*3.05*
Misc. and general expense	*1.96*	*1.88*	
Road maintenance and construction	*2.00*		
Overhead[d] *(stump to truck)*			*4.76*
Fire protection[d]			*.14*
Depreciation[d]			*1.14*
Stumpage			*.21*
Transportation/50 miles	*6.84*	*6.40*	*6.18*[e]
Totals	*21.82*	*21.50*	*29.08*

[a]*Rust Engineering Company, May 1979. "Advanced Systems Demonstration for Utilization on Biomass as an Energy Source in Westbrook, Maine: Alternative Harvesting Strategies."*

[b]*Biltonen, Frank E., and others, USDA Forest Service Research Paper NC-137.*

[c]*From R-6 Amendment No. 305 (Aug. 1979) to FSH 2409.22 at 220 cu ft/MBF; .82143 tons, o.d./cunit for lodgepole pine.*

[d]*Includes overhead, fire protection, and depreciation if not shown separately.*

[e]*Based on 4 trips/day @ 156 min. each; $.595/min., 8 hrs; $.694/min, 2.4 hrs. Comparative labor costs: N. E. Oregon, $14.24/hr; Maine and Michigan, $7.37/hr.*

raw material or fuel because of its attractive price. Prices at the mill range from about $2.00 per ton O.D. for wet hogged fuel to $10.00 per ton for dry planer shavings suitable for particleboard production. These costs of manufacturing residue are much lower than the cost of logging residue loaded for transport and the transportation costs are normally less.

The availability of manufacturing residue has discouraged the recovery of logging residue to this time, but the potential value of wood residue for fuel increases as oil prices rise. Nevertheless, it appears that the most likely utilization of logging residue will occur through more complete wood recovery in the initial logging; that sorting of all material will be made for highest value use; and that the portion used for fuel will be that least suited for other uses. This hierarchy of raw material disposition will, however, permit high-valued products to continue paying the major share of delivered wood costs, and that used for fuel may be charged a proportionately smaller amount.

Such an integrated system of harvesting and utilization permits maximum flexibility in response to market shifts or the need for energy, and eases concerns about the long-term availability of wood for energy. Estimates of the domestic pulp industry's need for wood fuel by the year 2000 (Table VI) may amount to one-half of its wood requirements for fiber, or 35% of its total wood needs.

IX. CONCLUSIONS

1. The nearly 50 million tons of wood and bark residue in pieces 4 inches and larger in diameter, that remain annually after logging operations in the United States is potentially an important source of additional energy. Logging residue in pieces less than 4 inches in diameter is estimated to amount to somewhat more than the tonnage of larger material but is more expensive to collect in most situations.

2. Whole tree harvesting and chipping operations are expected to increase as a means of more efficiently recovering maximum wood and bark volume per acre. The practice of whole tree chipping, however, has two disadvantages, namely: (1) it reduces the opportunities to sort wood for maximum value use, i.e., poles, sawlogs, veneer logs, bark-free pulp chips, etc., and (2) it increases the risk of removing too much fine material that is essential to maintain the timber-growing capacity of the land.

3. As the practice of sorting wood for maximum value use is extended to currently unused material, the fraction

remaining for fuel will be most readily available to those doing the sorting, namely forest products manufacturers.

4. The forest products industries, particularly the pulp and paper industry, are the most likely users of the fuel fraction of logging residue for additional reasons. These added reasons include such considerations as: (1) forest industry plants have logistic advantages because they are located close to wood supplies; (2) the in-plant energy requirements of forest industry plants make it possible to produce energy efficiently in amounts as small as 2 megawatts, and (3) the experience of forest industry firms in handling and storing wood in many forms make them less concerned than utilities of others with the problem of maintaining adequate fuel supplies.

5. Wood supplies available as logging residue can be supplemented by wood available in low-value hardwood stands that have been uneconomical to harvest; and in trees now lost through cumulative annual mortality or through catastrophic losses to insect epidemics, windstorms, major fires and widespread diseases. It is expected that such material, like logging residue, also will be used selectively to obtain maximum value, but that substantial tonnages will be available for use as fuel.

6. The expected practice of sorting currently unused wood for a variety of products may appear to limit the quantities available for fuel. On the other hand, selective use to obtain maximum value may permit the recovery of wood and bark that is not economically available for use as fuel only.

GLOSSARY: FOREST RESOURCE TERMS

COMMERCIAL FOREST LAND -- Land at least 10% stocked by forest trees of any kind, is capable of producing crops of industrial wood in excess of 20 cubic feet per acre per year and is available for such use.

CULL TREES -- Live trees that do not contain one merchantable log because of defect (rotten cull) or because of roughness, poor form or noncommercial species (sound cull).

DIAMETER BREAST HEIGHT D.B.H. -- Tree size as measured outside the bark at breast height (4-1/2 feet above ground).

GROWING-STOCK TREES -- Live sawtimber, poletimber, saplings or seedlings of commercial species that are now, or may be expected to become, suitable for use as industrial wood; excludes cull trees.

LIMBWOOD -- That part of the tree above the stump which does not meet the requirement for logs, including all live, sound branches to a minimum of 4 inches outside bark.

LOGGING RESIDUE -- The unused portion of wood and bark left on the ground after harvesting. The material may include tops or broken segments of growing stock, or material contained in live or dead, standing or down trees that were not included in the timber inventory because of rot, roughness, seasoning checks, small size or of a species considered unmerchantable.

MANUFACTURING (OR MILL) RESIDUE -- Wood materials other than the primary products (such as lumber), manufactured from roundwood; materials other than wood pulp produced in making that product.

MORTALITY -- The volume of sound wood in sawtimber or poletimber trees dying from natural causes during a specific period.

NET ANNUAL GROWTH -- The gross annual change in volume of sound wood in live trees less deductions for rot or other defects.

NONCOMMERCIAL FOREST LANDS -- Unproductive forest land incapable of yielding crops of industrial wood because of adverse site conditions, and productive forest land withdrawn from commercial use through statute or administrative regulation.

NONSTOCKED AREAS -- Commercial forest land less than 10% stocked with growing-stock trees.

POLETIMBER TREES -- Live trees 5.0 to 10.9 inches, d.b.h.

REMOVALS -- Volumes of timber removed from the growing-stock inventory, including timber products, logging residues, and other removals such as land clearing.

ROUNDWOOD -- Logs, bolts, or other round sections cut from trees.

SALVABLE DEAD TREES -- Standing or down dead trees that are considered merchantable by regional standards.

SAWTIMBER TREES -- Live trees, 11.0 inches or larger in d.b.h., containing at least one merchantable log.

TIMBER DEMAND -- The volume of timber that would be purchased at specified prices at a specified point in time under specified or implied assumptions relating to population, income, and other technological or institutional factors.

TIMBER SUPPLY (OR TIMBER HARVEST) -- Net volume of roundwood products available to forest industries from all sources at specified or implied price levels.

VOLUME OF GROWING STOCK -- The cubic volume of sound wood in the trunk or main stem of noncull sawtimber and poletimber trees of commercial species from a 1.0 foot stump to a minimum 4.0 inch top outside bark or to the point where the central stem breaks into limbs.

WOOD OUTPUT -- Includes roundwood products such as logs, poles, pulpwood plus material removed as chips, split posts, fuelwood, etc. This could in some instances exceed the estimated volume of growing stock because output may be in part from rough or rotten trees, windfalls, or species not regarded as having commercial value when timber inventories were made.

BONE DRY UNIT (BDU) -- 2400 pounds of moisture-free wood, unless otherwise stated.

CORD -- 128 cubic feet bulk measure, containing 75-90 cubic feet solid measure, usually taken as 80 c.f., solid measure.

CUBIC METER (m^3) -- 35.31 cubic feet.

CUNIT (CCF) -- 100 cubic feet, usually taken as solid measure, without voids.

THOUSAND BOARD FEET (MBF) -- In log or tree form this is the estimated volume of bark-free wood needed to produce a thousand board feet of lumber (1 thousand square feet, 1 inch thick, or 83.8 cubic feet). In log form a thousand board feet may be equivalent to 150-200 cubic feet of solid wood depending on log diameter; generally considered equivalent to 2 cords; but in small diameter timber it is equivalent to about 2 cunits.

TON -- short ton, 2000 pounds, green (gr.) or oven-dry (o.d.) with zero moisture.

TONNE -- Metric ton. Approximately 2204.6 pounds.

UNIT -- 200 cubic feet, bulk measure, used to measure hogged fuel, pulp chips, etc. Contains varying amounts of solid material depending on the amount of compaction. It is now more customary to weigh material correct for moisture and calculate the number of bone-dry units.

REFERENCES

Biltonen, Frank E., Hillstrom, William A., Steinhilb, Helmuth M., and Godmen, Richard M. 1975. Mechanized thinning of northern hardwood pole stands: Methods and economics. USDA For. Serv. Res. Pap. NC-137, St. Paul, Minn.

Clark, Alexander, III. 1978. *Forest Prod. J. 28*(10), 47-52.

Davidson, David A. 1978. *In* "Complete Tree Utilization of Southern Pine" (C. W. McMillin, ed.). FPRS, Madison, WI.

Folkema, M. P. 1977. Evaluation of Kockums 880 Tree King Feller-Buncher. Forest Eng. Research Institute of Canada, Tech. Rpt. TR-13.

Johnson, Leonard R., Simons, George, and Peterson, James. 1977. Unconventional Energy Sources Study Module IIIB of the Northwest Energy Policy Project for the Pacific Northwest Regional Commission, Vancouver, WA.

Mattson, James A., Arola, Rodger A., and Hillstrom, William A. 1978. *In* "Complete Tree Utilization of Southern Pine" (C. W. McMillin, ed.). FPRS, Madison, WI.

Mifflin, Ronald W., and Lysons, Hilton H. 1975. Skyline yarding cost estimating guide. USDA Forest Service Research Note PNW-325.

O'Dair, James R. 1978. *In* "Complete Tree Utilization of Southern Pine" (C. W. McMillin, ed.). FPRS, Madison, WI.

Phelps, Robert B. 1979. Consumption and demand for forest products in the United States. USDA For. Serv. For. Econ. Research Paper delivered at FPRS Conference on Timber Supply: Issues and Options.

Rust Engineering Co. 1979. Advanced system demonstration for utilization of biomass as an energy source in Westbrook, Maine: Alternative harvesting strategies. May.

Spurr, Stephen H., and Vaux, Henry J. 1976. Timber: Biological and economic potential. *Science 191*(4228). USDA, Forest Service.

U.S. Dept. Agric., For. Serv. 1978. Structural flakeboard from forest residues. USDA For. Serv. Gen. Tech. Rep. WO-5.

U.S. Dept. Agric., For. Serv. 1978. Forest statistics in the United States, 1977. (Review Draft)

Wahlgren, Harold E., and Ellis, Thomas H. 1978. Resource availability for energy uses with whole-tree utilization TAPPI.

GENETIC IMPROVEMENT OF FOREST TREES FOR BIOMASS PRODUCTION

Bruce Zobel

School of Forest Resources
North Carolina State University
Raleigh, North Carolina

ISBN 0-12-535902-0

I. INTRODUCTION AND GENERAL

When growing trees for biomass for energy and chemicals, the foresters' outlook and attitude must change somewhat from those when growing cellulose only for conventional fiber or solid wood products. Especially for energy, the objective must change from cellulose to Btu production; in fact, net Btu per acre at as reasonable as possible as quickly as possible is the product sought. This means that the total Btu produced per acre must be adjusted by the energy inputs and the cost of producing the biomass. Neither yields measured as total cubic volume nor green weight can be used to directly estimate success when growing biomass; in fact, they can be quite misleading. Even dry weight yields sometimes are not too precise an indication of the relative productivity and value of a species or acre of land when biomass is produced to be used for nonconventional forest products. To do a good job of growing wood for energy or chemicals, it is necessary to understand more about wood properties and growth potentials than is currently known or practiced.

A. Hardwood Emphasis

Most emphasis on biomass production in forestry has been with hardwoods, although there is real potential with some conifers. With the exception of a few genera such as *Populus* and *Eucalyptus*, we know less about management and genetic manipulation of the hardwoods than is known about most of the conifers. Although not always true, it is the general case that the hardwoods are more difficult and costly to grow in the nursery and to establish in field plantings than are the conifers. They frequently require intensive site preparation, fertilization and postplanting cultivation or release from competition. All these add up to heavy initial establishment and plantation costs, making fast growth and short rotations mandatory if a favorable economic return is to be obtained.

Hardwoods have some distinct advantages for biomass production. Even though less is known about management and genetic manipulation than for the conifers, the information currently available indicates greater potential gains from intensive forest management and breeding in the hardwood group than from the conifers. One major advantage of several of the hardwood species is their ability to sprout following logging, making it possible to grow successive crops without the need for a new planting each rotation. Sprouts grow at a rapid rate and anywhere from two to as many as seven sprout (coppice) rotations can be harvested before replanting is necessary. Additionally, some species can be reproduced using rooted cuttings, which results in maximum genetic gains through utilization of both the additive and dominance variance components; only the additive genetic component of variation is captured when using seed and seedlings.

Some hardwoods, especially the eucalypts, are already widely used and developed for energy and chemicals, such as in the state of Minas Gerais in Brazil, where over 200,000 acres per year are currently being planted for charcoal production. Now there is a move to also use eucalypt wood to produce alcohol to mix with gasoline. Most of the alcohol used in the mandatory 20% alcohol-80% gasoline mix in Brazil has come from excess sugar but that source is not large enough to fill all the needs. Manioc and wood are viable alternatives for alcohol production. The biomass productivity by the eucalypts in parts of Brazil is unbelievably large, being four to five times greater than wood production from hardwoods or pines in the southern United States. In the eucalypts it has been possible to double the biomass production per acre per year through better forest management and better genetic stock in just a few years.

B. Land Availability

One primary consideration, too often overlooked in the enthusiasm of discussing an increase in biomass production, is the availability of suitable operable land in large enough blocks to make economic energy or chemical production programs feasible. Some wildly unrealistic predictions have been made and even published about the biomass productive potential that can never be met because operable size forestry units with suitable soils are not available in the regions involved. Sometimes farmland is included in the calculations of available area and at other times steep or rocky lands are assumed to be available for intensive cultivation and mechanical harvesting. It is "terrifying" when energy or chemical productive potentials are predicted without a realistic understanding of the availability of the quantity and quality of forest land necessary to fulfill the predictions.

C. *Forestry and Agriculture*

There is a continuing debate about the relative potential and advantages of producing biomass by forest trees or through using agricultural crops. The values quoted as "tons per acre" often appear to favor biomass production by crops, but when all aspects are weighed the advantage is nearly always in favor of forest trees. Some reasons why this is so are:

(1) Trees are storable "on the stump" until the biomass is needed and can be harvested nearly any time of the year. Storage and deterioration are major problems with biomass produced from agricultural crops just as it can be with "whole tree" chips from forest trees. Weight and extractive losses occur in chip piles, with cellulose degradation and discoloration taking place very rapidly after harvest. The piles of green chips, leaves and bark heat and the chips turn black as they start to disintegrate.

(2) The net energy balance strongly favors forest trees. Annual crops require harvesting, site preparation, fertilization and planting each year, all of which takes considerable energy input and cost. Trees do not require such high-energy usage, and management activities are on a less intensive basis and are needed only a time or two during the tree's rotation (harvest age), which may vary from 5 to 20 years. This very large advantage of trees relative to energy input-output is all too frequently overlooked. Net energy balance required to produce a unit of biomass must be a key determining factor as to which is the best source of biomass.

(3) Trees can be grown on lands marginal or unusable for efficient agricultural production so there is minimal competition for lands to grow biomass or to grow food. Such competition over land between food and fiber production is a serious problem when growing fiber-producing plants such as kenaf. Despite its reported large tonnage yields, kenaf must be grown on good sites, must be fertilized and cultivated, and can be grown successfully only on soils that are not infested with nematodes. The yields from growing kenaf cited in the literature, and by some of its more avid proponents, are often more than double those actually obtained when kenaf is grown on an operational basis over extensive acreages.

D. Total Tree Harvesting

One often hears about the advantages of "total tree" harvest for forest biomass production. For some types of natural forest stands such as open-grown hardwoods or overdense conifers, yields are indeed increased greatly when the large, heavy-limbed trees of the numerous too-small trees that occupy the site are harvested. For open-grown hardwoods, volume and dry weight yields can sometimes be doubled over the conventional wood harvest by using total tree chips. For managed forest stands containing well-formed, rapidly growing trees, the advantage of total tree harvest is much less; some studies have shown for pine or well-formed hardwoods the added volume from total tree harvesting will not greatly exceed 10% of the normal merchantable volume. In addition, the volume added by total tree chipping contains 25% or more bark and leaves, which produce a smaller proportion of the desired product than does wood.

Total tree harvest always carries with it a possibility of creating soil problems through removal of excessive amounts of nutrients. Conventional logging removes few nutrients, since the bulk of the nutrients in the aboveground parts of the tree are in the inner bark, small limbs and leaves. For some sites, if very short rotations are used resulting in a heavy leaf and small branch harvest, it is mandatory to monitor for any possible nutrient imbalance in the soil following total tree harvests. For example, in some areas of Brazil, harvest age has been reduced to five or six years for stands that have an exceedingly fast growth rate. The forester must have knowledge of what is happening and adjust the value of the added biomass obtained from total tree harvest against any nutrient changes and the cost of reestablishing the needed balance in the soil. It is my opinion that in the future the heavy nutrient-containing portions of the tree (the small tops, leaves and small branches) will be returned to the site rather than being harvested for biomass. Yields of desired products from such material usually are low, while moisture content is high. Although I have no data to support this idea, it may well be that leaves and small branches are worth more for their nutrient content when placed back on the site than is their value for biomass if fertilizers are needed to replenish the nutrients removed in total tree harvest.

II. WHAT ARE THE POTENTIALS--HOW CAN GENETICS HELP?

As a generalization, an acre of land can only produce a given amount of biomass. It is the job of the forester to determine and control the form and size of the plants on which this biomass will be produced, within the most suitable time frame and cost structure. The ultimate skill of the forester is to have the land produce trees of the desired type at the earliest date and at the lowest cost. This silvicultural task sounds easy but is very difficult to achieve.

A. *The Genetic Approach*

Because of the complexity of species and the environments in which they are grown, there certainly is no general or universal genetic approach that should be applied to increase biomass production. Methods must be developed for each species and condition; in this chapter I will illustrate methods and potential, using a few chosen case examples. *For biomass production, the major objective is to obtain the maximum growth of the most desired wood in the shortest possible time at as low a cost as possible.* This does not mean that tree quality can be ignored, because crooked trees or heavy-limbed trees will produce quite different woods than will straight, small-limbed trees because of the presence of reaction wood and knotwood. Also, the cost and ease of handling and harvesting will be greatly affected by tree form.

To get the desired gains using genetically improved trees, it is essential to also use improved methods of forest management. Trees that are planted and grown under adverse environmental conditions will never have a chance to express their full potential, no matter how good they are genetically. It is mandatory, therefore, that there be a "marriage" between good cultural practices and genetic improvement. Forest trees are just like improved crop plants such as corn. If, for example, corn is planted but not cared for or fertilized, one would obtain the best yields by using the old unimproved varieties that are adpated to growing under adverse conditions. But if the best land and good cultural care are used, the old varieties are capable of producing only a small fraction of the yield available when improved varieties are sown. Only when the best culture is combined with the best genetic potential can the gains be maximized. An understanding of this interrelationship is of key importance, because in forestry all too commonly there is willingness to invest in the development of genetically superior trees, but when these are planted there often is a reluctance to expend the effort and money necessary to manage the improved trees intensively.

Genetic manipulation can help increase production of biomass in three ways:

(1) Increase the yield on already good sites.

(2) Develop trees capable of growing a merchantable crop of wood on marginal or nonproductive areas which currently will not support an economical forest enterprise.

(3) Develop trees with wood that has more Btu and less moisture or a higher chemical content.

B. Increasing Yields on Good Sites

Even on the best sites, most forests are producing considerably less biomass than their potential. The first and most obvious improvement is to establish or change the stand to proper stocking so that the total environmental potential of soil, sunlight, moisture, and nutrients can be converted to biomass in the desired form. Understocked or overstocked stands never produce the potential of which a site is capable. One aspect of the environment always comes into short supply before the others and thus becomes the limiting factor. To be fully productive, the stand must be so manipulated that the time at which any factor becomes limiting is increased, an easy statement to make but not an easy thing for the forester to accomplish. One of the most common limiting factors that must be overcome is attack by pests, whether disease, insects or other. Breeding for pest resistance or resistance to specific environmental limitations such as drought, excess moisture or cold has been most successful and has resulted in millions of acres being made operable that could not have been accomplished without genetic pest control or improve adaptability to environmental limiting factors.

After stand structure and the resultant utilization of the environmental factors are optimized, how is it possible for improved genetic strains to produce additional biomass? Basically it can be either by (1) producing more biomass at a given time through a more complete utilization of nutrients, moisture and sunlight, or (2) by producing an amount of biomass equal to the unimproved stock but in a lesser period of time than was formerly required.

Generally benefits from a *reduction* of the time period to full productivity are preferred by most foresters because of their massive effect upon economic returns. Most people are aware of the shape of the compound interest curve with its very steep rise over time. Therefore the large site preparation and establishment costs necessary for maximum yields of biomass result in a very large increase of the cost of the biomass as time increases. The effect of rapidly increasing interest with time, and thus the greatly increased costs, becomes controlling in determining what forest practices can be used.

A quicker capture of the site production potential can be obtained by planting more trees per acre just as well as by using trees that grow faster. However, this results in many small stems; it is more costly to regenerate and to harvest large numbers of trees per acre. Planting and nursery costs greatly increase when more seedlings are needed per acre; the small stems harvested have an inordinate amount of less desired biomass such as bark. In the southern pines, for example, 5-inch diameter trees will average about 30% bark while 10-inch diameter trees only have 10 or 11% bark. The best and most efficient gain will therefore result from planting genetically faster-growing plants at wider spacing, with the resultant full utilization of the yield capability of the site in a shorter time period from trees of larger diameter.

Therefore, shortening the time to full productivity of forest land is usually the major objective of any genetics program designed to increase biomass yields. According to Lambeth as cited in Zobel (1979), the following would be obtained.:

33-year rotation, no genetic improvement	10% profit
33-year rotation, 10% gain in yield at age 33	21% profit
Same volume yield at age 30 as formerly obtained at age 33	39% profit

As a result, the catchwords in biomass production are "short rotation." This will not be of maximum utility until fast-growing strains are produced that will result in larger trees in less time at initially wider spacing.

C. *Increasing Yields on Marginal Sites*

Genetic improvement to develop strains of trees able to produce economically profitable crops on marginal sites is becoming more emphasized in some areas than is developing faster-growing trees on good sites. Forestry operations are rapidly being "pushed off" the better sites which are suitable for the production of agricultural crops. When there is competition for use of land between food or fiber, the food production usage always predominates. Therefore, if the acreage of forest production is to be increased it must be on the submarginal sites which currently are not economical to operate. The *only* way to make such sites reasonably productive is to combine good management with planting stock bred especially to overcome the limiting factor(s) (including pests) that have rendered the sites marginal in the past.

An example of currently submarginal sites in the southeastern United States are those referred to as pocosins that contain excess moisture. We have developed trees with root systems and physiological adaptations that now enable us to grow trees reasonably well on the deep peat in the pocosin despite the limiting factors of excess moisture and low soil pH. The roots of the improved strain have the ability to elongate and function with a level of oxygen much below that normally considered to be necessary for normal root functioning. Morphologically the roots develop a stiltlike system rather than a regular taproot that enables the plant to be stable and grow normally in water-saturated peat soils. The stiltlike roots consist of a number of major roots that develop as anchor roots rather than the conventional single taproot; this makes the tree perfectly stable when growing in peat over 10 feet deep before mineral soil is encountered. In the same conditions normal loblolly pine with a taproot grows a few years, then "wallows with the wind" and falls over or leans badly. We do not yet know how the roots develop and grow through completely water-saturated peat, a pattern not expected based on the physiological need for roots to have oxygen to develop; current research may clarify this puzzle somewhat. Thus through selective breeding we have successfully developed a strain of loblolly pine that is capable of growing well and apparently normally and with stability, in continuously water-saturated, deep peat soils.

Many kinds of limiting factors occur on the submarginal sites for which new strains of forest trees can be bred. Well known are the food results from breeding trees for drought tolerance and resistance to certain insects and diseases. It is becoming more critical that we develop trees that can grow in soils with excessively high or low acidity; equally important, we need to find or develop trees that will grow reasonably well with a nutrient deficiency of elements such as boron or phosphorus or where there is an excess of undesired chemicals such as aluminum.

Breeding trees adapted to grow and produce an economic crop on problem sites has been much more successful than any of us had ever expected or even hoped. We have been able to combine good growth, form, and desired wood qualities along with the improved adaptability. I feel the greatest future overall gains from the application of forest genetics, especially in biomass production, will be through improving adaptability to adverse environments or pests. The large land areas on which foresters will be forced to operate in the foreseeable future if biomass needs are to be met require emphasis on the silviculture and genetics required to improve adaptability.

D. Increasing Yields by Improving Wood Qualities

Most wood properties are strongly inherited, especially the key property of wood density (specific gravity). When specific gravity of wood is increased, the result is more solid wood substance per unit volume. Most fortunately, moisture content and wood density are negatively correlated, and within a species, high density wood has a low moisture content.

One major relationship that affects wood specific gravity of conifers such as the pines is tree age. Near the center of the tree, juvenile wood is produced. Depending on the species, juvenile wood formation occurs from 5 to 15 rings outward from the pith at tree center. Juvenile wood is formed at all heights in the tree. Thus in a given year, a 30-year-old tree produces mature wood at its base while producing juvenile wood near the top of the tree. Since juvenile wood has low specific gravity, high moisture content, and cellulose differences from mature wood (such as a high hemicellulose content), a young tree (or topwood or plywood cores) will have wood that will give relatively low net Btu yields or low chemical production per unit volume or unit green weight. Thus age and location of wood has a major effect on yield of biomass and products obtained therefrom.

Hardwoods have a pattern of growth and wood development which produces only "mild" juvenile wood. Because of this, short rotations are of greater advantage to hardwoods than to most conifers which usually produce distinct juvenile wood having low specific gravity and high moisture content. Juvenile wood of hardwoods usually has a higher moisture content than mature wood and its wood density may be a little less, but overall the negative quality characteristics of juvenile wood of hardwoods are quite small compared to conifers.

The magnitude of the relationship between specific gravity, age and moisture content is indicated in Tables I and II for pine and in Table III for several species of hardwoods. Note that ash has more dry wood substance per cubic foot than does sweetgum and that it has wood with a very low moisture content.

It is of key importance that specific gravity in conifers is inherited essentially independently of growth rate. With care in selection and breeding, it is therefore feasible to develop a fast-grown, high-gravity strain of forest tree. Additionally, because of the high negative correlation

TABLE I. Dry Wood Weight for Young Loblolly Pine in the Piedmont of South Carolina, with Comparisons for the 25- and 35-Year Age Class

		Weight--Pounds/Cubic Ft. Inside Bark	
Age	Specific Gravity	Oven Dry	Green
10	0.385	24.0	60.1
11	0.391	24.4	60.0
12	0.396	24.7	59.8
13	0.400	25.0	59.7
14	0.405	25.3	59.5
15	0.409	25.5	59.3
25	0.438	27.4	57.0
35	0.458	28.6	54.6

TABLE II. Generalized Yields from 100 Solid Cubic Feet of Green, Peeled Loblolly Pine Wood of Different Ages[a]

Stand Age (Yrs.)	Dry Clear Wood (Lbs.)	Dry Clear Wood (%)	Water (Lbs.)	Water (%)	Resin Extractives (Lbs.)	Resin Extractives (%)	Other[b]	Total
25	2500	42.2	3600	56.3	100	1.5	?	6400
30	3000	56.8	3300	51.4	120	1.8	?	6420
35	3100	49.7	3000	48.1	140	2.2	?	6240

[a]Data are generalized for loblolly pine in the eastern portion of its range. Note the data are for 100 solid cubic feet of wood, not for a cord.

[b]Includes knots, resin associated with knots, "included" bark, etc. This material constitutes between 3 and 7% of the volume of woods-run logs.

TABLE III. Wood Weight and Moisture Content of Hardwoods

Species	*Specific Gravity*	*Percent Moisture Content*[a]	*Dry Wt. in Lbs./ Cu. Ft.*	*Green Wt. in Lbs./ Cu. Ft.*
Sweetgum	*.467*	*119.4*	*29.1*	*63.9*
Tupelo	*.478*	*91.3*	*29.8*	*57.0*
Red Maple	*.461*	*72.3*	*28.7*	*49.5*
Yellow-Poplar	*.377*	*110.5*	*23.5*	*49.5*
Cottonwood	*.380*	*124.0*	*23.7*	*53.1*
Red Oaks (Cut Leaf)	*.604*	*72.7*	*37.6*	*65.1*
Red Oaks (Entire Leaf)	*.596*	*77.2*	*37.1*	*65.9*
Sycamore	*.460*	*111.0*	*28.7*	*60.5*
Birch	*.484*	*94.2*	*30.2*	*58.6*
Ash	*.551*	*50.4*	*34.3*	*51.7*
Black Alder (3-yr-old)	*.358*	*125.6*	*22.3*	*50.4*

[a]*Based upon dry weight of wood.*

between specific gravity and moisture content, it is possible to develop a fast-growing tree with dense wood that has a low moisture content which is ideal for energy production. The relationship between growth rate, wood density and moisture content is important for any biomass usage. If manipulated properly it can result in greater yields from a given volume of wood and less moisture must be deposed of in developing the final product, be it Btu or chemicals.

Because of the large individual tree differences in wood qualities within a given species on a given site, and because of their general high heritability, it is very feasiable and profitable to breed for desired wood qualities for improving biomass production. Once the genetically improved trees have been obtained, the gains will be permanent from one generation to another and will not require repeated treatment and investment as is needed for things such as fertilization.

III. HOW IS IMPROVED GENETIC MATERIAL OBTAINED?

It is not appropriate in this paper to go into great detail as to the techniques needed to obtain genetic improvement. However, the broad concepts and methodology should be of interest and will be described briefly. Genetic improvement consists of (1) locating and using the correct species; (2) using the best geographic sources within the best species; (3) selecting and breeding the best individuals within the best sources of the best species. The only way to make maximum gains through seed production is to combine all three methods of genetic improvement listed above. When vegetative propagation is feasible a fourth method of mass-producing the very best trees to obtain maximum gains is available. An example is *Eucalyptus* in Brazil. There, in a few years, it has been possible to double the volume yields while at the same time improving tree form, wood density, and probably cellulose yield. Improvement in biomass production has been dramatic. I understand that growth rates of *Eucalyptus* in Brazil were nearly doubled early in the century through selective breeding and now it has been possible through selection and breeding and vegetative propagation to approximately double growth again along with better wood and tree form.

A. *Correct Species*

The choice of species is sometimes limited, sometimes broad, depending upon the area to be planted, the economic need, and the products desired. Hardwoods are usually preferred for energy production and for chemicals from wood. Many hardwood species have a high wood density and relatively low moisture content when compared to the conifers of the same age. The chemical composition of the wood of hardwoods is usually most desirable as a base for organic chemical production; for example, hardwoods contain several times more of the "chemical building block" furfural than do the pines. However, if maximum product gains are to be obtained from biomass at the most reasonable prices, especially in the more temperate areas, wood quality desirability, volume growth and the cost of establishing plantations must all be assessed. Sometimes conifers will be favored despite their less desirable wood. Conifers are especially useful when resinous byproducts are specifically desired.

In the warmer climates, members of the Genus *Eucalyptus* are usually the major group of trees planted for biomass production. Many species have inherently rapid growth, and those planted most widely have reasonably dense wood which does not contain too high a moisture content. The advantage of the eucalypts is that they are particularly responsive to genetic manipulation and to silvicultural treatment. Genetic gain can be dramatic and swift, as mentioned earlier. Along with their desirable growth and wood characteristics, recent results show that species such as *E. grandis* can be rooted on an operational basis with costs for plantlets that do not differ greatly from those of seedlings. Use of rooted cuttings enables rapid and maximal genetic gains because all the genetic variability can be captured in the rooted plantlets rather than only a part, which is the situation when sexual reproduction with its attendant genetic recombinations is used. An additional advantage, quoted by many people, is that some species of the eucalypts sprout (coppice) after the tree is felled. This capability potentially results in some unique operational and cost advantages. The Genus *Eucalyptus* contains numerous species with wide adaptability; along with their good genetic and silvicultural response, the eucalypts are an ideal group to work with for biomass production.

Many other hardwood species can be used to produce biomass. As in the southern United States, these often fall into two groups, (1) trees with relatively fast growth, low wood density and high moisture content such as sweetgum and sycamore (*Liquidambar styraciflua* and *Platanus occidentalis*); (2) species with generally slower growth but which have denser wood and low moisture content such as ash (*Fraxinus*) or the oaks (*Quercus*). Both these groups of trees are widely utilized for biomass production. In order to plan satisfactorily for the desired species to use, it is necessary to weigh volume growth potential against wood qualities, using criterion such as net Btu production or total yield of the chemicals desired. Studies over the past twenty years have shown that all the species mentioned are responsive to genetic manipulation.

In the northern areas of the United States and in Canada, several hardwood species are widespread. Most prevalent is aspen (*Populus tremuloides*), a fast-growing species which generally sprouts profusely following logging. Aspen wood has many good qualities although moisture content is somewhat high and certain chemical substances in the wood of aspen are considered undesirable for some products. There is a very large inventory of standing timber of this species, along with maple and birch. I foresee aspen developing into a species which will be much more widely used for several products including biomass production.

Another group of hardwoods that are fast-growing and suitable for biomass production are the alders. In the northern part of the South and the southern part of the North in the eastern United States, the European black alder (*Alnus glutinosa*) is proving to have considerable biomass potential. This species is fast-growing, sprouts well, has a reasonably wide distribution, usable wood, seeds early and seems to be well adapted for several different sites. In the western United States and Canada, *Alnus rubra*, the red alder, grows very rapidly on the better sites. This species was commonly categorized as a weed by many foresters in the past but now is being more widely used and could supply a great deal of biomass in a short period of time. Limited genetic work with the alders in the United States and more intensive work in northern Europe shows this genus to be responsive to genetic manipulation. Along with the alders, the birches (*Betula* sp.) have been found to be fast producers of biomass and genetically tractable.

One very large group of hardwoods often associated with biomass production but only sparsely used as of now are the mixed tropical hardwoods. This group of heterogeneous species contains by far the largest inventory of biomass. However, management of the difficult-to-handle tropical forests is just starting and very little has been done with them genetically, with the exception of a few species such as *Gmelina*. The tropical hardwoods with their vast acreages and potential are really the "sleeping giant" of hardwood biomass production. The contribution that genetic manipulation will make to increasing the volume and quality of biomass production of the many species in a tropical hardwood forest is not now known, but in my opinion it will be considerable, certainly more than most foresters now consider, many of whom have mentally "written them off."

It would take too much space to discuss the potentials of the numerous coniferous species for biomass production. Genetically we know a lot about many conifers. They have the advantage of growing rapidly with minimal culture and are easy to handle silviculturally. Since so much is known about them genetically as well as silviculturally, it is possible to move into massive regeneration programs in a very short period of time. The conifers are more tolerant of adverse sites than are many hardwoods, making them especially desirable in currently submarginal forest areas. Most species do not coppice well but they are generally easy to grow in the nursery and to plant. If conifers are desired in the tropical areas, one usually thinks about the potentials of the so-called tropical and subtropical pines for biomass production. However, in the colder areas, trees such as the larches (*Larix* sp.) have an excellent biomass production potential. The fast-grown conifers that are harvested on short rotations have low yields, with the bulk of the wood produced being juvenile wood with low specific gravity, high moisture content, low cellulose and high hemicellulose content. Despite some of these adverse characteristics of the wood for energy and chemicals, I expect to see an upsurge in the use of the wood of the conifers for biomass production.

B. Best Sources Within a Species

Most tree species vary a great deal in adaptability and growth potential, depending upon the geographic location of their origin. Geographic source variation (or provenance variation as it is often called) is of key importance; it takes a full rotation to be sure the correct source within a species is being used. All species show differences depending on their geographic source, but in some it is dramatic indeed; a couple of examples are loblolly pine (*P. taeda)* and *Eucalyptus grandis*. Geographic differences are so great within these species that they sometimes appear to the casual observer to be different species.

A species cannot be categorized as to utility for biomass production without reference to provenance, since sometimes differences among seed sources are so large that one geographic source may be ideal for an area or specific product while another provenance of the same species may be quite useless. Ignoring geographic source or categorizing the value of a given species based on tests of only one or two sources is the most common error made in determining the correct trees to use for biomass production. Aside from growth potential, wood properities can vary greatly, both by geographic source of origin and by the new environment in which the tree is grown. Within its natural range, *P. caribaea* has wood that is relatively but uniformly moderately dense. When grown in certain "ideal environments," such as in parts of South Africa, the species fails to produce much summerwood resulting in a soft, white, low-density wood of poor utility for biomass. In the wet Tropics this species quite often has a core of extreme juvenile wood surrounded by very dense mature wood. Knowing the response to provenance and the effect of plantation location for both growth and wood properties is essential if efficient biomass production is to be achieved.

Other than by using the correct species, the largest and easiest genetic gains in biomass volume and quality can be made by properly determining the best seed source within a species. Additional large and quick gains can be added by using the *land race* most suitable to a region. All exotics are only partially adapted to the new site on which they are planted; the new environment and the genetic potential of the individuals grown there interact, with the result that certain trees grow better, faster, and have more desirable wood than do the bulk of the trees within the exotic plantations. When these best performers within exotic plantations are selected and are then brought together so they can interbreed, the resulting seedlings are referred to as a *land race*. In almost all situations in which I have been involved, the land race will outperform any new seed sources that may be brought in from the natural range of a species if the original exotic seed was reasonably well adapted to the new environment. *Development and use of land races have been ignored* all too often when biomass is desired, and thus one of the easiest and most effective tools to obtain large and efficient genetic gains has not been used widely enough. On the average, the best and certainly the most rapid gains in adaptability characteristics will be obtained by combining the best provenance with the land race of outstandingly adapted individuals selected from it.

C. Best Individual Trees Within the Best Source of the Best Species

For certain characteristics such as wood specific gravity, tree form, disease resistance and others, the greatest genetic improvement can be made when individual tree selection is used within a provenance. Development of a land race automatically includes individual tree selection for the most adaptable individuals. When combined with a selection for tree quality and wood characteristics, maximum improvement will be obtained from using the best individuals from the plantations of a provenance.

Individual tree selection consists of two parts: (1) If the family structure is available (*i.e.*, if the parentage of the trees being assessed is known) the selection process consists first of choosing the outstanding families[1] and then the best individual trees within these best families. Certain characteristics such as growth rate, adaptability and disease resistance may show the greatest gain from family selection, while many quality characteristics that are strongly inherited, such as tree form or wood specific gravity, will respond and produce the greatest gain when individual tree selection is used. There is really no debate as to whether family or individual selection should be used singly because both are necessary if maximum gains are to be obtained. (2) If, as often happens, family structure is not known, then only individual tree selection can be employed and gains from genetic manipulation will be less than when family and individual tree selection can be combined.

[1] *Family refers to progeny from a given parent tree.*

D. Alternative Methods

There are methods other than selection that can be used to obtain genetic gain in biomass production. If trees can be vegetatively propagated, genetic simularity between parent and progeny will be greater than through seed regeneration because of a more complete capture and use of the genetic potential of the mother tree. Vegetative propagation enables mass plantings of hybrids which is difficult through seed production because when two hybrids are crossed their progeny are variable, grading from one parent through the intermediate hybrid form to the other parent. First-generation or F_1 hybrids can rarely be easily produced on a large scale and thus they are expensive; often control, hand-pollination is required. Vegetative propagation allows the characteristics of the hybrid to be transferred directly from an outstanding cross to the propagules of plantlets that will be used in mass planting. In theory at least, the vegetative propagules will be genetically identical to the tree from which the cuttings were taken, whether it be a pure species or hybrid.

There is danger in utilization of vegetative propagules without good planning; thus problems are already observable in some tree improvement programs. Since each vegetatively produced tree is genetically identical to its donor parent, it is essential to use a number of different clones (mother trees) in operational planting programs, using vegetative propagules, if a sufficiently broad genetic base is to be maintained. A broad base is necessary to avoid catastrophe from pests or adverse environments and to get full production from the site. Seed produced through sexual means has a built-in variability produced by recombination of characteristics from both parents, so each plant is genetically different from its sister or brother. On the other hand, vegetative propagules are like identical twins, with little variability evident among the propagules from a given parent.

Hybridization has the potential to create something new by combining characteristics of both parents into a unique tree not currently found in nature. Most economically important characteristics are inherited in an intermediate manner between the two parents of a cross. It is therefore possible, with good planning, to develop a hybrid with a combination of characters desired for a given site or product. Because of the difficulty of producing hybrids on a mass scale, forest tree hybrids have not been much used operationally except for easy-to-root species like some poplars. Vegetative propagation will become much more important as techniques are developed and perfected for more species.

IV. WHAT GAINS ARE POSSIBLE, USING GENETICS?

The question related to gains cannot be answered without a whole series of qualifications and "if's" about species, product, intensity of selection, and breeding methods used. But some generalities can be made that will be useful as a guide.

A. *Gain vs. Risk*

There is always some dispute about what is the best genetic approach to use. However, all genetic improvement involves a weighing of gain against risk. Control of pests, control of the environment, rotation age, and variability of the species to be used, all play a major role in making the decision as to how much risk can be taken to obtain a given amount of gain. The shorter the rotation age, the greater can be the risk to obtain additional gain. Greater gains from genetic manipulation will be possible in biomass production relative to conventional forest products because rotation ages can be short. Because of environmental pressures and the inherent conservatism of foresters, the risk factor has tended to be overemphasized and gains are often smaller than possible. The inherent variability within a tree species is usually underestimated, and too often tree populations are assumed to have the great genetic uniformity present in homozygous crop plants. In fact, numerous studies have shown that forest trees have the largest genetic variability of any major group of plants or animals. Each tree must have a greatly variable and adaptable genetic makeup to survive and reproduce over many years, growing on different sites in vastly differing environments that can occur within and between years. As rotation ages are shortened, the danger from too great a risk from genetic uniformity becomes less for the forest manager. If biomass production is to be fully successful, the risk factor must be increased somewhat to obtain as much gain as possible and the monoculture bugaboo so often cited as a major problem must be looked at as the the true biological risk rather than an emotional pseudo-fear based on homozygous crops that currently often is the basis of decision-making.

B. *Gains Possible*

For biomass production on good sites I estimate 60% of the gains from genetic improvement will be in the form of volume and 40% will be from improved quality characteristics such as wood. For pulp and solid wood products, the ratio of growth and quality is closer to 50-50. The magnitude of the gains can be very large, especially in the initial stages of selection and breeding. For example, in the eucalypts it appears that when combined with good management it will be easy to double volume yields in one or two generations. We have experience in pine that volume improvement obtainable is 10% to 20% the first generation and 35% to 50% the second one. Quality gains will be large the first generation but may fall off somewhat in succeeding generations. For example, in loblolly pine we have been able to improve straightness of tree bole enough in the first generation so that there is little need for additional straightening in future generations of breeding.

The characteristics of biomass production that should be concentrated upon on the good sites are volume growth and the wood qualities of specific gravity and moisture content; these must be always combined with pest resistance. On the poor or marginal sites, better adaptability must be a paramount objective of the breeder. Within the adaptable strains, growth and quality characteristics can be improved; we have been remarkably successful in such a combination. It is not possible to quote gains from breeding for adaptability; how does one express an improvement from unproductivity to economic productivity?

My estimate of total gains in biomass production over current yields on good sites made possible through combining good management and genetics is as follows for the first generation:

Soft hardwoods (sweetgum, sycamore, etc.)	75%
Hard hardwoods (oak, ash, and others)	50%
Eucalypts	100%
Pine	20%

Total volumes or tonnage production per acre are not possible to generally quantify because they vary so with site, species, and location. For example, pine in the southeastern United States with the best care on good sites can grow 2 to 2.5 cords/acre/year in a 25-year rotation if good genetic stock is used. The same pine species will produce 6 cords/acre/year in 20 years in southern Brazil. In the drier areas of Minas Gerais, Brazil, 6 cords of eucalypts per acre per year in 6 years are usual, and gains to 10 cords through genetics are probably possible. In comparison, in the very best coastal areas of Brazil, 6 to 8 cords are being produced from earlier moderately improved stands, but it now appears that 12 to perhaps even 14 cords/acre/year can be obtained in 8 years, using the best genetic stock, the best management techniques and vegetative propagation. This outstanding volume growth can be translated to an excess of 16 tons dry weight per acre per year.

C. *Assessing Biomass*

Estimating production of biomass by green weight can be most misleading. Many publications report tonnage yields but some do not specify whether it is green or dry weight. Moisture content of wood varies widely with species, environment, age of tree, and location in the tree from which the wood is obtained. For example, 30-year-old loblolly pine will have a moisture content of about 90% (calculated on a dry wood weight basis) while 15-year-old trees will have a moisture percent close to 150% (*i.e.*, 1.5 lbs. water for every lb. of dry wood). To be meaningful for biomass estimates, therefore, *all weight values should be expressed as dry weight*. Young trees, such as those harvested using the "silage concept," contain a considerable volume of leaves and small branches so the moisture content is very high.

Dry weight gains in biomass production are meaningless without working out an energy balance of "energy in, biomass out." Such determinations are complex and vary from area to area and among species. Studies that have been done, mostly still unpublished, show that the energy balance using tree species per unit of biomass produced is generally better than that for agricultural crops.

V. SUMMARY

Without the use of genetics, maximum gains in biomass will never be realized. There is nothing magic about genetics; it is only one of several silvicultural tools that enable the plants to make maximum use of the environmental factors affecting growth and quality. It does allow development of plants that can bypass or overcome elements of the environment that limit growth and production. Genetics is a most useful tool in maximizing biomass production because, once a gain is obtained, it is permanent. For example, with fertilization, each crop must be treated but, once developed, genetically improved stock maintains its superiority for succeeding generations. However, a program of genetic improvement cannot be static and ongoing breeding programs are needed to continuously upgrade the material for even better biomass yields.

REFERENCE

Zobel, B. J. 1979. Trends in forest management as influenced by tree improvement. 15th South. For. Tree Impr. Conf., Miss. State Univ., pp. 73-77.

WOOD FUELS CONSUMPTION METHODOLOGY AND 1978 RESULTS[1]

G. F. Schreuder
D. A. Tillman

College of Forest Resources
University of Washington
Seattle, Washington

[1]*This paper presents the results of research sponsored by the Forest Products Laboratory, U. S. Forest Service.*

ISBN 0-12-535902-0

I. SUMMARY AND INTRODUCTION

Total wood fuel consumption in 1978 was 1.7×10^{15} Btu (quads) ± 0.2 quads as shown in Table I. Of this total 59% was consumed by the pulp and paper industry; 20% by lumber, plywood, and related board manufacturing plants; 16% by residential users; and 5% by all other wood fuel users including furniture and fixtures manufacturers.

These estimates represent an 0.1 quad increase from wood fuel consumption in 1976 (Bethel *et al.*, 1979; Tillman, 1978). They confirm the continued dominant influence of the forest products industry (FPI) in total wood fuels consumption, as the FPI consumes almost 80% of all wood fuels burned.

Because 80% of the industrial wood fuels consumed are used in the forest products industry, both the volumes of fuel generated and consumed can be treated as a direct function of lumber, plywood, and pulp production. Annual wood fuel consumption, therefore, will be somewhat sensitive to new housing starts as well as more general economic conditions.

TABLE I. Estimate of Total Wood Fuels Consumed, 1978 (in 10^{12} Btu)

	Case		
Economic Sector	*Low*	*Medium*	*High*
Residential	*200*	*270*	*340*
Pulp and paper	*960*	*1010*	*1070*
Lumber, plywood, and related board prod.	*320*	*340*	*350*
Furniture and fixtures	*20*	*40*	*60*
Utilities and all other	*40*	*50*	*60*
Total	*1540*	*1710*	*1880*

II. BACKGROUND AND METHODOLOGY

The problems associated with estimating wood fuels consumption are the result of several factors: (1) residential wood fuels are produced and consumed in local markets, rather than national markets; (2) most industrial wood fuels are mill wastes which are consumed at the same point of production, never entering a marketplace; and (3) wood fuels are produced in many forms including residential cordwood and charcoal, spent pulping liquor, hogged bark and wood, and dry planer shavings and sanderdust.

Because estimation of wood fuels consumption is compounded by several factors, national statistics are not developed on a regular basis. Further, previous estimates (e.g., Tillman, 1977; Bethel *et al.*, 1979) were developed using divergent methodologies. Thus it is important here to identify not only the estimates made, but also the methodology employed in deriving the estimates.

A. *Background*

Official statistics on wood fuels consumption, kept by the U. S. Bureau of Mines, were discontinued in 1950. At that time consumption had declined to 1.2 quads (Enzer, Dupree and Miller, 1975); oil, natural gas, and hydroelectric power were making increasing contributions to U. S. energy supplies; and nuclear power was on the horizon. In 1950, wood fuel supplied 3.3% of the nation's energy needs, down from over 90% in 1850.

In 1950 no one could foresee the geopolitical events of the 1970s which, combined with declines in domestic oil and gas reserves, and production, created enormous interest in renewable resource fuels. In the years since 1973, when this interest was rising, no time series data were available to correlate interest with activity.

Two estimates were developed for 1976 consumption: Tillman (1977) and Bethel *et al.* (1979). The first study was conducted for the Federal Energy Administration, and the second was part of a larger study for the Office of Technology Assessment, U. S. Congress. Different approaches were taken, however both yielded national estimates of 1.6 quads. Both studies recognized the difficulties of measuring fuels which are gathered from dispersed sources rather than produced, in large quantity, in a centralized operation such as a coal mine or oil field; gathered, concentrated, and used frequently without ever entering the commercial marketplace; produced by many economic units as a secondary activity; and very diverse in form, with particle sizes ranging from micron size to cordwood and moisture content ranging from 2% to 200% (O.D. basis).

To address these problems a series of methodologies were devised to estimate wood fuels consumption. These methodologies were developed to make duplication of results, and updating possible

B. Methodology

The general approach taken in this study is as follows:

(1) Establish the point of measurement as fuel supplied to the point of use--the woodstove, the industrial boiler, or the utility power plant;

(2) Establish the Btu as the principal unit of measurement in order to provide comparability between estimates of such fuels as spent pulping liquor and cordwood; and

(3) Disaggregate utilization into the following sectors for individual methodology development: residential, forest products industry, and other industry.

1. Residential Fuel Wood Methodologies. Two methodologies were employed to develop residential wood fuels consumption data. The first relates residential wood fuels consumption to total energy consumption in rural areas. The second relates wood fuels consumption to energy requirements.

In the first case total 1978 residential fuel consumption for heating and cooking is established, along with rural residential fuels consumption. These values are then multiplied by the fractions of homes using wood for heating and cooking. Finally an efficiency correction is made to arrive at input Btu measurement.

In the second case fuel requirements are established as a function of climate--specifically heating degree days. Using a datum temperature of 70°F, heat losses are estimated for average rural homes, and wood fuel consumption is then calculated.

2. Energy Consumption in the Forest Products Industry. Direct wood fuels consumption data are available for the forest products industry. Data for the pulp and paper industry were obtained for the years 1971-1977. A one year extrapolation was made for 1978. These data came from the American Paper Institute.

Similarly the National Association of Manufacturers published wood fuels and fossil fuels consumption in sawmills, plywood mills and other forest industry plants for the year 1971 (Jamison, Methven, and Shade, 1978). That ratio provides a basis for using Bureau of Census data to update the wood fuels consumption estimates in the lumber and board product plants.

Given the large percentage of wood fuels consumed by these industries, the availability of such data lends considerable credence to the estimates thus developed.

3. Energy Consumption in Other Industries. Estimates of wood fuel utilization in the electric utility industry were developed by applying a heat rate of 15×10^3 Btu/kW-h to electricity production from wood. The heat rate is consistent with estimates by Evans (1974) and Hill (1977). The electricity production is reported by the Federal Power Commission.

Estimates of wood fuel used in all other industries were obtained by applying efficiency and operating hours assumptions to the national wood and bark boiler inventory (Ultrasystems, 1978). Discrepancies in that inventory preclude its use where more reliable data are available.

4. Data Aggregation. In all cases a range of values is estimated. These ranges are then added together for a final set of values. Despite the presentation of a 1.7 ± 0.2 quad estimate, the high and low estimates are not considered in any way to be confidence limits. The variety of techniques and sources used precludes positing values with such measures of precision. These values are more akin to scenario analysis results.

III. ENERGY CONSUMPTION BY SECTOR

A. *Energy Consumption in American Households*

Wood may be either the primary or secondary source of energy for space heating, water heating, or cooking. The extent to which it is used among households, as well as the efficiency with which it is used will vary among households. Further, nothing is known about the number of rooms that are heated in households that use wood. Another question is posed by the differences in quality of house insulation by energy user groups. Answers to these questions are given by a survey in New Hampshire (Dalton, 1977) suggesting that there are differences in insulation quality between wood burning households and nonwood burning households. The same is suggested for differences between houses in urban areas with houses in rural areas (Housing Survey, 1976).

All these factors will have their effect on the energy needs of a household and it is possible that firewood using households can be distinguished from other fuel user groups.

Two approaches are taken to find the total energy consumption by all American firewood using households:

(1) To estimate the energy consumption of wood fuels by end-use and to correct this with an estimate for efficiency.

(2) To estimate the energy needs of wood users.

1. Estimate by Energy Consumption. This estimate is based on the energy use of all the households. The latest refined data of total energy consumption by households is from 1973 (Ross and Williams, 1975). Later data are not as refined but they can be used to update the figures from 1973. The 1978 data for fuel consumption generally is presented in Table II. The typical commercial/residential consumption ratio is 43:57 (Doe, 1979). Thus residential consumption for 1978 is estimated at 16.7 quads. Linear extrapolation of column one, presented in Table III, shows space heat at 8.50 quads and cooking at 0.75 quads. Water heating would be estimated at 2.53 quads. These total to 11.78 quads.

The current Housing Report (1976) supplies data on the use of wood as a source for heating and cooking in urban and rural areas. In urban areas, wood can be neglected as a primary or secondary energy source. Some 0.16 cords per year are used on average, in 25 million households for esthetic use (Ellis, 1976).

The figures for wood use in rural areas are consistently small. The standard error of the sample is 8.3% and therefore there is a considerable variation in the figures of firewood usage.

The figures that are given by the Housing Report presented in Table IV, are considered to be the minimum numbers of rural households that use firewood as a primary source for heating and cooking.

Three percent of all rural households use wood as a source for space heating, water heating and cooking (or 3% of 35% of all households). Two percent of the rural households use wood as the primary source, one percent as the additional source. This amounts to the following consumption per year for primary use:

$$11.78 \times 10^{15}\ \text{Btu}^{(a)} \times 0.35^{(b)} \times 0.02^{(c)} \quad \frac{0.8^{(d)}}{0.3^{(e)}} = 0.22 \times 10^{15}\ \text{Btu} \tag{1}$$

where

(a) total residential energy consumption,
(b) fraction of homes in rural areas,
(c) fraction of rural homes using wood for primary energy source,
(d) assumed fossil conversion efficiency, and
(e) assumed maximum average wood conversion efficiency.

TABLE II. Fuel Utilization in the U.S., 1978 (including electricity)

Sector	Energy Consumption (Quads)[a]
Industrial	28.1
Commercial/Residential	29.3
Transportation	20.6
Total	78.0

[a]Source: Monthly Energy Review, DOE, March 1979.

TABLE III. Primary Energy Consumption by End-Use in American Households in 1973 and 1978.

Sector	Direct fuel	Elec- tricity	Total 1973	Total 1978
Residential Total	7.88	1.97	14.07	16.7
Space heat	6.16	0.32	7.19	8.5
Water heat	1.26	0.28	2.13	2.53
Air conditioning	-	0.32	1.00	-
Refrigeration	-	0.38	1.18	-
Cooking	0.38	0.08	0.63	0.75
Lighting	-	0.26	0.82	-
Drying	0.09	0.08	0.34	-
Other electrical	-	0.25	0.78	-

TABLE IV. Number of Households Where Firewood Is Used for Secondary or Primary Heating Purposes

Region	No. of Rural Households	Percentage as Wood Users		No. of Wood Consumers	
		Heating	Cooking	Heating	Cooking
Northeast	3,425,316	1.54	0.59	52,750	20,209
North Central	5,746,751	1.42	0.03	81,603	17,240
South	7,948,359	4.4	1.00	349,728	79,484
West	2,124,929	5.1	0.78	108,371	16,574
Total	19,245,355	3.1	0.69	592,452	133,507

Source: Housing Report, 1976.

Secondary or supplemental energy consumption can be estimated as follows:

$$8.5 \times 10^{15}\ \text{Btu}^{(a)} \times 0.35^{(b)} \times .01^{(c)} \times .25^{(d)} \times \frac{0.8^{(e)}}{0.15^{(f)}} = 0.04 \times 10^{15}\ \text{Btu} \quad (2)$$

where

(a) total residential heating load,
(b) fraction of U.S. homes in rural areas,
(c) fraction of rural homes using wood heat for supplementary purposes,
(d) assumed % of heating load supplied by wood,
(e) efficiency of fossil conversion, and
(f) maximum average efficiency assumed for wood systems.

The consumption for primary and secondary heating and cooking purposes in rural areas is 0.26 quads/year. Consumption for primary and secondary use in urban areas is neglected in these estimates.

These estimates, however, are derived from data on energy needs per household in 1978, and an estimate of the number of firewood consuming households in 1976. Since that year many stoves have been sold and wood fuels have become increasingly competitive when expressed in cost per million Btu. Studies of elasticities of demand for different fuels cannot be applied to wood fuels because of the unknown effect of inconveniences that are involved in burning wood fuel in comparison to gas, oil, or electricity (Leis *et al.*, 1978). Thus it is not clear how much higher consumption in 1978 will be over the 1976 consumption. The estimates presented, therefore, are conservative.

2. *An Estimate of Residential Energy Needs.* The energy needs of the household, given insulation characteristics and outside temperatures, can also be computed. The consumption is done here for the average rural American house. The flux of energy is necessary to make up for heat losses to the environment.

The effectiveness of space heating can be measured by the heat loss to the surrounding atmosphere. Heat is lost from dwellings in two ways.

(1) Heat transmission through walls, ceilings, floors, doors and windows.

(2) Air infiltration into the apartment through cracks around windows and doors and as a result of opening doors.

The perceived comfort zone for room temperature has been established at ~70°F. This becomes the datum temperature for heat loss analysis. Temerature extremes between the outside temperature and the desired inside temperature require that the inside temperature be heated or cooled. The Housing Survey (1976) states that houses in rural areas, the ones of primary concern for our study, are generally better insulated than houses in urban areas. This is not further quantified in that report but it is assumed here that houses in rural areas are insulated to conform to standards set by the Federal Housing Authority (FHA pamphlet No. 51a). The FHA has set the maximum allowable heat loss at 35 Btu/ft^2/hour, over the expected temperature range Δ t.

Estimates are made assuming a standard average single family dwelling of 1500 square feet. The allowable heat loss for a 1500 square feet house is expressed as: The total heat loss through ceilings and walls and must be less than Q_t, where Q_t equals the maximum allowable heat loss × ft^2 living space of the average rural household. $Q_t = 35 \times 1500 =$ 52,500 Btu/hour. The heat loss rate for a 1500 ft^2 house is as follows:

$$Q_t \div t = \frac{1500 \times 35}{\Delta t_{max}} \quad \text{(see Table V).} \tag{3}$$

When Q_t, the heat loss taken is multiplied by Degree Days Heating one then obtains the annual heat loss in Btu/year per average single family dwelling for that particular region (Gordon 1977). In Table V, total annual heat losses for average houses in the four regions are shown as well as for Vermont and New Hampshire units. These are heat losses of houses of 1500 ft^2 living space and where all the rooms are heated.

When fuelwood is used as main or only source of energy, it is likely that the inhabitants are more aware of energy use and are more frugal than other households and will have:

(1) better insulation than the average house;
(2) only those places heated that are frequently used, such as kitchen and living room.

New Hampshire and Vermont data on residential firewood usage are available (Dalton 1977; Action Research 1976). A comparison is made possible by this data between the energy needs of an average household and one where firewood is the primary heating source. The ratio between the two is the constant here, introduced to correct the heat loss values of Table V.

TABLE V. Regional Heat Loss Rates and Energy Needs for Average Fixed Single Family Dwellings

Region	T_{min}[a] °F	Δt °F[b]	$Q_t/\Delta t$ Btu/hr/°F[c]	DDH	MM Btu/yr/house[d]
Northeast	-15	85	617.6	6926	100
South	5	65	807.7	3020	85
North Central	-30	100	525	6774	78.2
West	-15	85	617.6	5276	58.5
New Hampshire	-30	100	525	10600	133.5
Vermont	-25	95	552.5	8269	109

[a] Source: Climatic Atlas of the U.S.

[b] Δt: Is the difference between minimum expected outside temperature and inside room temperature of 70°F.

[c] Q_t = (heat loss rate) × (Degree Days Heating/year) × (24 hours)

[d] Averages for each region or state weighted by size of the rural population in the area.

The energy supplied by firewood is then calculated by the following formula:

$$\frac{AHL \times K_i \times RH}{E_w} = C \qquad (4)$$

where:

AHL = annual heat loss (Btu)
K_i = insulation constant
RH = number of rural households using wood
E_w = wood efficiency
C = energy consumed (Btu)

K_i is defined as insulative effectiveness and is calculated as follows:

$$K_i = \frac{HL_w}{HL_a} \qquad (5)$$

where HL_w = heat loss rate of homes heated by wood
HL_a = heat loss rates of average homes

For New Hampshire the value is .63 and for Vermont it is .77. Estimates for these states are presented in Table VI.

While these data are area specific, it may be used generally to forecast wood energy consumption, recognizing that the project is, at best, imprecise. Table VII presents such estimates by region using formula (4) as previously presented.

Ellis estimates the total use of firewood for aesthetic use as .16 cords per household, an average for 25 million households (Ellis, 1978). Table VIII summarizes the total residential wood fuel consumption in the U.S., combining heating, cooking, and aesthetic consumption. Residential wood fuels use in the U.S. is at least 0.2 quads/year. This figure is based on the energy need for heating, cooking, and aesthetic purposes. It also could approach 0.34 Quads/year. The range, then is 9.5 - 16.2 $\times 10^6$ cords/year at 21 $\times 10^6$ Btu/cord. Table IX presents the estimated contribution of the U.S. Forest Service permit program to this total as of 1977.

It seems reasonable from the above data that total U.S. residential wood fuel consumption was between 0.20 and 0.34 quads, or 9.5 $\times$ 16.2 $\times 10^6$ air dried cords per year in 1978. Our estimated probable consumption is 0.3 quads, or 14.3 $\times 10^6$ cords. National forests are supplying 15 - 25% of the firewood consumed.

TABLE VI. Estimates of Firewood Use in New Hampshire and Vermont[a]

	New Hampshire	Vermont
1. Number of households using wood	120,000	300,000
2. Amount of wood used on average for primary heating	4 cords[b]	4 cords
3. Average amount in all houses including secondary and cooking fuel	2.5 cords	1.3 cords
4. Total number of cords per year[c]	300,000	390,000

[a] Sources: Dalton, 1977; Action Research 1976.

[b] A heating value of 21 × 10^6 Btu per cord is assumed.

[c] Line 1 × line 3 = line 4.

TABLE VII. *Energy Demands of U.S. Households, 1978*

Region	*Annual energy demand per house (10^6 Btu)*	*Insulation constr.*	*Number of rural households using wood*	*Assumed stove efficiency*	*Range of wood energy used (10^{12} Btu)*
Northeast	*100.0*	*.63-.77*	*52,750*	*0.3*	*11.1-13.5*
North Central	*78.2*	*.63-.77*	*81,603*	*0.3*	*13.4-16.4*
South[a]	*85.0*	*.63-.77*	*349,728*	*0.3*	*62.4-76.3*
West	*58.5*	*.63-.77*	*108,371*	*0.3*	*13.3-16.3*
Total	-	*.63-.77*	*592,452*	*0.3*	*100.2-122.5*[b]

[a] *Sample calculation:* $\frac{85 \times 10^6 \times .77 \times 349{,}728}{.03} = 76.3 \times 10^{12}$

[b] *At 21 × 10^6 Btu/air dried cord, this represents 4.8 - 5.8 × 10^6 cords.*

TABLE VIII. Estimates of Wood Fuel Use in American Households, 1978 (in 1×10^{15} Btu)

Wood fuels for	*By energy use*	*By energy demand*
Heating and cooking (Btu)	*0.26*	*0.12*
Aesthetic (Btu)	*0.08*	*0.08*
Total	*0.34*	*0.20*

B. *Energy Consumption in the Pulp and Paper Industry*

The forest products industry includes manufacturing operations producing pulp and paper, lumber, plywood, particleboard, and other secondary products. Typically manufacturing complexes often include pulp, lumber, and plywood mills in integrated fashion. Typically the pulp mills consume more Btu than they produce as wood fuels, while sawmills produce a surplus of fuel. Despite the integrated nature of many mills, energy accounting is handled separately here.

The pulp and paper industry consumes more wood based fuel than all other economic sectors combined. Because the pulp and paper industry is one of the five largest energy consuming manufacturing industries (Conference Board) detailed statistics are kept by the American Paper Institute. Table X presents detailed estimates of energy consumption in pulp and paper manufacture for 1977.

Projections for 1978 may be made by following one of the three trend lines in Figure 1, a tracing of wood fuels consumption in the pulp and paper industry. The highest projection is derived by employing the 1975-1977 trend; the lowest is obtained by employing the (statistically insignificant) 1971-1977 trend. The median estimate used here follows a 1971-1977 trend line but disregards the years 1972 and 1975. The range of estimates varies by 110×10^{12} Btu (10.2% of the 1978 estimate), lending considerable confidence to the value posited. The probable value selected is 1.01×10^{15} Btu with high and low boundaries being 1069 and 959×10^{12} Btu, and hogged bark and wood supplied $\sim 187 \times 10^{12}$ Btu.

Regionalization of wood consumption in the pulp and paper industry assumes continuation of the pattern extant in 1975 when, as Table XI shows, the south dominated pulp and paper capacity, and wood fuel utilization. From these factors

TABLE IX. *Firewood Survey of the National Forest in Fiscal Year 1977*

Regions	Number of firewood permits issued fiscal year 1977	Estimated volume of fuel collected by all permit holders, etc.		
		10^6 bd. ft. equivalent roundwood	10^6 Cords	Energy equivalent if air-dried (10^{12} Btu)
All regions	436,000	1030	2.4[a]	50.4[b]
Region 1	27,000	99	0.2	4.2
Region 2	61,000	46	0.1	2.1
Region 3	71,000	125	0.3	6.3
Region 4	77,000	157	0.4	8.4
Region 5	56,000	165	0.4	8.4
Region 6	81,000	245	0.6	12.6
Region 8	41,000	114	0.3	6.3
Region 9	21,000	78	0.2	4.2
Region 10	1,000	1	Negl.	Negl.

[a] $$1030 \times 10^6 \text{ bd ft} \times \frac{1 \text{ ft}^3 \text{ W}}{6 \text{ bd ft}^*} \times \frac{1.12 \text{ W} + \text{B}^{**}}{1.0 \text{ W}} \times \frac{1 \text{ cord}}{80 \text{ ft}^3} = 2.4 \times 10^6 \text{ cords}$$

*6 bd ft accounts for losses in lumber processing.

**1.12 wood and bark accounts for bark used in fuel (W = wood; W + B = wood plus bark).

TABLE X. 1977 Energy Consumption in the Pulp and Paper Industry

Fuel	*Contribution (10^{12} Btu)*
Hog fuel	*86.1*
Bark	*94.4*
Spent liquors	*797.0*
Total wood fuel	*977.5*
Fossil fuels, electricity, and other	*1197.2*
Total all fuels	*2174.7*

Source: Duke (1978).

TABLE XI. Regional Patterns of Capacity and Fuel Utilization

Region	*% of paper and board capacity*	*% of pulp capacity*	*% of industry fuels supplied by wood*
Northeast	*18.0*	*8.7*	*20.5*
North Central	*19.4*	*8.9*	*14.2*
South	*48.6*	*64.0*	*50.0*
West	*14.0*	*18.4*	*50.4*
Total	*100.0*	*100.0*	*45.0*

Source: Duke and Fudali (1976).

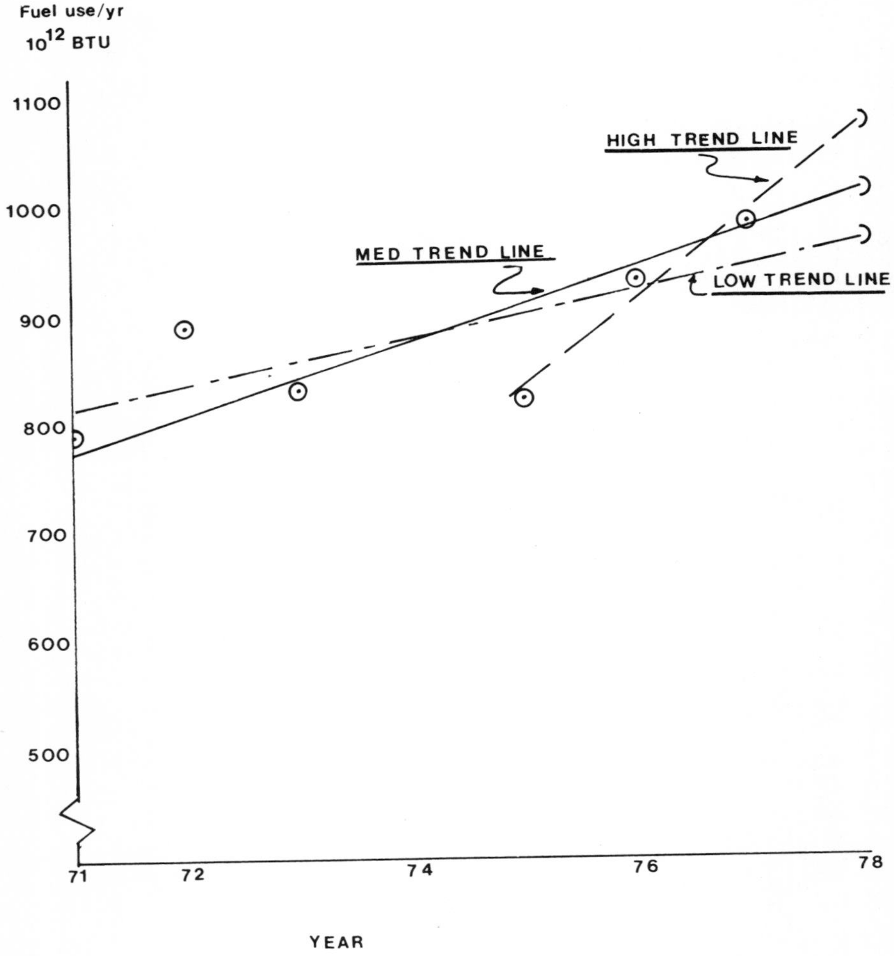

FIGURE 1. Projection trends for wood fuels in pulp and paper manufacturing.

crude regional estimates of wood fuels consumption are as follows: Northeast, 56.1×10^{12} Btu; North Central, 50.2×10^{12} Btu; South, 700.0×10^{12} Btu; and West, 203.2×10^{12} Btu.[1]

C. *Energy Consumption in Lumber, Plywood, Particleboard, and Related Products*

While data for pulp and paper manufacturing is well developed, data for lumber, plywood and related products is less adequate. The U.S. Census Bureau does collect purchased energy statistics, however; and its data is within 6.9% of the API data for pulp and paper, with the Census Bureau having the higher estimates. Table XII presents the 20-year trend of purchased energy consumption in the wood products industry (not including logging). The trend is reproduced in Figure 2. Significant to note, Jamison, Methven, and Shade (1978) present fossil purchases of 115×10^{12} Btu for 1971--2.2% higher than those in the census data.

Jamison, Methven, and Shade (1978) estimate wood fuel use in the wood products industry at 250×10^{12} Btu, for a purchased fuel/wood ratio of 1:2.17 in 1971. For two reasons this is considered to be a conservative ratio. Purchased fossil fuel prices were low at that time, discouraging large investments in wood fired equipment. Domestic crude oil prices ranged from \$3.39/bbl (well-head), and natural gas sold, at the well-head, for $18.2¢/10^3$ ft^3 (or $\$0.182/10^6$ Btu) in that year (Bureau of Mines, 1974). Further, sawmills produce more fuel than they can consume (Boyd *et al.*, 1976).

Given that ratio, estimates for wood fuel consumed in 1976 are 325×10^{12} Btu. A high 1978 estimate is developed by a linear extrapolation of the fossil and wood fuel trends. It assumes no energy conservation. A low estimate is obtained by using the 1976 estimates unchanged. The median estimate is the arithmetic average of the two. These are presented in Table XIII. Based on the 1976 census the regional consumption is as follows: Northeast, 20.6×10^{12} Btu; North Central, 38.8×10^{12} Btu; South, 144.9×10^{12} Btu; and West, 132×10^{12} Btu. This assumes mid-case totals.

[1]*Sample calculation:* $\frac{180 + .087}{2} \times .205 \times 2050 = 56.1$ *where* 2.05×10^{15}.

Btu is the assumed total fuels consumption of the industry exclusive of self generated hydro. Data from the American Paper Institute (1977) shows a relatively even distribution of energy between pulping and paper making. This is confirmed by Gyltopoulos (1976).

TABLE XII. Purchased Fuels in Forest Products Industry by Type and Year (in 10^{12} Btu)

Year	Fuel oil[a]	Coal[b]	Natural gas[c]	Total
1976	NC[d]	NC	NC	150.1
1971	37.0	4.4	71.1	112.5
1967	20.8	6.4	54.4	81.6
1962	22.3	7.8	24.5	54.6
1958	13.3	6.1	6.9	26.3

[a]Assumes 5.8×10^6 Btu/bbl.

[b]Assumes 24×10^6 Btu/ton.

[c]Assumes 1×10^3 Btu/ft^3.

[d]Not calculated due to incomplete data published by U.S. Bureau of Cenus.

TABLE XIII. Wood Fuel and Purchased Fuel Estimates for 1978 in the Wood Products Industry (10^{12} Btu)

Fuel type	Low estimate	Medium estimate	High estimate
Wood	325	337	350
Purchased fossil fuel	150	115.5	161

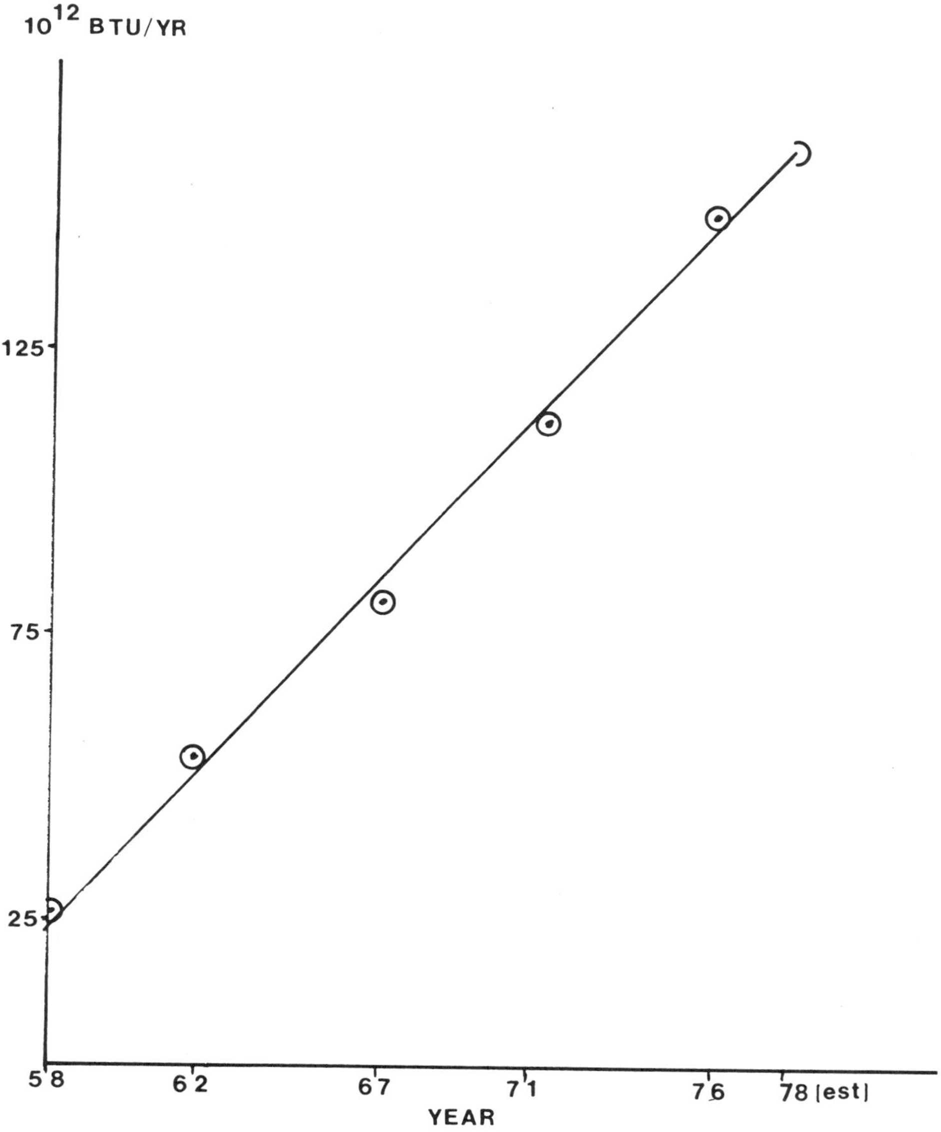

FIGURE 2. Annual fossil fuel purchases in lumber and board manufacturing.

D. Wood Fuel in Electric Utilities

Estimates of wood fuel utilization in electric utilities are based on data from the Federal Power Commission on power produced in 1978. Power plants fueled by wood are assumed to have a heat rate of 15×10^3 Btu/kW·h. That in turn assumes steam at about 850 psia/850°F throttle, condensing to 2" Hg, and a water rate of 8.1 lb/kW·h. This assumption is consistent with Evans (1974) in studying fuel farms for Canada. Further it is consistent with efficiencies quoted by Hill (1977) after adjusting for hog fuel boiler efficiencies relative to fossil fuels.

Six wood fired power plants in Vermont, Minnesota, Montana, Oregon, and Wisconsin generated a total of 66×10^6 kW·h in 1978. This represents the consumption of 0.99×10^{12} Btu in wood. That trillion Btu is equal to 124×10^3 green tons (50% moisture) or 61.9×10^3 oven dry (O.D.) tons of hog fuel. This estimate does not include the cogeneration facility of the Eugene (Oregon) Water and Electric Board, operated with Weyerhaeuser, and producing 131×10^6 kW·h in 1978. The fuel for that facility was included in the pulp mill calculations presented previously. Thus the 61.9×10^3 O.D. ton estimate is achieved by the following calculations:

$$\frac{66 \times 10^6 \text{ kW·h}}{} \times \frac{15 \times 10^3 \text{ Btu}}{\text{kW·h}} \times \frac{1 \text{ O.D. ton}}{16 \times 10^6 \text{ Btu}} = 61.9 \times 10^3 \text{ O.D. tons}$$

This estimate represents a lower bound. An upper bound is obtained by employing the heat rate cited by Methven (1978): 20×10^3 Btu/kW·h. Assuming the less efficient heat rate, some 1.32×10^{12} Btu or 82.5×10^3 O.D. tons, annually, are used. In either case the quantity of wood fuel consumed in utility applications is quite small.

E. Wood Fuel Use in Other Industrial/Commercial Applications

Data on wood fuel consumption in secondary wood products manufacturing (e.g., furniture and fixtures) and other nonwood industries are sketchy at best. The Ultrasystems (1978) boiler inventory is used here. Table XIV is a listing of such boilers by SIC code and capacity range. SIC code 25 is the Furniture and Fixture industry.

Any number of assumptions may be made regarding thermal efficiency, operating hours, and the actual capacity of these boilers. Certainly furniture industry boilers operate at

TABLE XIV. Wood and Bark Fired Boilers by Industry and Capacity

Industry (SIC)	Boiler capacity (in 10^6 Btu/hr)			
	0-50	51-100	101-150	151-200
13	1	0	0	0
20	12	3	1	0
22	1	0	0	0
23	2	0	0	0
25	235	17	0	2
28	7	0	3	0
30	2	1	0	0
31	4	0	0	0
32	2	0	0	0
33	4	0	0	0
34	3	2	0	0
35	2	0	0	0
37	4	0	0	0
39	11	1	0	0
45-97	22	0	0	0
Total	312	24	4	2

higher efficiencies than hog fuel boilers under optimum conditions. The fuel is dry. However, as waste disposal mechanisms, they may not be operated in optimum modes. Permitting these boilers to operate at near-fossil efficiencies while providing renewable fuels to other industries is one of the few useful applications of pelletization technology. In order to make estimates the following assumptions were made:

(1) That the mid point of each range represents the actual capacity estimate. Thus these boilers have a combined total capacity of 10.5×10^9 Btu/hr output. Of this total, the furniture and fixture industry has a total capacity of 7.50×10^9 Btu/hr.

(2) The thermal efficiency of nonfurniture plant boilers is 65%; the efficiency of furniture plant boilers is 80%. On this basis the hourly fuel consumption totals 4.62×10^9 Btu for all but the furniture industry; and 9.38×10^9 Btu/hr for furniture manufacturers.

(3) For a lower bound it is assumed that all boilers are operated at 2000 hrs/yr. Total fuel consumed is 28.0×10^{12} Btu/yr (1.75×10^6 O.D. tons) under this assumption. Of this total, 18.8×10^{12} Btu were consumed in the furniture industry, and 9.2×10^{12} Btu in other firms.

(4) For an upper bound, the boilers were all assumed to operate 6000 hrs/yr (3 shifts/day). The total consumption was 84×10^{12} Btu, with 56×10^{12} Btu being used in the furniture and fixture industry.

To the totals, 27×10^{12} Btu may be added in order to include ferroalloy uses of hog fuel, and charcoal use in residential and restaurant applications (see Tillman, 1978; Bethel *et al.*, 1979).

F. Consumption Technologies

It is important to note that this 1.7 quad consumption rate represents virtually complete utilization of mill residues. The quantity of wood fuel consumed can increase modestly over time, as lumber and pulp production increase. For dramatic changes in wood fuels consumption to occur, however, the more expensive forest residues must be utilized.

Given this position on the marginal cost or supply curve, it is also important to consider the technologies currently being employed. To date, various direct combustion systems have been deployed raising process steam and heat; and some quantities of electricity are generated in both stand-alone condensing power plants and cogeneration facilities. The deployment of additional condensing power and cogeneration systems, gasifiers fueling retrofitted gas fired boilers and other facilities, and large scale liquefaction facilities will cause a shift in the demand function. This shift will again imply the use of higher cost forestry residues, raising the price of wood fuel to all industrial consumers in the local market area. Within that context it is important to note that the cogeneration systems, and secondarily the process heat and steam systems, can afford to pay more for wood fuels than can the other conversion facilities (Bethel *et al.*, 1979).

IV. CONCLUSIONS

Table XV is a summary of the estimated 1978 wood fuels consumption by industry and by estimate class. The middle estimate is an arithmetic average of the upper and lower bound, with the exception of the pulp and paper estimate. The middle estimate is some 1.7 quads or 1.7×10^{15} Btu/yr of wood fuel consumed ($\pm 1.2 \times 10^{15}$ Btu). This is equivalent to 106×10^6 O.D. tons of wood; however, its form includes green and dry cordwood, hog fuel, black liquor, and such dry fuels as planer shavings and sander dust, and charcoal. If one converts the total number to cords at 21×10^6 Btu/cord, annual consumption is equivalent to 81×10^6 A.D. cords; however this number is also misleading.

A. *Relation to Previous Estimates*

Both the University of Washington College of Forest Resources (Bethel *et al.*, 1979) and Tillman (1978) put 1976 consumption at 1.6×10^{15} Btu. The Bethel estimate put hog fuel consumption at 463×10^{12} Btu, secondary manufacturing wood fuel utilization at 50×10^{12} Btu, spent liquor use at 979×10^{12} Btu, household fuelwood consumption at 260×10^{12} Btu, and charcoal use at 15×10^{12} Btu. It is essentially comparable to the above estimate, and from a comparison of the two, certain conclusions can be drawn:

(1) There are very small increases in wood fuel use generally;
(2) Gains in residential, furniture, and nonwood manufacturing sectors have been small indeed; and
(3) The bulk of the 0.1 quad gain has been in the forest products industry, about evenly split between hog fuel and spent liquor.

This indicates that any growth in the use of wood fuel is bound to be slow. This low rate of growth can be attributed to the near 100% use of hog fuel from mill residues now, and the high cost of obtaining forest residues (Bethel *et al.*, 1979). Growth in the use of mill residues, at this point, has to be driven by the demand for lumber; sawmills produce more hog fuel than they consume (Boyd *et al.*, 1976). Growth in use of pulping liquors is driven by pulp production.

Recent price increases in oil and national policy making coal the marginal fuel for industrial applications may drive wood fuel utilization past the 2×10^{15} Btu level in the near

TABLE XV. Estimate of Total Wood Fuels Consumed, 1978 (in 10^{12} Btu)

Economic sector	Case		
	Low	Medium	High
Residential	200	270	340
Pulp and paper	960	1010	1070
Lumber, plywood, and related board prod.	320	340	350
Furniture and fixture	20	40	60
Utilities and all other	40	50	60
Total	1540	1710	1880

future. This would be caused, in no small part, by the energy cost of lumber and paper relative to the energy cost of their substitutes.

B. *Future Consumption Forecasts*

In order to forecast adequately wood fuels consumption (hence resource needs), localized marginal cost curves are needed for forest residues. Only one such curve has been reported to date, and it is for the Oklahoma-Arkansas area (Jamison, Methven, and Shade, 1978). Similar curves are needed for such areas as Mississippi-Louisiana, North Carolina, Washington-Oregon (west side and east side), and northern New England. Forecasts of industrial activity are also necessary.

Equally important to energy forecasting are techniques to improve efficiency of wood fuel use, industrially and residentially. Within the industrial sector this means acquiring more fundamental knowledge concerning particle size distributions and moisture content distributions within the mass of hog fuel. It further requires the development of a body of knowledge concerning how such processes as sawlog debarking or breakdown influence the fuel characteristics.

Already it can be seen that such technologies as cogeneration are making gains in the forest products industry. These are efficiency improving systems. More detailed technical knowledge concerning the physical and chemical nature of various wood fuel fractions, and the amenability of those fractions to fuel preparation processes, would facilitate forecasting fuel wood needs.

Finally forecasting technological advances in processes for manufacturing lumber, plywood, pulp, and paper would enhance our ability to forecast wood energy consumption and fuel wood (resource) needs.

REFERENCES

Residential Wood Fuel

Action Research, Inc. 1976. Final report on the use of wood as a heat source and the quality of insulation in Vermont households. Prepared for the State of Vermont Energy Office, Burlington, Vermont.

Dalton, M., and others. 1977. Household Fuel-wood Use and Procurement in New Hampshire. New Hampshire Agric. Exp. Sta. Univ. of New Hampshire, Durham. New Hampshire Research Rep. Mc. 5g. Ock.

De Angelis, D. G. and Renwith, R. B. 1978. Source Assessment, Residential Combustion of Wood. Monsanto Research, Dayton, Ohio. Prepared by the U.S. Environmental Protection Agency. July.

Dell, J. 1978. Utilization of Forest Residues for Firewood on the National Forest of Oregon and Washington. Forest Services Region 6. Fuel Management Notes, February 16.

Department of Energy. 1976. Energy Data Reports. Annual Fuel Oil Sales. Edison Electric Institute, 1975. Statistical textbook of the electric industry for 1974. Mo 75-39 November.

Ellis, T. H. 1975. The Role of Wood Residue in the National Energy Picture, Wood Residues as an Energy Resource. Forest Products Research Society, Madison, Wisconsin.

Gordon, H. 1977. Perspectives on the Energy Crises. Ann Arbor Science, Michigan.

Keays, Y. L. 1975. Biomass of Forest Residuals. Am. Inst. of Chem. Eng. Symp. Series 71 (146).

Kennel, R. P. 1977. Wood Energy-commercial Application Outside the Wood Industry. MASA Conference Publication 2042.

Leis, W., Hirst, E., and Cohn, S. 1976. Fuel Choices in the Household Sector. Oak Ridge Mak. Lab., Tennessee. Energy Research and Development Administration.

Livingston, R. S., and McNeill, B., eds. 1975. Beyond Petroleum. Institute for Energy Studies, Stanford.

Meyer, W. W. 1979. Personal communication with W. W. Meyer from Meyer Engineers, Kirkland, Washington.

Ross, M., and Williams, R. H. 1975. Assessing the Potential for Fuel Conservation. Institute for Public Policy Alternatives, State University of New York, Buffalo, New York.

Sliven, R. Y. 1972. A Survey of Fuel and Energy Use in the U.S. Pulp and Paper Industry, 1971 through 1975. American Paper Insitute, New York.

U.S. Department of Commerce. 1970. Characteristics of the Population, Vol. 7. Social and Economic Statistics Admin. Bureau of the Census.

U.S. Department of Commerce. 1976. Current Housing Reports. Series H-150-76. Annual Housing Survey, Part E.

U.S. Department of Commerce. 1970. Climatic Atlas of the U.S. U.S. Government Printing Office.

U.S. Department of Energy. 1979. Monthly Energy Review, March.

U.S.D.A. Forest Service. 1973. The Outlook for Timber in the U.S. Forest Resource Report No. 20. Washington, D.C.

Watson, D. 1977. Designing and Building a Solar Home. Garden Way Publ., Charlotte, Vermont.

Industrial Wood Fuel

Bethel, J. S., and others. 1979. Energy from Wood. A Report to the Office of Technology Assessment. College of Forest Resources, University of Washington, Seattle, Washington.

Boyd, C. W., and others. 1976. Wood for Structural and Architectural Purposes. Wood and Fiber 8(1). Special CORRIM Panel II Report.

Bureau of Mines, U.S. Department of the Interior. 1975. Minerals Yearbook, Vol. 1: Metals, Minerals and Fuels. U.S. Government Printing Office, Washington, D.C.

Bureau of the Census. 1972. Fuels and Electric Energy Consumed, 1972. Census of Manufacturers. U.S. Government Printing Office, Washington, D.C.

Bureau of the Census. 1976. Fuels and Electric Energy Consumed, Annual Survey of Manufacturers, 1976. U.S. Government Printing Office, Washington, D.C.

Duke, J. M., and Fudali, M. J. 1976. Report on the pulp and paper industry's energy savings and changing fuel mix. American Paper Institute, New York.

Duke, J. M. 1978. Report to the Department of Energy on Energy Conservation: Progress in the Pulp, Paper and Paperboard Industry in 1977. American Paper Institute, New York.

Enzer, H., Dupree, W., and Miller S. 1975. Energy Perspectives. U.S. Department of the Interior, Washington, D.C.

Evans, R. S. 1974. Energy Plantations: Should We Grow Trees for Power Plant Fuel. Western Forest Products Laboratory, Vancouver, B.C.

Federal Power Commission Form 4, Monthly Power Plant Report, Summary for 1978.

Gyftopoulos, E., and others. 1976. A Study of Improved Fuel Effectiveness in the Iron and Steel and Paper and Pulp Industries. Thermo Electron Cor., Waltham, Mass. (NSF Grant 7422046).

Hill, P. G. 1977. Power Generation: Resources, Hazards, Technology, and Costs. MIT Press, Cambridge, Mass.

Jamison, R. L., Methven, N. E., and Shade, R. A. 1978. Energy from Forest Biomass. National Assoc. of Manufacturers, Washington, D.C.

Methven, N. E. 1978. Existing Cogeneration Facilities. Presented at Increasing Energy from Biomass: 1985 Possibilities and Problems, the Pacific Northwest Bioconversion Workshop, October 24-26, Portland, Oregon.

The Conference Board. 1974. Energy Consumption in Manufacturing. Ballinger Publishing, Cambridge, Mass.

Tillman, D. A. 1977. The Contribution of Nonfossil Organic Materials to U.S. Energy Supply. Materials Associates, Washington, D.C. (FEA Contract No. P-03-77-4426-0).

Tillman, D. A. 1978. "Wood as an Energy Resource." Academic Press, New York.

Ultrasystems Incorporated. 1978. Feasibility Study for a National Wood Energy Data Base. Final Report to U.S. Department of Energy, Contract E1-78-X-01-4951. McLean, Virginia.

SUGAR STALK CROPS FOR FUELS AND CHEMICALS

E. S. Lipinsky
S. Kresovich

Battelle's Columbus Laboratories
Columbus, Ohio

ISBN 0-12-535902-0

I. INTRODUCTION

The need to supplement or supplant fossil resources with renewable resources for the production of fuels and chemicals now is well established. The major issues concern the choice of renewable resource, the desirable end products, and the technology to proceed from the resources to the end products. The sugar stalk crops merit discussion because they produce high yields of simple carbohydrates that are directly fermentable at low nutrient uptake, and the stalk residue (bagasse) has considerable potential as a fuel and/or chemical feedstock in itself.

It is the purpose of this chapter to place the sugar stalk crops into perspective as sources of fuels and chemicals so that those contemplating development of renewable resources can determine in which geographical areas sugar stalk crops may be the most desirable and which agricultural and chemical process technologies may be applied to convert these resources into fuels and chemicals.

The scope of this chapter is confined to sugarcane and sweet sorghum and their close relatives which are collectively designated as sugar stalk crops. The major sugar crop omitted thereby is the *Beta vulgaris* (sugar beets and fodder beet). Beets deserve separate treatment because of two important differences from sugar stalk crops--they are much more storable (McGinnis *et al.*, 1971) but lack the built-in fuel that the stalk provides (Lipinsky *et al.*, 1977).

The chemicals that are within the scope of this chapter are those that have been made from fossil resources. Therefore, citric acid, monosodium glutamate and similar chemicals are not discussed. Although economic parameters and requirements are discussed in this chapter, quantitative estimates of the cost of producing fuels and chemicals from sugar crops is so dynamic that newspapers rather than books are the appropriate medium for discussion of that topic. The problem is not meeting a specific cost target but one of achieving a position in an overall system.

This chapter is organized as follows. A systems overview is presented to put the sugar stalk crops into perspective as biomass resources. Then these crops are described and their processing and conversion to chemicals and fuels are discussed. The fuels and chemicals possibilities then are described. The chapter closes with a few conclusions, emphasizing policy implications.

II. SYSTEMS OVERVIEW

The factors of production are transformed into the products demanded by the economy by the application of appropriate technologies, as shown in Figure 1. Although this model has some serious deficiencies (see below), it does highlight some important considerations. For example, it is clear that renewable resources are not free goods; they are created from factors of production at costs that may be far from negligible. It is also clear that the degree to which these resources can be created and used effectively and economically depends on the state of development of diverse technologies. This chain of technologies is only as strong as its weakest link and the weakest link may not be a popular technology for support by industry or government. Capital is formed when the system functions effectively.

The major weakness of the classical view of resource systems is that insufficient attention is given to the feedback attributes of such systems. Figure 2 is an illustration of the application of feedback principles to sugar stalk resource systems. Unlike large scale corn grain conversion facilities which use unit trains of grain brought in over long distances, it would be rare for the sugar stalk raw material to be transported more than 32 km to reach a processing facility (Irvine, 1977). This makes possible the back hauling to the farm of unneeded trash obtained in the separation process and of potash fertilizers obtained by stillage processing. The steam and electricity for operation of the mill and the distillery are generated at the site from bagasse or trash. Production of the steam and electricity from the biomass resource makes this system much more energy independent than is conventional corn grain conversion (Jenkins *et al.*, 1979). If small scale ammonia plants could be developed, nitrogen fertilizer might be obtained from this source at the site also.

Up to now, the assumption has been made that the sugar stalk crop system is exclusively a fuels and chemicals system. This assumption is not warranted because these crops are grown for food uses now and any use for fuel or chemicals must take into account alternative opportunities in other end use products. The price of raw sugar derived from sugarcane and sugar beets has fluctuated widely in recent years, reflecting weather conditions, competition with corn sugars, and other factors. A viable fuels/chemicals sugar stalk crop system must deal effectively with this issue. An effective approach is to build an adaptive system (Lipinsky, 1978) in which variable quantities of food, and nonfood products can be made, depending on the demand of the end use markets (Figure 3). The Brazilian sugar-alcohol industry illustrates how the technology evolves

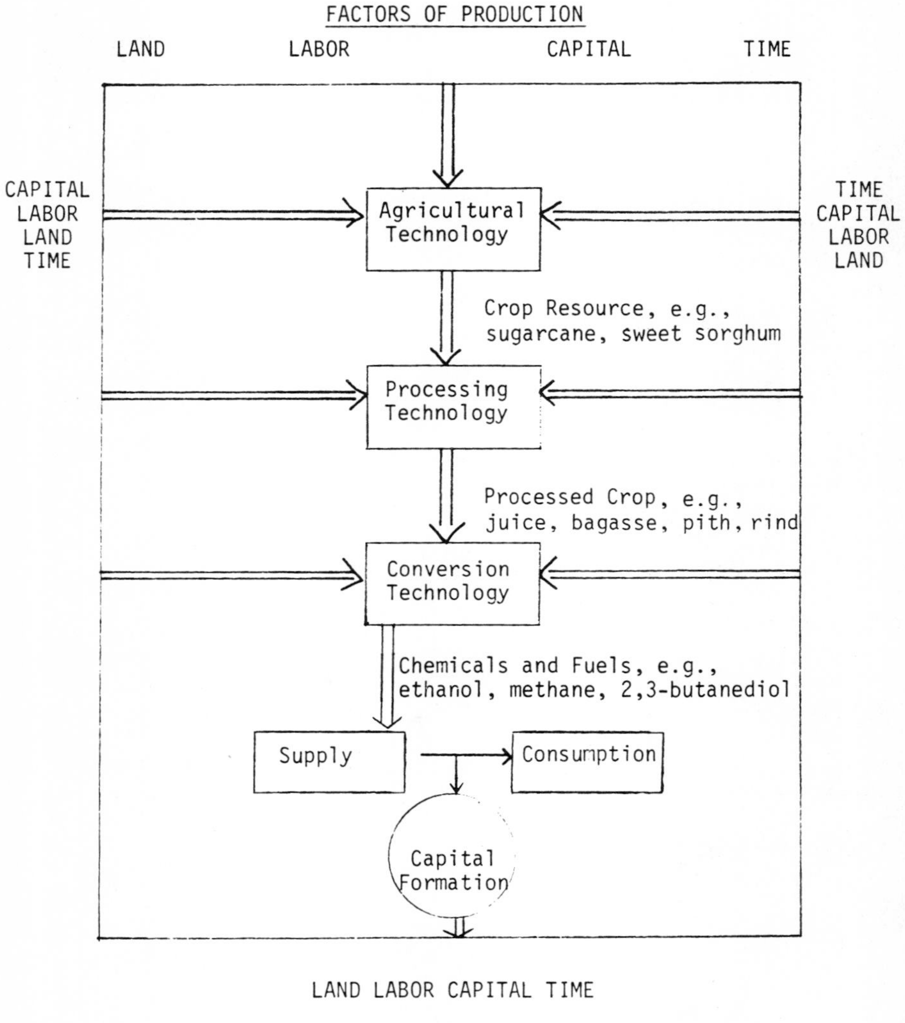

FIGURE 1. Transformation of factors of production into chemicals and fuels.

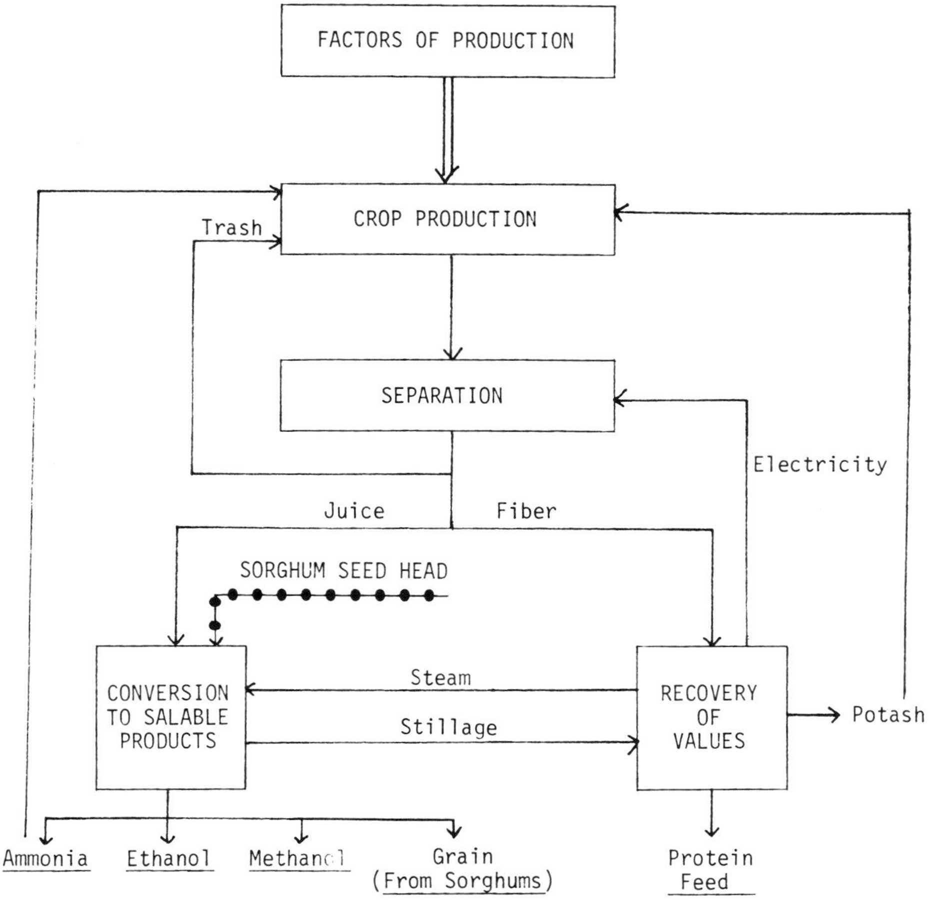

FIGURE 2. Feedback in a sugar stalk crop system.

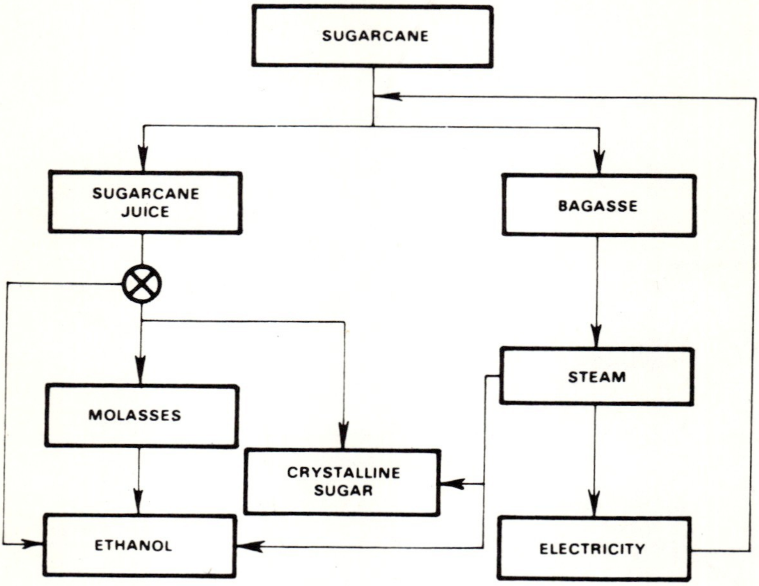

Figure 3. Use of ethanol production to control crystalline sugar supply. Source: Lipinsky, 1978.

in such a situation. Conventional sugar technology uses the percentage of sucrose recovered as a measure of industry results. Energy intensive processes are employed to re-evaporate and recombine the residues from early crystallizations to maximize yields, as described in the Cane Sugar Handbook (1977). In contrast, some Brazilian sugar factories now take the most easily crystallized sugar (known in the trade as "A" sugar) and send the rest of the solution to the fermenters for conversion to ethanol (Chenu, 1977). This system saves on capital investment, energy, and skill requirements of operators. The amount of crystalline sugar removed depends on the market demand for food and for fuel.

III. SUGAR STALK CROP RESOURCE DESCRIPTION

A. *Sugarcane (Saccharum officianarum L.)*

1. Botanical Description and Composition. The basic structure of the sugarcane (see Fig. 4) is closely related to that of other members of the family Gramineae, of which it is a giant member (Barnes, 1974). The major structure of this perennial grass is the culm, stem, or stalk (the principal organ of sucrose storage). The stem is solid, 2.5 - 6m in height, usually erect and unbranched except for tillers (secondary stems) at the base. The stem is composed of a series of joints, 5 - 25 cm long and 1.5 - 6 cm in diameter consisting of a node and internode. The length, diameter, shape, and color of the joints varies with the cultivar but can also be influenced by climatic and other factors, particularly light and nitrogen status (Purseglove, 1972).

The leaves are two-ranked as in other grasses, alternating on opposite sides. The number of mature green leaves during the "grand growth" period is about 10. As new leaves emerge, the older lower leaves dry and die and may drop off or be retained (Purseglove, 1972).

The root system is fibrous and composed of two components. The sugarcane plant is generally propagated by cuttings, therefore the first component of the system is a group of thin,

Figure 4. Sugarcane in Louisiana.

highly branched roots which develop from the cuttings. The secondary roots, which develop from the secondary shoots as the plant matures, are thicker than the primary roots and penetrate the soil to a depth of 2 m or more. However, the roots most active in uptakes of water and nutrients function in the upper 50 cm of the soil.

The flower of the sugarcane plant is a loose terminal panicle 25 - 50 cm long and silky in appearance. The extent of flowering varies greatly with cultivars and climate (photoperiod). In fact, most sugarcane produced in the mainland United States is harvested prior to flowering.

In the United States, the proportion of total millable stalk to the total wet biomass varies greatly, ranging from 50% in 12-month old Hawaiian sugarcane to approximately 70% in 9-month old Louisiana cane (Irvine, 1977). The composition of the millable stalk is given in Table I. The amount of each of three components is primarily genetically determined; however, the environment and cultural practices can affect the percentages of the components to a slight degree. The constituents found in the extracted juice of the stalk (Table I) are primarily three sugars, of which sucrose is by far the major component.

2. *Production*. Table II highlights the culture of sugarcane.

3. *Harvesting*. In the United States, sugarcane is harvested both manually and with machines. In Florida, approximately 50% of the sugarcane is hand-harvested, while in Louisiana, Texas, and Hawaii machines are utilized. Louisiana's erect, low yielding crop favored the development of the soldier harvester, whose gathering arms snap leaning stalks to attention and a base cutter and topper cut the upright stalks, and a carrier marches the stalks in line through the moving machine and trips them across a leap row (Irvine, in press). Following burning (to reduce the quantity of trash), the stalks are loaded into wagons by a self-propelled grab loader. In future biomass production systems, it appears likely that field trash burning will be discontinued; firstly, due to environmental regulations and secondly, and more importantly, biomass (or trash) is a valuable by-product that deserves collection and utilization as an energy source for the processing facility.

Elsewhere, the combine harvester is utilized. Following upright burning, the combine harvester pushes the cane over after topping, then cuts the base of the stalk. The stalks are then cut into billets which are elevated past forced air cleaners and then dropped into a tractor drawn bin traveling synchronously with the harvester (Irvine, in press).

TABLE I. Composition of Sugarcane and Juice Solids

	Percent
Millable stalk	
Water	73 - 76
Solids	24 - 27
Soluble solids	10 - 16
Fiber (dry)	11 - 16
Juice constituents	Soluble solids
Sugars	75 - 92
Sucrose	70 - 88
Glucose	2 - 4
Fructose	2 - 4
Salts	3.0 - 4.5
Inorganic acids	1.5 - 4.5
Organic acids	1.0 - 3.0
Organic acids	1.5 - 5.5
Carboxylic acids	1.1 - 3.0
Amino acids	0.5 - 2.5
Other organic nonsugars	
Protein	0.5 - 0.6
Starch	0.001 - 0.050
Gums	0.30 - 0.60
Waxes, fats, phosphatides	0.05 - 0.15
Other	3.0 - 5.0

Source: Irvine, 1977.

TABLE II. Summary of Sugarcane Production in the United States

Where grown	The major areas of production include Louisiana, Florida, Hawaii, and Texas. Minimal cane production in Alabama, Georgia, and Mississippi.
Area harvested	351,000 hectares (1976).
Yield	Average U.S. 65 MT ha^{-1}; range from 45 MT ha^{-1} in Georgia (9-month crop) to 112 MT ha^{-1} in Hawaii (24-month crop).
Typical farm size	Ranges from 100 to 1000 ha varying widely throughout the United States.
Planting date(s)	August-December on the mainland United, continuous (year-round) in Hawaii.
Harvest season	October through February on the mainland U.S.; virtually continuous in Hawaii.
Length of growing cycle	7-12 months on the mainland and 24 months in Hawaii.
Number of ratoon crops	2-4.
Rotation	Either continuous or rotated with soybeans or vegetables.
Soil requirements	Grown on a diversity of soils, ranging from sand to muck.
Fertilizer application	Varies greatly with soil, environment, and yield expected. A crop of 74 MT ha^{-1} is said to remove 110 kg N, 60 kg P_2O_5, and 300 kg K_2O.
Water management	Both irrigated and unirrigated. A general rule is 30 mm of water per MT of stalks.

(Continued)

TABLE II. (Continued)

Disease problems/ control	*Major disease include mosaic virus, smut, and ratoon stunting disease. Controlled by cultivar selection.*
Pest problems/control	*The major pest is the sugarcane stalk borer; minor problems with wireworm, white grubs, and sugarcane beetle. Control through insecticide application, use of resistant cultivars, and clean cultivation.*
Weed problems/control	*The major weed is johnson grass; minor weeds include bermudagrass and guineagrass. Weeds are controlled with mechanical cultivation and herbicide use.*

4. Potential Geographical Range. When considering the potential geographical range of sugarcane, both biological and economic factors must be used in the selection process. Furthermore, it is usually the economic return and not the biological potential which will play the major role in the development of production areas (Lipinsky *et al.*, 1976). But currently we find ourselves in a period when the economics of "energy farming" fluctuate greatly from week to week, depending on the use and type of production inputs and new developments in biomass processing and conversion. Thus, this analysis will address primarily the biological potential of sugarcane while recognizing the potential economic constraints will limit the eventual usage.

The two primary environmental parameters which determine the geographical range of any crop are temperature and moisture. Vital temperatures must not only include the potential of frost or freeze periods but also include temperatures which allow sugarcane growth to occur (>20C). Other important temperature-related factors include the length of the period above 20 C and how high above 20 C the temperature rises.

In the absence of irrigation, sugarcane requires a minimum effective rainfall of 1500 mm (Barnes, 1974) for good growth. However, growth will occur with less moisture. With these considerations, it has been estimated that roughly 10 million hectares within the southeastern United States have sufficient moisture and appropriate temperatures to successfully produce sugarcane (Lipinsky and McClure, 1977). This quantity of land represents the biological potential of sugarcane; the next production criterion required is the economic analysis of the land potential of sugarcane. Possibly if a critical need arose, roughly an additional 100,000 hectares in Florida, Louisiana, and Texas could be put into production by 1985 (Lipinsky *et al.*, 1976).

5. *Yield Potential.* The theoretical maximum yield of sugarcane is 280 MT ha^{-1} yr^{-1} (Bull and Tovey, 1974) while the average yield within the United States is approximately 65 MT ha^{-1} yr^{-1}. Thus, there is considerable room for yield improvement despite the climatic restraints of the temperate United States. Extended research in temperate areas shows that closer spacing between rows (from 1.5 m to 0.6 m) results in higher yields. The average yield of small plots in four harvests on 0.6 m rows in Louisiana was 160 MT ha^{-1} yr^{-1}, over 60% of the theoretical maximum; on large plots of close spaced sugarcane planted with plantation scale equipment, yields of 110 MT ha^{-1} were obtained (Irvine, in press).

In addition to the yield increases demonstrated by the use of improved cultural practices, yields may be improved through breeding. As mentioned previously, the fiber content of commercial sugarcane cultivars ranges from 11 to 16% with sucrose content ranging from 10 to 16%. Since these cultivars were developed through selection for low fiber, a change in selection pressure for high fiber and high total sugars would likely result in genotypes with a higher energy content both per ton and per unit area (James, in press). Irvine (1980) reported 28% fiber for a hybrid between a commercial cultivar and *S. spontaneum*.

6. *Implications for Usefulness in Energy Crop Systems.* Acclaimed as the most efficient bioconverter of solar energy, sugarcane owes this reputation partly to high rates of photosynthesis, and more importantly, to a long growing season (Loomis and Gerakis, 1975; Irvine, 1980). Maximum short-term crop growth rates of sugarcane range from 50 to 54 g M^{-2} ground day^{-1}, while maximum growth rates of other crop plants, i.e., wheat, barley, and soybeans, range from 34 - 39 g m^{-2} ground day^{-1}. The growing season of sugarcane can last throughout a year, thereby maintaining an efficient solar converter in the field.

Development of sugarcane as an energy crop could result in expansion of the crop to higher latitudes. Cultivars for energy developed by intercrossing sugarcanes and related genera offer potential in the United States. It is likely that frosts and freezes would affect sugarcane less for energy production than for crystalline sugar production (James, in press). However, other than being an extensive energy crop like sorghum or corn, sugarcane has the greatest potential to be a viable energy crop within a selected region like southern Florida, Louisiana, and Texas.

B. Sweet Sorghum and Sweet-Stemmed Grain Sorghum (Sorghum bicolor L. Moench)

1. Botanical Description and Composition. Like sugarcane, the sorghums shown in Figures 5 and 6 are members of the grass family and are often grown in cultivation as a single-stemmed type, but also show great variation in tillering capacity, determined by both cultivar and plant population (Doggett, 1970). The stem of the plant varies in height from 0.5 - 4 m in height and, like sugarcane, can accumulate and store sugar within it. The stem diameter ranges from 0.5 - 3 cm with the general size tapering from the base to the seedhead.

The number of leaves varies from 10-30, according to the cultivar and appear to alternate in two ranks. Leaves of the sorghum plant have long blades ranging from 30 - 135 cm (Artschwager, 1948) with a width of 1.5 - 13 cm.

The roots of sorghum are divided into a temporary and a permanent system (Freeman, 1970). A single radicle is produced by the seedling, followed by adventitious fibrous roots from the lowest nodes of the stem. The entire system may penetrate to a depth of 1 m width a spread of 1.5 m, depending on soil conditions and cultural practices.

The sorghum inflorescence, or "head," is a somewhat compact or loose panicle. A well-developed panicle may contain as many as 1,000 - 2,000 seeds. As an indication of the variation in seed size, sorghum cultivars range from 44,000 to 60,000 seeds per kilogram (Martin *et al.*, 1976).

The proportion of total wet stalk to total wet biomass is highly dependent on the type of sorghum (whether sweet or sweet-stemmed grain) and the cultural practices employed, particularly the plant population and the row spacing. A general range for this value is 60 - 80%, with the sweet sorghums being at the higher end of the range and the sweet-stemmed grain sorghums at the lower. Grain yields will range from 500 kg ha^{-1} with the sweet sorghum to approximately 6000 kg ha^{-1} with sweet-stemmed grain sorghum. Table III highlights the composition of the stalk and grain. Like sugarcane, the component percentages shown in Table IV are genetically determined; however, environment and cultural practices do influence the composition more with sorghum than with sugarcane.

The sorghum plant, whether sweet or sweet-stemmed grain, has a number of distinctive physiological and agronomic characteristics, which increase its potential as a viable multiuse crop over a wide geographic range (Lipinsky and Kresovich,

Figure 5. Sweet sorghum in Ohio.

Figure 6. Sweet-stemmed grain sorghum in Texas.

TABLE III. Composition of the Sorghum Stalk

	Percent
Sweet	
Water	*67 - 80*
Solids	*20 - 33*
Soluble solids	*6 - 18*
Fiber (dry)	*12 - 20*
Sweet-stemmed grain	
Water	*65 - 75*
Solids	*25 - 35*
Soluble solids	*4 - 8*
Fiber (dry)	*20 - 30*
Juice constituents	*Soluble solids*
Sweet	
Sugar (sucrose, fructose, and glucose)	*67 - 82*
Salts	*18 - 33*
Sweet-stemmed grain	
Sugar (sucrose, fructose, and glucose)	*40 - 75*
Salts	*25 - 60*

in press). Firstly, sorghum exhibits the C_4 (Hatch-Slack) photosynthetic pathway and is therefore quite efficient in assimilating carbon dioxide. Also, sorghum lacks the process of photorespiration and is highly productive, achieving maximum short-term crop growth rates of approximately 51 g m^{-2} ground day^{-1} (Loomis and Williams, 1963).

Secondly, sorghum utilizes water efficiently. The root system is fibrous and extensive. Ponnaiya (1960) demonstrated the existence of heavy silica deposits in the endodermis of the root, forming a complete silica cylinder in the mature roots. This mechanical strength is of great importance in preventing collapse of the system during drought stress. A waxy cuticle covers the above-ground structure to retard drying. Unlike corn, sorghum has the ability to remain dormant during

TABLE IV. Average Composition of Sorghum Grain

Constituent	*Percent*
Starch	*60.4 - 76.6*
Pentosans	*1.8 - 4.9*
Sugar	*0.5 - 2.5*
Protein	*8.7 - 16.8*
Fats and waxes	*1.4 - 6.1*
Fiber	*0.4 - 13.4*
Ash	*1.2 - 7.1*

Source: Miller, in press.

a drought period and then to become active rapidly following moisture reintroduction. The water requirement to produce one kilogram of sorghum dry matter ranges from 250 - 350 kilograms, whereas the requirement for wheat and soybean is approximately 500 and 700, respectively.

The nutrient requirements for sorghum are also low. A number of experiments in Texas have shown that the grain in a sorghum crop yielding 6,175 kg ha^{-1} removes an average of approximately 110 kg of nitrogen and 16 kg of phosphorus and potassium. For sweet sorghum in the southeastern United States, the normal fertilizer recommendation is 45, 45, and 35 of nitrogen, phosphorus (P_2O_5), and potassium (K_2O), respectively (Freeman *et al.*, 1973).

2. *Production.* Table V highlights the culture of sweet sorghum and sweet-stemmed grain sorghum.

3. *Harvesting.* Because most sweet sorghum is grown on a small scale, it is harvested both by hand and by mechanical harvesters. Some growers utilize modified silage harvesters to cut stalks into 10 - 20 cm billets. The stalk billets and leaves are then separated by air with the billets then conveyed to a trailer body (Freeman *et al.*, 1973). Modified sugarcane harvesters may also be utilized for harvesting sweet sorghum (Ruff *et al.*, in press). This method, however, is insufficient if one desires to collect the seedhead. In general, there is a lack of suitable equipment; therefore, if sweet sorghum is to become a major crop, efficient equipment and methods will need to be designed (Kresovich, in press).

TABLE V. *Summary of Sweet Sorghum and Sweet-Stemmed Grain Sorghum Production in the United States*

Where grown:	The major production areas are the southeastern United States. Production occurs in over 25 states from Florida to Minnesota.
Area harvested	Probably less than 10,000 hectares (no official production statistics available).
Yield	Sweet sorghum yield ranges from 60 MT ha^{-1} in North Dakota (4-month crop) to 120 MT ha^{-1} in Texas (6-month crop) under experimental conditions. Sweet-stemmed grain sorghum yield has ranged approximately 40 - 60 MT ha^{-1} composed of 2 - 6 MT ha^{-1} of starch and 0.8 - 1.2 MT ha^{-1} of fermentable sugar under experimental conditions.
Typical farm size	Average traditional producer grows less than 4 hectares. Commercial operations have grown as much as 400 hectares.
Planting date(s)	March - May in the southern United States, April - June in the Midwest and Great Plains.
Harvest season	July - December in the southern U.S., August - October in the Midwest and Great Plains.
Length of growing cycle	3.5 - 6 months.
Rotation	Normally would be rotated with cotton, sugarcane, corn, and soybeans.
Soil requirements	Grown on a diversity of soils; loam and sandy soils preferred.

(Continued)

TABLE V. (Continued

Fertilizer application	*Varies greatly with soil, environment, and yield expected. For sweet sorghum, a general recommendation of 45, 45, and 34 - 45 kg ha^{-1} of N, P_2O_5, and K_2O, respectively. For sweet-stemmed grain sorghum, 100, 45, 34 - 45 ka ha^{-1} of N, P_2O_5, and K_2O, respectively.*
Water management	*Usually no irrigation is required. Some water management necessary in Florida and Texas.*
Disease problems/ control	*Major diseases include downy mildew, head smut, mosaic virus, and antdracnose. Controlled by cultivar selection.*
Pest problems/control	*The pests include the sorghum midge, greenbug, lesser cornstalk borer, sugarcane stalk borer, corn leaf aphid, wireworm, and chinchbug. Control through insecticide application, use of resistant cultivars, and clean cultivation.*
Weed problems/control	*Weed problems vary greatly but can be controlled with herbicide use, mechanical cultivation, and crop rotation.*

Sweet-stemmed grain sorghum is a new crop and no work has been performed concerning harvesting procedures and equipment. Because of its intermediate plant height, approximately 1.5 - 2.0 m (compared to the height of 2.5 - 4.0 m for sweet sorghum, the grain of the sweet-stemmed grain sorghum can be harvested with a mechanical combine. Unfortunately this method is insufficient to collect the sugar-rich stalk. Conversely, the modified sugarcane harvester is equipped to handle the stalk without the grain. Thus a two-step harvest operation must be performed using modified, existing equipment or a "hybrid" harvester, which combines the useful attributes of a grain combine and a stalk harvester, must be developed. Another solution involves the development of processing/conversion technologies which utilize the entire plant. A high-powered forage harvester could then be useful for whole-plant removal.

4. *Potential Geographical Range.* Sorghum is adapted to a wider range of environmental conditions and potentially can be grown anywhere that corn, soybeans, cotton, and sugarcane can grow. It seems reasonable to assume that sweet sorghum will initally be integrated into the sugarcane growing regions whereas the sweet-stemmed grain sorghums will be utilized in the grain sorghum regions of the Great Plains.

Because of the tremendous diversity within the species *Sorghum bicolor*--currently there are over 17,000 accessions within the World Collection--its potential appears unlimited. In the future, as the increasing population of the United States causes agriculture to move into more marginal lands for the production of food and energy, sweet sorghum and sweet-stemmed grain sorghum will be able to produce useful yields under conditions which are unfavorable for most other crops.

5. *Yield Potential.* Current commercial yields of sweet sorghum range from 33 to 44 MT of millable stalks ha^{-1} (Kresovich, in press). These yields are achieved with the use of older, early maturing cultivars and row spacings of one meter. Also, no hybridization of sweet sorghum has been used commercially; therefore, if an analogy can be drawn between sweet sorghum and corn improvement, sweet sorghum is at the same stage of development as corn was in the 1930s.

The development of sweet-stemmed grain sorghum has been conducted within only the past few years. Breeding efforts have been carried out at a low level; however, results have been quite encouraging, ranging from 2-6 MT ha^{-1} of starch and 0.8 - 1.2 MT ha^{-1} of fermentable sugar under experimental conditions. It is quite apparent that, if sweet sorghum and sweet-stemmed grain sorghums received the same amount of funding and research interest as corn, sugar beets, or sugarcane, they could well become useful crops in the United States' agricultural system.

6. *Implications for Usefulness in Energy Crop Systems.* Sorghums (both sweet and sweet-stemmed grain) have potential, not only as energy crops, but as multiple-use crops in an adaptive system in which variable quantities of food and nonfood products can be made. Current efforts in breeding and selection have barely scratched the surface of the potential of the sorghum germplasm. Further development could result in commercialization of sweet and sweet-stemmed sorghum in the Midwest and Great Plains. Sorghum, like corn, has the potential to become an extensively grown feedstock for future energy production systems.

IV. ALTERNATIVE PROCESSING AND CONVERSION TECHNOLOGIES

Processing is defined as the physical separation and beneficiation that is employed prior to conversion by chemical means to obtain fuels and/or chemicals. Some systems have a clean break between processing and conversion. Others integrate the steps, with considerable potential for savings in energy and capital investment. The sugar industry has built up a conventional technology for processing that cleans the sugarcane stalks, crushes, grinds, and mills to obtain a juice that is clarified. The clarified juice then is evaporated and crystallized to obtain raw sugar (Meade *et al.*, 1978). If fuels and chemicals are the major markets for the sugar crop output, numerous shortcuts can be considered. The investment in in-place processing equipment will have a major impact on selection of processing technology. Therefore, it is likely that two separate stalk processing systems will develop (at least)--one an evolution of the current technology, perhaps along the lines of the Brazilian system described above, and the other a more radical system that is not equipped to compete in food markets.

Many alternatives to conventional sugar stalk crop processing and conversion are under development (Table VI). The Tilby process (Tilby, 1976, 1971a, 1971b) could be employed to obtain table sugar from sugar stalk crops by juice extraction from the pith and crystallization. The pith also would be an excellent fermentable sugar source.

The Tilby process (Figures 7 and 8) involves separation of the pith and rind of sugar stalk crops by means of a slitting and scraping operation. High quality plywood substitutes can be made from the rind, after it has been rendered sugarfree. The sugar-containing pith can be dried for prolonged storage, the sugars can be extracted relatively readily by diffusion processes, or the EX-FERM process can be employed. This process appears especially attractive for use in geographical areas with little forest product resources for lumber and plywood production.

The Envirogenics process involves hammer milling and steaming of sugar stalk crops to obtain a sugar solution that can be fermented directly, without removal of the fiber (Bruschke, 1978). The juice would be too dirty and inverted for crystalline sugar production but is quite acceptable for fermentation.

The EX-FERM process is still more radical in that little or no effort is made to rupture the sugar crop's sugar-containing cells (Rolz, 1979). This process (Figure 9) permits the buildup of a relatively concentrated ethanol solution by recycling the yeast through several fixed-bed reactors

TABLE VI. Alternative Processing and Conversion Technologies

Technology	*Characteristics*
Tilby	*Separation only, no in-situ converion* *Separates pith and rind by slitting and scraping* *Requires clean aligned stalks*
Envirogenics	*Separation only, no in-situ conversion* *Hammermills and steams to obtain fermentable sugar solution/fiber*
EX-FERM	*In-situ conversion is emphasized* *Ferments crushed stalk or pith, relying on osmosis to extract sugars*
Dry biomass	*Burn sugar crop biomass for steam generation* *No attempt to remove sugars from original site in stalk*
Maximum EtOH	*In-situ conversion is emphasized* *Use cellulose and hemicellulose in stalk for fermentation to ethanol.*

containing chopped sugar crop stalk or sugar stalk piths. From a mechanical handling viewpoint, sugar crop pith would be preferable to chopped stalk, but the additional cost may not be tolerable. Osmosis transfers the fermentable sugars to the yeast that is present in the fermentation medium.

Alexander's concept (1979) is to sun dry and/or flue-gas dry sugar crop biomass to the point that it can be used to generate steam and electricity in a boiler-turbine facility. In this instance, the concept of the sugar crop as a source of sugar is replaced with the concept of growing the crop solely for its heat of combustion value.

Significant recent advances in the microbiological conversion of cellulose and hemicellulose into ethanol improve the prospects for sugar stalk crops. The delivered cost of lignocellulose to an ethanol producer located in an area with a

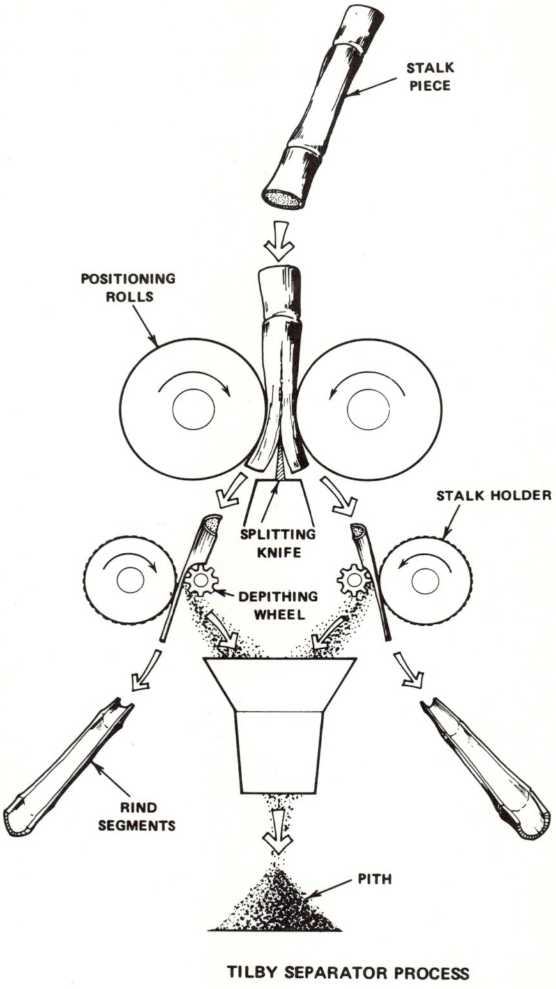

FIGURE 7. Simplified schematic of the Tilby Separator Process.

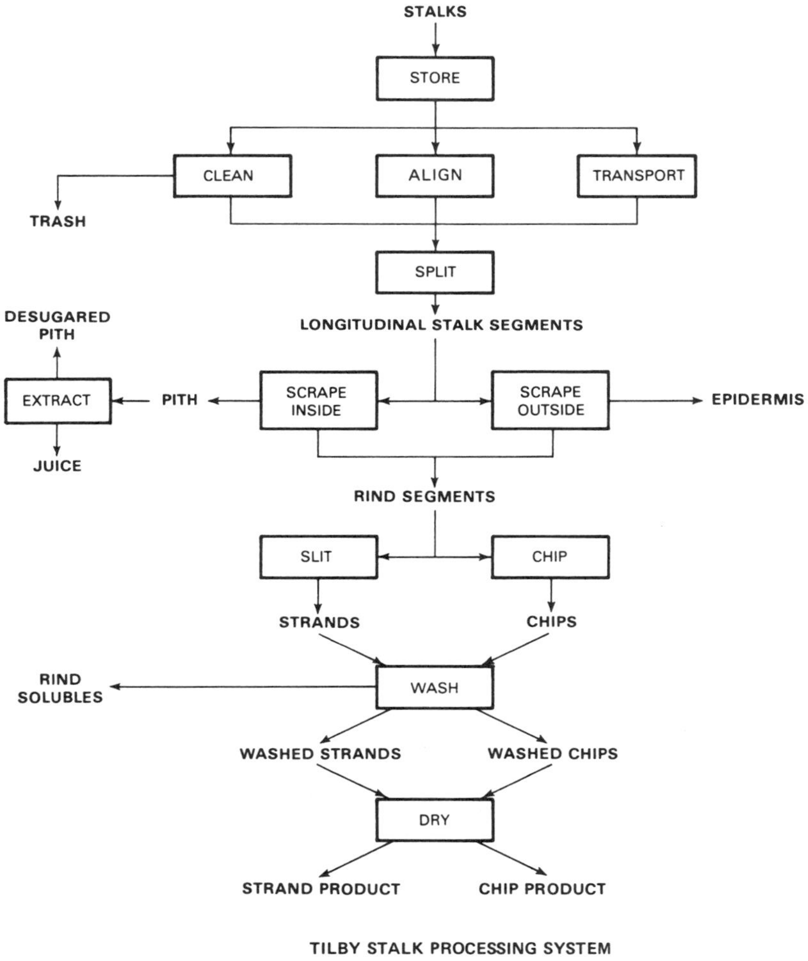

FIGURE 8. Tilby Stalk Processing operations.

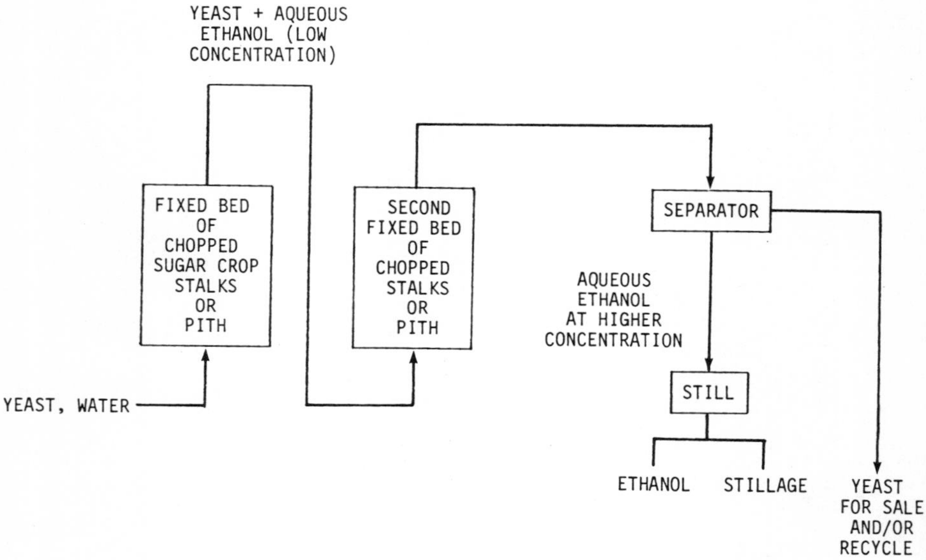

FIGURE 9. Principle of the Ex-Ferm Process. Source: Rolz, 1979.

long harvesting seasons is likely to be as low as for forestry residues. The stalks are delivered with the intent of sugar extraction either for crystalline sugar production or for ethanol production.

The front-running process for conversion of lignocellulose into ethanol is the one under development at MIT by Wang and coworkers (Wang *et al.*, 1980). Two thermophilic bacteria are employed. One (Clostridium thermocellum) hydrolyzes cellulose to glucose and converts glucose to ethanol; the other (Clostridium thermosaccharolyticum) hydrolyzes hemicellulose to 5-carbon sugars and ferments both glucose and 5-carbon sugars to ethanol. The yields with Solka Floc are approaching those achieved in the fermentation of starch to ethanol. It is necessary to select strains that minimize the production of by-products--acetic acid and lactic acid. Early results with stalk material resembling those obtained from sugar stalks indicate that the fermentation can be adversely affected by constituents of stalks that are not present in purified Solka Floc. These development problems are not considered insurmountable by the MIT Group. The process appears to be similar to a conventional fermentation (Jenkins *et al.*, 1979).

Recognition that the lignocellulose of sugar stalk crops contains more pentose polymers than glucose polymers centers attention on this substrate as the key to increasing ethanol yields from the sugar stalk crops. Even if the MIT process were not employed, the hemicellulose of the stalk could be extracted with hot water/dilute acid and fermented to ethanol. The residual lignin-cellulose complex could be burned or anaerobically digested.

The pentose fraction could be converted to furfural (Paturau, 1969) by strong acid digestion. This process has been commercially practiced in the United States and elsewhere on bagasse for many years (e.g., the Quaker Oats facility at Belle Glade, Florida). The year-round availability of rice hulls appears to make possible more economical production of furfural from that source than from seasonal production of sugarcane bagasse.

V. CONVERSION STRATEGIES

Conversion of raw materials derived from sugar crop stalks usually is considered to be a choice between biochemical and thermochemical options (Ward, 1980). Sugar stalk crops are over 70% moisture prior to processing and most processing techniques leave the product at more than 40% moisture (usually about 50%). In addition, the dry matter of carbohydrates is more than 50% oxygen. Therefore, conversion technology for making fuels concentrates on removing oxygen and on accommodating to the presence of water. For example, fermentation or anaerobic digestion processes operate in an aqueous environment so that the initial water content is not a disadvantage. These fermentations yield such products as ethanol or methane which represent the loss of most or all of the oxygen in the original sugar crop biomass. The conversion of sugar stalk crop biomass into chemicals is not limited to oxygen removal. Glycerine which has as much oxygen as the original carbohydrates but which sells for perhaps ten times as much can be made by conversion processes in which oxygen removal is not a factor.

Because carbohydrate crop conversion antedates the formal sciences by many centuries, familiarity renders rigorous analysis difficult. As shown in Figure 10, carbohydrate conversion to energy-rich chemicals and fuels can be achieved either by using biomass or by reaction of carbohydrates with nonbiomass materials. Both approaches merit discussion, but only the former will be discussed now. Carbohydrates can undergo dehydration or oxidation-reduction reactions (internal disproportionations) in which no outside reactants are needed. Examples of dehydration are the conversion of pentoses into furfural by acid digestion and the formation of levoglucosan from hexoses via pyrolysis. Familiar oxidation-reduction reactions include the fermentation of sugars to produce ethanol and the formation of methane by pyrolysis.

A convenient means of visualizing the effectiveness of conversion processes in concentrating energy is illustrated in Figure 11. The Y-axis shows directly the degree of concentration that is achieved, the X-axis shows the comparison of heats of combustion, and the areas of the rectangles described by the points (X,Y) are comparisons of the energy stored by the alternative fuels for release through combustion. For example, glucose is relatively dilute as a fuel, as shown by its heat of combustion of 15,200 kJ per kg. One mole (180 kg) of glucose can be converted either into about 90 kg of ethanol ($- \Delta H_c$ = 28,600 kJ per kg) or 48 kg of methane ($- \Delta H_c$ = 53,400 kJ per kg). Each is theoretically capable of storing about 95% of glucose combustible energy in a fraction of the weight. In contrast, conversion of glucose to acetic acid provides no concentration (Figure 11) and reduces the energy content by 1100 kJ per kg.

Dehydration yields energy-rich chemicals by removing a very poor fuel (water) from the carbohydrate. The oxidation-reduction reactions do not violate the laws of thermodynamics--to the extent that the energy-rich compound is high in energy, the rest of the carbohydrate molecule must have been converted to something of very low energy. Thus, ethanol which contains

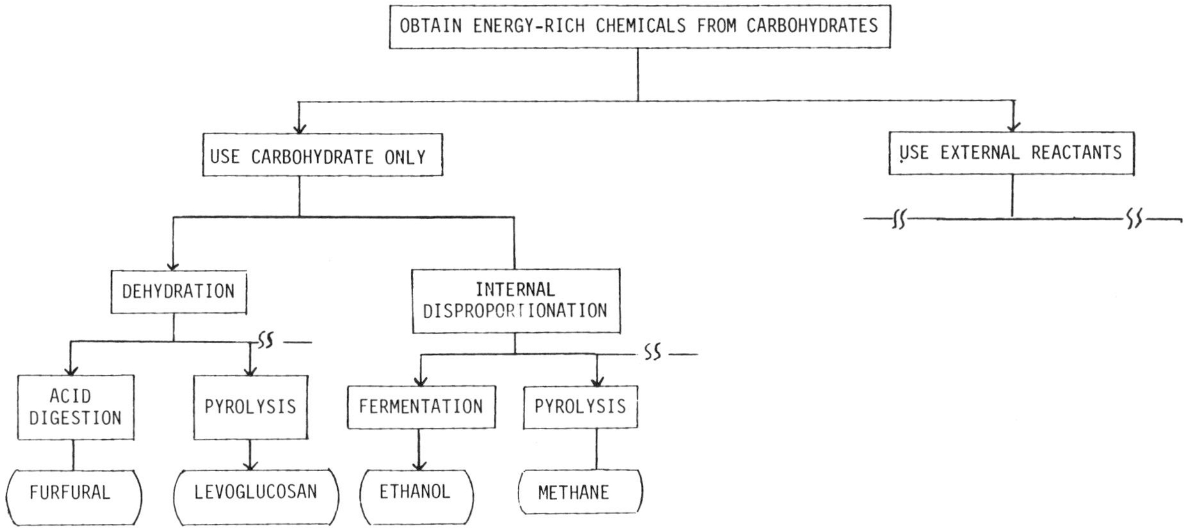

FIGURE 10. Alternative methods to obtain energy-rich chemicals from carbohydrates.

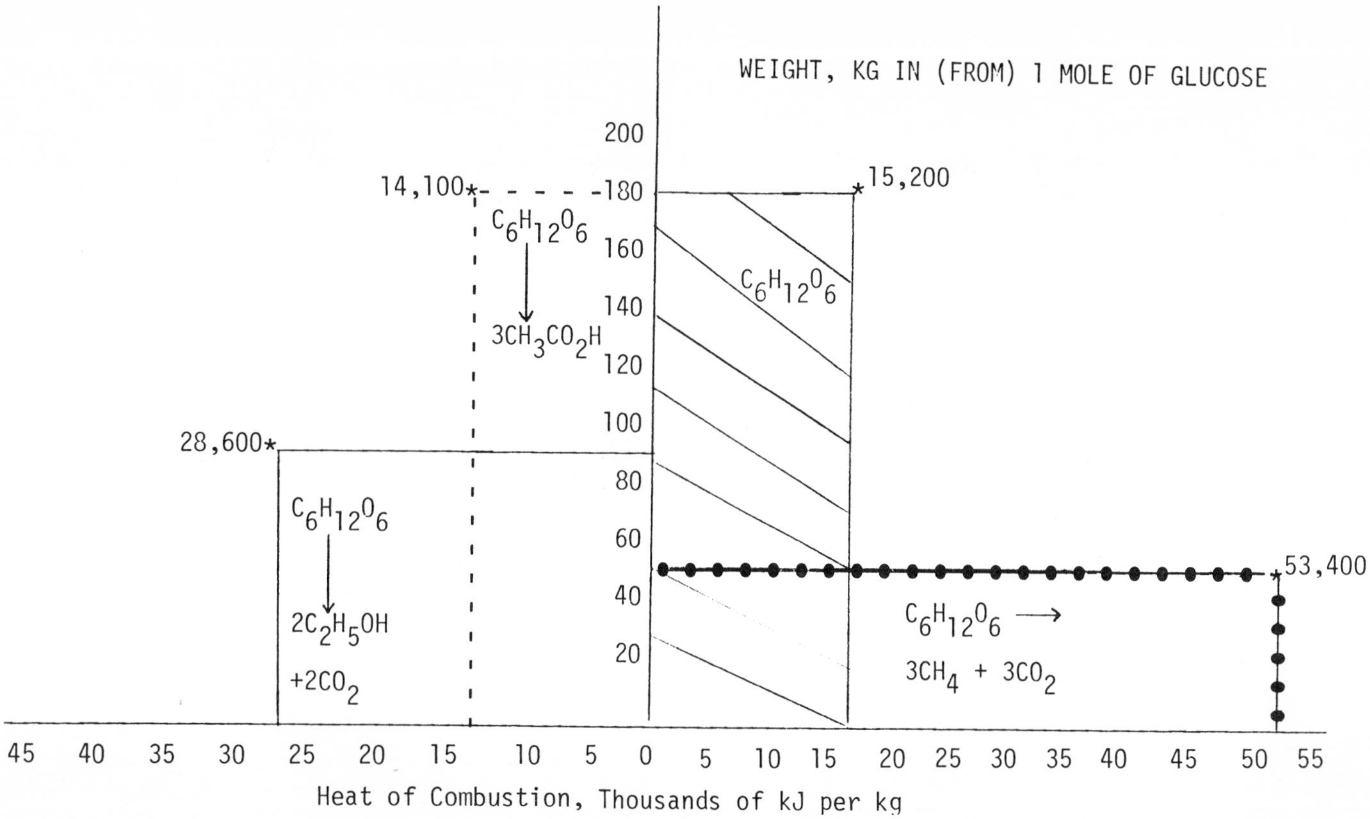

FIGURE 11. Graphical comparison of the heat of combustion of 1 mole of glucose with the heat or combustion of the ethanol or methane that could be derived from it.

a high percentage of glucose's energy is accompanied by non-oxidizable carbon dioxide in fermentation reactions.

The intense interest in fermentation ethanol as a source of motor fuel has obscured the fact that there are many opportunities for carbohydrates to be converted to energy-rich compounds. As shown in Table VII, many formal reactions can be expressed in which dehydration and/or disproportionation provide chemicals from carbohydrates without need for external reactants. Some disproportionations yield very high energy products (methane via anaerobic digestion or thermochemical gasification); others produce low energy compounds that have functional groups that are valuable for the purpose of chemical synthesis (e.g., lactic acid and acetic acid). It is possible to write some equations for which no known carbohydrate reaction has been reported to our knowledge (e.g., the formation of oxalic acid and 2,3-butanediol from glucose).

Both dehydration or disproportionation processes may operate at once. The formation of levulinic acid and formic acid is a dehydration-disproportionation reaction. Pyrolysis of carbohydrates to obtain methane, carbon monoxide, and hydrogen represents a complex series of dehydration and disproportionation reactions.

VI. SELECTION OF FUELS AND CHEMICALS

Future needs of the marketplace play a major role in selection of chemicals and fuels to be made from sugar stalk crop resources. The other major determinant is the relative effectiveness of these resources versus other biomass resources and other resources in general. These considerations involve raw material costs, yield, availability on both an annual basis and a daily basis, scale of enterprise, and state of the art. In addition, numerous institutional regulatory, and political considerations play important roles.

The considerations presented above are too complex for rigorous selection of the most desirable chemicals and fuels to be derived from sugar stalk crops. However, some guidelines can be presented. The marketplace will want chemicals now derived from imported petroleum to be made from renewable resources. In some instances, such as the aromatic chemicals, the competition between motor fuel uses and chemical uses will be intense (Storck, 1980). New routes to chemicals derived from benzene, toluene, and xylenes are the ones that would be most welcomed. From the world-wide viewpoint, chemicals now made from ethylene, propylene, and the butanes also are desirable. In the United States, the availability of natural gas liquids as a source of these chemicals makes the pressure in

TABLE VII. *Examples of Dehydration and Disproportionation Reactions of Hexoses and Pentoses*

Formal reaction	Type	Example
$C_6H_{12}O_6 \rightarrow 3CH_4 + 3CO_2$	Disproportionation	Methane via anaerobic digestion or thermochemical gasification
$C_6H_{12}O_6 \rightarrow 2C_2H_6O + 2CO_2$	Disproportionation	Fermentation for EtOH production
$C_6H_{12}O_6 \rightarrow 3C_2H_4O_2$	Disproportionation	Acetic acid via fermentation
$C_6H_{12}O_6 \rightarrow 2C_3H_6O_3$	Disproportionation	Lactic acid via fermentation or alkali digestion; dihydroxyacetone via fermentation
$C_6H_{12}O_6 \rightarrow C_4H_{10}O_2 + 2CO_2 + H_2$	Disproportionation	2,3-Butanediol production via fermentation
$C_6H_{12}O_6 \rightarrow C_5H_8O_3 + H_2CO_2 + H_2O$	Dehydration, disproportionation	Levulinic acid via acid digestion
$C_5H_{10}O_5 \rightarrow C_5H_4O_2 + 3H_2O$	Dehydration	Furfural via acid digestion
$C_6H_{12}O_6 \rightarrow 6CO + 6H_2$	Disproportionation	Gasification
$C_6H_{12}O_6 \rightarrow 6C + 6H_2O$	Dehydration	Charcoal production
$C_6H_{12}O_6 \rightarrow C_6H_6O_3 + 3H_2O$	Dehydration	HMF by acid digestion

$C_6H_{12}O_6 \longrightarrow C_6H_{10}O_5 + H_2O$	*Dehydration*	*Levoglucosan via thermal treatment*
$C_6H_{12}O_6 \longrightarrow C_2O_4H_2 + C_4H_{10}O_2$	*Disproportionation*	*Unknown*
$C_6H_{12}O_6 \longrightarrow C_4H_{10}O_2 + 2CH_2O_2$	*Disproportionation*	*2,3-Butanediol via fermentation*

this direction somewhat less intense. Chemicals that are normally derived from methane (methanol, formaldehyde, and ammonia) have large markets but are not under an intense pressure as are the aromatic chemicals.

Where does biomass, and in particular sugar stalk crops, stand in relation to these chemical needs? Carbohydrates appear to be very far removed from aromatic chemicals, so far as duplicating their chemical structure is concerned. However, the aromatics are in gasoline to raise its octane rating and such products as ethanol and methanol from biomass also can achieve this function. However, ethanol can be dehydrated to ethylene and Brazil is proceeding with plans to install facilities for this purpose (Yang, 1979 and Tsao *et al.*, 1978). The stoichiometry of the reaction dictates that approximately 1.5 pounds of ethanol must be used to make each pound of ethylene. This leads to high raw material costs (Lipinsky *et al.*, 1977).

The stoichiometry of the production of acetic acid from carbohydrates appears more promising in that theoretically one pound of simple sugar can produce one pound of acetic acid. This route would have to compete with the highly efficient carbonylation of methanol which exploits the low cost of carbon monoxide. The carbohydrate route may have good prospects when effective means to recover acetic acid from dilute solutions are developed or when means to ferment carbohydrates to relatively concentrated solutions of acetic acid are found.

Some of the best prospects for carbohydrate conversion to traditional petrochemicals are in the area of 4-carbon chemicals. Maleic anhydride which is derived from benzene or butenes might be made from 2,3-butanediol. Conversion of 2,3-butanediol to methyl ethyl ketone in high yield has been reported (Neish *et al.*, 1945).

Routes from lactic acid to acrylates were explored in previous feedstock shortages. Acrylates and acrylonitrile are made now from propylene in high yield by processes that have very favorable stoichiometry in that the oxidation is adding substantially to the molecular weight of the starting material with cheap reagents (oxygen or ammonia). Where propylene is not available, carbohydrate-based processes might be considered in the future.

Although ethanol is considered by many to be the primary goal of sugar crop systems, agricultural energy independence will depend on the future availability of diesel fuel. It is a tribute to the versatility of the diesel engine that it will burn even ethanol which is a high octane fuel but surely better diesel fuels can be made from sugar stalk crops. The chemical composition of lipids is much more conducive to good diesel fuel performance. The motivation for converting sugar stalk crop carbohydrates into lipids rather than harvesting soybeans or sunflowers and extracting their oils is that the yields per acre can be increased. Thus, availability would be enhanced, but cost probably will be high.

VII. CONCLUSIONS AND POLICY IMPLICATIONS

Although the sugar stalk crops could make a strong contribution to energy independence, perhaps equivalent to 1 billion to 10 billion gallons per year of ethanol ultimately, they do not represent a panacea for the production of chemicals and fuels to replace imported petroleum. The contributions of forest biomass and starch crop biomass are likely to be greater than sugar stalk crop biomass' contribution because they are not plagued with seasonality and stillage problems. Abandonment of the sugar stalk crops, however, would be as big a mistake as over-reliance on this biomass resource. The favorable energy balance, ease of fermentation, and ease of integration with both lignocellulose and starch crops render the sugar stalk crops a valuable member of a renewable resource team in countries with tropical climates and in countries with large temperate areas (e.g., the United States).

The political importance of substituting renewable resources for imported petroleum resources places heavy burdens of responsibility on the public policy makers and the legislators who decide which resources and processes will receive support and incentives. For example, the utilization of the factors of production for sugar stalk crop production, processing, conversion, and use for chemicals and fuels applications is a seamless system. Such adaptive systems grow by evolutionary processes or even by metamorphosis in which feedback concerning technology, prices, and competition from other resources interact (Jantsch, 1975). Dismemberment of such a system by setting one government agency in charge of crop production system, another in charge of processing and conversion, and a third agency in charge of environmental impacts with little or no means of communication among the agencies is clearly not an optimal approach.

The dominant roles of lignocellulose and starch must be taken into account in planning for sugar stalk utilization system development. Sugar stalk crops contain lignocellulose and can be genetically designed to contain considerable starch. Means to make these other resources allies need to be developed.

The selection of chemicals and fuels to be made from sugar stalk crop resources should not be narrowed arbitrarily to ethanol. Nor should it be assumed that sugar crop juice is the product that is to be converted to a fuel or chemical.

For companies already manufacturing crystalline sugar or other sweetener products, conversion of lignocellulose or starch in the grain of sweet stemmed grain sorghum may be the most profitable opportunity. Diesel fuel or chemicals may be preferred over gasoline additives, if the free market is allowed to choose. Arbitrary subsidies of ethanol could be quite counterproductive for the U. S. economy.

In developing sugar stalk crop systems, realism in establishing time and cost parameters needs to be applied. These systems are likely to require 10 to 20 years to reach fruition when breeding of new cultivars, process development, technology transfer to thousands of farmers, and many other system needs are taken into account. Other resources will require at least as much time for full development. What is required is a sustained effort that facilitates the harmonious evolution and metamorphosis of the system. Experience in commercial development indicates that systems grow best on reinvested profits; therefore, there should not be a long research and development period prior to initial commercialization. On the other hand, a subsidized demonstration plant that is clearly uneconomic retards real progress.

The prospects for sugar stalk crops as renewable resources for the production of chemicals and fuels depend on institutional and political factors that presently surpass the agronomic and technical factors in importance.

REFERENCES

Alexander, A. 1979. Proceedings of the Conference "Alternative Uses of Sugarcane." San Juan, Puerto Rico, March 27.

Artschwager, E. 1948. U.S.D.A. Tech. Bull, 957, Washington, D.C.

Barnes, A. C. 1974. "The Sugar Cane." Wiley, New York.

Bruschke, H. 1978. Proceedings of the International Symposium on Alcohol Fuel Technology Methanol and Ethanol. Wolfsburg (November 21-23, 1977). U.S. Department of Energy Document CONF-771175, July, Springfield, VA.

Bull, T. A., and Torrey, D. A. 1974. Proc. Int. Soc. Sugar Cane Techol. 15, 1021-1032.

Chenu, P. M.A.M. 1977. Proceedings of the 16th Congress of the International Society of Sugarcane Technologies. Sao Paulo 3, 3241-3250.

Doggett, H. 1970. "Sorghum." Longmans, London.

Freeman, J. E. 1970. *In* "Sorghum Production and Utilization" (J. S. Wall and W. M. Ross, eds.), pp. 28-72. Avi Publishing Co., Westport, CT.

Freeman, K. C., Broadhead, D. M., and Zummo, N. 1973. U.S.D.A. Ag. Handbook No. 441, Washington, D.C.

Irvine, J. E. 1977. *In* "Cane Sugar Handbook" (G. P. Meade and J. C. P. Chen, eds.), pp. 15-29. Wiley, New York.

Irvine, J. E. (in press). *In* "Handbook of Biosolar Resources" (E. S. Lipinsky and T. A. McClure, eds.), Vol. II. CRC Press, West Palm Beach, FL.

James, N. I. (in press). Proc. BioEnergy '80, Atlanta, GA.

Jantsch, E. 1975. "Design for Evolution." George Braziller, New York.

Jenkins, D. M., McClure, T. A., and Reddy, T. S. 1979. Net Energy Analysis of Alcohol Fuels, Report No. 4312. American Petroleum Institute, Washington, D.C.

Jenkins, D. M. and Reddy, T. S. 1979. Economic Evaluation of the MIT Process for the Manufacture of Ethanol, Report 2. U. S. Department of Energy, June 28, 1979.

Kresovich, S. (in press). *In* "Handbook of Biosolar Resources" (E. S. Lipinsky and T. A. McClure, eds.), Vol. II. CRC Press, West Palm Beach, FL.

Lipinsky, E. S. 1978. *Science 199*, 644-651.

Lipinsky, E. S. and Kresovich, S. (in press). Proc. BioEnergy '80, Atlanta, GA.

Lipinsky, E. S. and McClure, T. A. 1977. *In* "Biological Solar Energy Conversion" (A. Mitsui, S. Mujachi, A. San Pietro, and S. Tamusa, eds.), pp. 397-410. Academic Press, New York.

Lipinsky, E. S., Nathan, R. A., Sheppard, W. J., McClure, T. A., Lawhon, W. T., and Otis, J. L. 1977. Systems study of fuels from Sugarcane, Sweet Sorghum, and Sugar Beets, Volume I: Comprehensive Evaluation, pp. 75-85, NTIS No. BMI-1957 (Vol. I), Washington.

Lipinsky, E. S., McClure, T. A., Nathan, R. A., Anderson, T. L., Sheppard, W. J., and Lawhon, W. T. 1976. E.R.D.A. Report BMI-1957, Vol. II. NTIS, Springfield, VA.

Loomis, R. S., and Gerakis, P. F. 1975. *In* "Photosynthesis and Productivity in Different Environments" (J. P. Cooper, ed.), pp. 145-172. Cambridge University Press, Cambridge.

Loomis, R. S., and Williams, W. A. 1963. *Crop Sci. 3*, 67.

Martin, J. H., Leonard, W. H., and Stamp, D. L. 1976. "Principles of Field Crops." Macmillan, New York.

McGinnis, R. A., and Sunderland, D. L. 1971. *In* "Sugar Beet Technology" (R. A. McGinnis, ed.), pp. 91-106. Beet Sugar Development Foundation, Ft. Collins.

Meade, G. P., and Chen, J. C. P. 1977. "Cane Sugar Handbook" Chapters 12 and 13. Wiley, New York

Miller, F. R. (in press). *In* "Handbook of Biosolar Resources" (E. S. Lipinsky and T. A. McClure, eds.), Vol. II. CRC Press, West Palm Beach, FL.

Neish, A. C., Haskell, V. C., and MacDonald, F. J. 1945. *Canadian Journal of Research 23*, 281-289.

Paturau, J. M. 1969. "By-Products of the Cane Sugar Industry," p. 97. Elsevier, New York.

Ponnaiya, B. W. X. 1960. *Madras Agric. J. 47*, 31.

Purseglove, J. W. 1972. "Tropical Crops: Monocotyledons." Wiley, New York

Rolz, C. 1979. *Biotechnology and Bioengineering 21*, 2347-2349.

Ruff, J. H., Dillon, R. C., and Coble, C. G. (in press). Proceedings of American Society Sugar Cane Techol., Fort Walton Beach, FL.

Storck. 1980. *Chemical and Engineering News 58*(18), 36.

Tilby, S. E. 1971a. U. S. Patent 3,567,510, March 2.

Tilby, S. E. 1971b. U. S. Patent 3,567,511, March 2.

Tilby, S. E. 1976. U. S. Patent 3,976,499, August 24.

Tsao, U., and Reilly, J. W. 1978. Hydrocarbon Processing, February, pp. 133-136.

Wang, D. I. C., Biocic, I., Fang, H. Y., and Wang, S. D. 1979. Proc. of 3rd Annual Biomass Energy Systems Conference, June 5-7, 1979, SERI/TP-33-285, pp. 61-68.

Wang, D. I. C., and Fang, H. Y. 1980. Ethanol Production by a Thermophilic and Anaerobic Bacterium Using Xylose Substrates, Paper presented at National Meeting of American Chemical Society, San Francisco, August 25-28, 1980.

Ward, R. F. 1980. *In* "Future Sources of Organic Raw Materials--Chemrawn I" (L. E. St-Pierre and G. R. Brown, eds.), pp. 333-342. Pergamon, Oxford.

Yang, V. 1979. The Prospects of Industries Based on Fermentation Ethanol in Brazil. Paper presented at the 72nd Annual Meeting of the AIChE, San Francisco, November 25-29, 1979.

ACID-CATALYZED DELIGNIFICATION OF LIGNOCELLULOSICS IN ORGANIC SOLVENTS

K. V. Sarkanen

College of Forest Resources
University of Washington
Seattle, Washington

ISBN 0-12-535902-0

I. INTRODUCTION

The renewable biomass materials of interest for conversion either to liquid fuels or chemical feedstocks consist of the following broad categories: (1) underutilized hardwoods such as aspen and other poplar species, eucalypts, southern hardwood species, etc.; (2) residues from logging both soft- and hardwoods; and (3) agricultural residues, such as wheat and rice straws, corn stalks and bagasse.

All these materials belong to the broad class of lignocellulosics and as such, consist of cellulose, hemicelluloses, lignin and varying amounts of extractive materials. The individual components possess characteristic, and quite variable responses towards chemical treatments, which circumstance is often the source of processing difficulties. Thus, enzymatic hydrolysis of polysaccharidic components to sugars is retarded by the presence of lignin. The alternative hydrolysis method by mineral acids tends to degenerate the lignin component and exerts an uneven hydrolytic action on polysaccharides, converting a substantial part of hemicelluloses to furfural derivatives. Consequently, the processing of biomass could be simplified substantially, if it would be preceded by separation to individual polymer components.

Of course, separation of the lignin component from wood biomass in order to convert it to papermaking fibers has been known for a long time, and is practiced by the pulp industry on a world-wide scale. The methods used do not, however, lend themselves advantageously for general biomass processing. Of known alternative biomass separation methods, delignification in organic solvents ("Organosolv pulping") has generated increasing interest. By this method, it is possible to separate biomass in the following components: (1) cellulosic fibers, suitable for either papermaking or enzymatic conversion to glucose; (2) solid low-molecular weight lignin, usable either as fuel or as feedstock for chemical conversions; and (3) hemicellulose sugars which may be used for various fermentation processes or as chemical feedstocks.

The organosolv process, together with the recovery of process chemicals, can be in principle simpler, less energy intensive and more economical than the conventional pulping processes, the dominating kraft process inclusive. Thus, organosolv delignification represents perhaps the most promising approach towards "total biomass utilization" in which all components of lignocellulosic material are effectively separated for appropriate utilization.

II. EARLIER STUDIES OF ORGANOSOLV PULPING

Delignification of wood in a large number of organic or aqueous-organic solvents containing various acidic catalysts has been known for a long time. The solvents used include lower alcohols, such as ethanol (Kleinert and Tayenthal, 1931, Kleinert, 1974) or butanol (Aronovsky and Gortner, 1936), ethylene glycol (Rassow and Gabriel, 1931), triethylene glycol (Grondal and Zenczak, 1950), dimethyl sulfoxide (Clermont and Bender, 1961) and tetrahydrofurfurylalcohol (Bobomolow *et al.*, 1979. Particularly in the earlier studies, small amounts of mineral acids--e.g., .05% HCl or 0.2% H_2SO_4--were used as catalysts. Later it has been found that such organic acids as oxalic (Paszner, 1979), acetylsalicyclic (Schwenzon, 1966) and salicyclic acid (Nelson, 1977) are probably more suitable catalysts. When sufficiently high pulping temperatures (180°C and above) are applied, no addition of catalyst is necessary for satisfactory delignification, at least in case of hardwoods (Aronovsky and Gortner, 1936; Kleinert, 1974). In the "uncatalyzed" process, organic acids released from wood probably act as acidic catalysts. Inorganic salts, such as aluminum chloride (Grondal and Zenczak, 1950) may also be used to furnish the desirable acidity to the medium in organosolv delignification.

The papermaking properties of organosolv pulps have been evaluated only by Kleinert (1974) and by Nelson (1977). The results obtained (Table I) suggest that, under favorable conditions, pulps may be obtained in higher yields than in the conventional kraft process, with strength properties equivalent to those of bisulfite pulps. While such hardwood species as poplars and eucalypts are pulped with relative ease at 170-180°C, softwoods are delignified with a great deal more difficulty.

The alternative use for cellulosic fibers obtained in organosolv delignification is their conversion to glucose by enzymatic hydrolysis. Humphrey and coworkers (Humphrey, 1979) have found that pulp produced from poplar by butanol delignification reacts unusually fast in enzymatic hydrolysis.

Since a bewildering variety of solvents, catalysts and reaction conditions have been applied in delignification by organic solvents, it is appropriate to discuss the relative advantages and disadvantages of the approaches taken so far.

TABLE I. *Strength Properties of Organosolv Pulps*

Species	Freeness of pulp, Csf, ml.	Strength properties			Reference
		Breaking length km	Burst factor	Tear factor	
Softwoods					
Spruce	500	10.8	98	105	Kleinert (1974)
Southern pine	500	10.4	102	112	Kleinert (1974)
Pine	500	8.0-8.3	49-60	75-111	Nelson (1977)
Hardwoods					
Poplar	500	7.1	39	78	Kleinert (1974)
Eucalypt	230	7.9-9.8	54	55-56	Nelson (1977)

III. PROCESS CONDITIONS

Of the process conditions meriting scrutiny, three are important here: the effect of catalyst, the effect of solvent, and the influence of species.

A. *Effect of Catalyst*

In a study conducted recently at the University of Washington, Seattle (Sarkanen *et al.*, 1978; Sarkanen and Assiz, unpubl. results), the selectivity of delignification of various catalysts was explored in ethanol-water systems (1:1 by volume) using Western cottonwood (*Populus trichocarpa*) wafers as substrate. In this study, an "uncatalyzed" system was compared with delignifications using H_2SO_4, HCl, aluminum salts, $NaHSO_4$ and various organic acids as catalysts.

Results on uncatalyzed delignifications are shown in Figure 1, depicting lignin-free yield as a function of lignin remaining associated with the fiber. For comparison, the selectivity curve for kraft delignification is shown in the same diagram.

It was found that, under otherwise similar pulping and conditions, the selectivity of delignification was strikingly dependent on the way the reaction mixture was brought to the peak temperature. If the chips and pulping liquor were heated rapidly to the pulping temperature (180-185°C), the selectivity was quite poor, as shown by experimental points along curve 1 of Figure 1. If, instead, the pulping system was first conditioned at approximately 100°C for a period of three hours prior to pulping, a much improved selectivity was the result, as shown by points along curve 2 of Figure 1.

In order to try to explain these rather unexpected observations, one must consider the physical factors involved. The penetration of ethanol-water in the cell wall causes selective swelling of the lignin component which, in turn, subjects the polysaccharidic components to a stress, under which the hydrolysis of glycocidic bonds may be accelerated. If this view is correct, preconditioning at lower temperatures would gradually relax internal stresses of this kind and the hydrolysis of polysaccharides would therefore proceed at a lower rate after reaching the pulping temperature.

Similar phenomena were observed in pulpings using acidic catalysts (Figure 2). Interestingly enough, the selectivity was found to be independent of the nature of the catalyst, as long as the amounts of catalysts were kept at sufficiently low levels. Curve 1 of Figure 2 represents selectivities obtained in rapid-heating pulpings with such varied catalysts

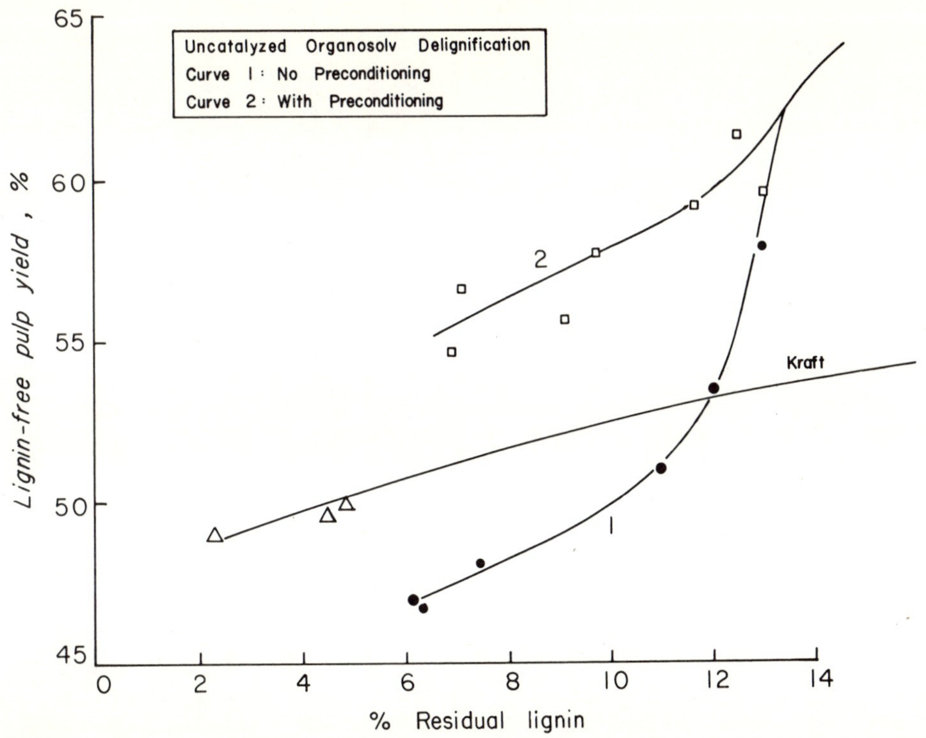

FIGURE 1. *Selectivity curves for the delignification of cottonwood wafers in uncatalyzed organosolv pulpings in ethanol-water (1:1 by volume) at 175 to 185°C. Curve 1: Temperature raised immediately to the pulping temperature. Curve 2: Temperature maintained at 100°C for 2 to 3 hours prior to pulping.*

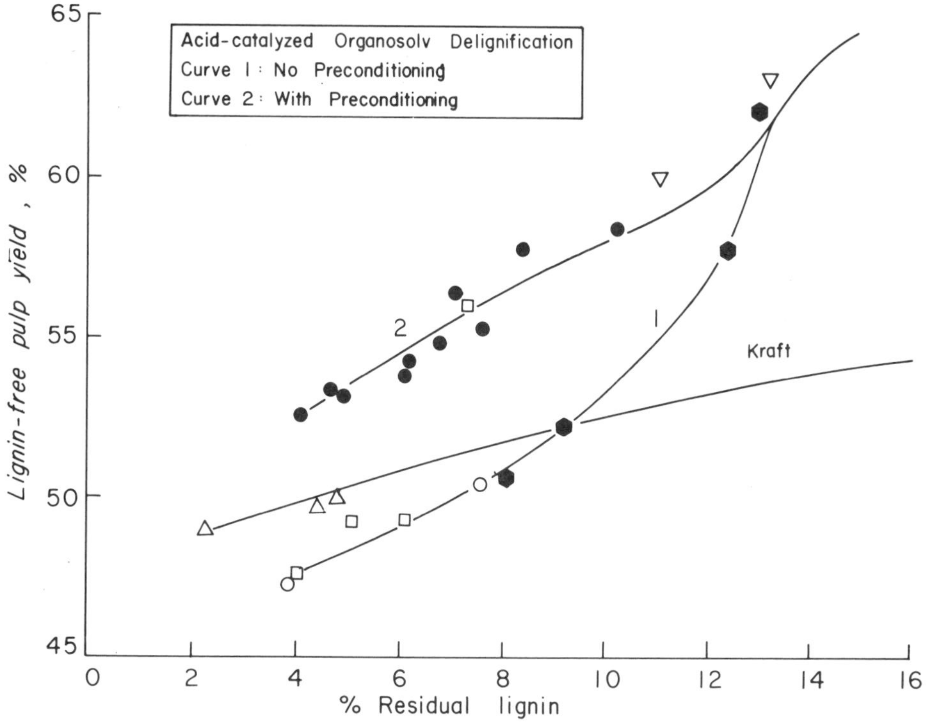

FIGURE 2. Selectivities of acid-catalyzed organosolv pulpings of cottonwood wafers. Curve 1: Immediate temperature raise. Curve 2: Preconditioning at 100°C (2-3 hours). Catalysts: (a) H_2SO_4, 155-165°, ∿ 10^{-3}M., solid hexagous. (b) .01-.03 M $NaHSO_4$, 175°, open and solid circles. (c) Oxalic acid, ∿ 10^{-2}M, 175°.

as H_2SO_4, $NaHSO_4$, oxalic-, salicylic- and 3,5-dinitrobenzoic acids, using pulping temperatures ranging from 145 to 175°C. The delignifications with sulfuric acid as catalyst were all conducted below 160°C varying the catalyst concentration in the range from .0008 to .009 M. Higher catalyst concentrations reduced the selectivity significantly. The actual pulping temperature appeared to have little effect on the selectivity. It should be noted that curve 1 of Figure 2 represents better selectivities than its counterpart in uncatalyzed pulpings.

By preconditioning at 100°C or, by simply extending the period of heat increase to two hours, the selectivities were again improved, as shown by curve 2 in Figure 2, which is nearly identical with the corresponding curve for uncatalyzed delignification. Therefore, no improvement in selectivity was achieved by the use of catalysts, but the rate of reaction and reproducibility were improved allowing the reaction to be carried out at lower temperatures. In ethanol-water systems, it appears not to be advisable to lower the reaction temperatures below 150°C, because of insufficient solubility of the lignin in the medium.

B. Effect of Solvent

Such solvent systems as triethyleneglycol-water (Orth and Orth, 1978) and dimethylsulfoxide (Clermont and Bender, 1961) are more effective in delignification than the ethanol-water system. These solvents have certain disadvantages. The high boiling point of triethyleneglycol (190°C) makes its recovery difficult and dimethylsulfoxide has been shown to react with lignin (Nahum and Pellegrini, 1970). Aqueous solutions of such volatile solvents as acetone and isopropanol offer no advantages in comparison with the ethanol-water system (Lee, 1979). Butanol and isoamylalcohol have been claimed by Aronovsky and Gortner (1936) to be more effective solvents than other volatile alcohols.

C. Effect of Species

Organosolv delignification gives satisfactory results with such hardwood species as poplars, alder, birch and eucalypts. Rice- and wheat-straw can be delignified with high selectivity (Sanwal *et al.*, 1978) while delignification of bagasse is combined with substantial polysaccharide losses (Resalati, 1979). Softwoods are delignified with a great deal more difficulty than hardwoods. Aronovsky and Gortner (1936) were unable to pulp jack pine in uncatalyzed butanol-water systems.

Nelson (1977) observed for acid-catalyzed pulpings in ethylene glycol that while eucalypt (E. regnans) could be delignified in a satisfactory manner within two hours at 170°C, the same degree of delification required 3.5 hours at 195°C in case of green (never dried) chips of Pinus radiata. Air-dried chips of the same species were delignified with more difficulty. Similar observations for other softwood species have been made by Paszner (1980).

IV. PROBABLE CHEMICAL MECHANISM OF ORGANOSOLV DELIGNIFICATION

Evidence accumulated so far suggest that lignicellulosic materials, with the probable exception of softwoods (gymnosperms), require only mild hydrolysis to be released from the cell wall matrix. This brings about the question: What types of linkages need to be hydrolyzed in order to accomplish the release of lignin?

Current knowledge on the structure of lignins has been summarized in an excellent review by Adler (1972). The hydrolyzable linkages and their estimated frequencies in softwood and hardwood lignins are presented in Table II. Among these linkages the β-aryl either (β-0-4) bonds predominate. They are, however, much more resistant towards hydrolysis than the α-aryl ether linkages. Model compound experiments suggest strongly (Miksche, 1972) that under mild hydrolytic conditions, the cleavage of ether bonds is exclusively restricted to those of the α-ether type.

TABLE II. Hydrolyzable Linkages in Lignins According to Adler (1977)

	Percent of total number of intermonomeric linkages	
Type of linkage	*Softwood (Spruce)*	*Hardwood (Birch)*
β-*Aryl either* (β-0-4)	*48*	*60*
α-*Aryl ether* (α-0-4)	*6-8*	*6-8 (?)*

The nature of chemical bonds anchoring lignins to hemicelluloses in the lignocellulosic framework has not been extensively characterized. Recent results (Erikson and Lindgren, 1977) are indicative of ether and 4-O-methylglucuronic acid ester bonds to the α-carbons of the lignin units. Although the hydrolysis rates of such linkages have not been characterized under acidic conditions, it can be surmised that they are more readily hydrolyzed than the β-O-4 bonds. These considerations lead to the proposition that the release of lignin in the organosolv process is essentially the consequence of the hydrolysis of α-aryl ether- and lignin-hemicellulose bonds. Observations made to date support this hypothesis. Isolated organosolv lignins are, in contrast to milled wood lignins, devoid of associated carbohydrates (Glasser, 1979). Monomeric hydrolysis products (Hibberts ketones), characteristic of β-O-4 hydrolysis, were found by McCarthy and coworkers in minute quantities only (S. Sarkanen *et al.*, 1980). Acidolysis of isolated organosolv lignins increases substantially the content of phenolic hydroxyl groups, indicating the presence of intact β-O-4 bonds (Goyal, 1980). The approximate activation energy of organosolv delignification, 8.4 kcal/mol (Lee, 1979), is much lower than the value 36 kcal/mol, obtained for the hydrolysis of β-aryl-ether bonds (Sarkanen and Hoo, 1980).

From the frequencies of the α-aryl ether linkages in Table II, it can be estimated that organosolv lignins should have an average $\overline{DP}_n$ of approximately 14, corresponding to a molecular weight range 2500 to 2800. Number average molecular weights of organosolv lignins (S. Sarkanen *et al.*, 1980) are of the correct order of magnitude, albeit somewhat lower (900-1200) than the predicted values. It should be noted, however, that the estimates in Table II for α-O-4 linkages are highly uncertain and may actually vary in lignins of different hardwood and other angiosperm species. Such differences are indicated by the solubilities of various lignins in 72% sulfuric acid in the first stage of the conventional Klason lignin determination, shown in Table III (Casey, 1970). It can be seen that while softwood lignins are virtually insoluble in 72% sulfuric acid, 44 to 78% of Klason lignin in hardwoods can dissolve in this medium. If it is assumed that α-aryl ether linkages are hydrolyzed in the treatment, the results would suggest a high frequency of these linkages in hardwood lignins, especially in those with high syringl to guaiacylpropane unit ratios.

As far as the softwood lignins are concerned, two alternative explanations should be considered in order to account for their resistance towards organosolv delignification. First, the frequency of α-O-4 linkages may not be sufficient to reduce the molecular size in hydrolysis to the level of dissolution. Consequently, partial hydrolysis of the more

TABLE III. Solubilities of Wood Lignins in 72% Sulfuric Acid (Casey, 1970)

Species	*Percent syringyl propane units*	*Klason lignin[a] percent*	*Lignin soluble in 72% H_2SO_4, percent of K.L.*
Western hemlock	*Negl.*	*29.5*	*0.7*
Douglas fir	*Negl.*	*28.4*	*2.3*
Ponderosa pine	*Negl.*	*27.9*	*1.4*
Eastern cottonwood	*20*	*22.4*	*43.8*
Red alder	*30*	*20.7*	*53.3*
Cascara buckthorn	*44*	*16.7*	*74.4*
Sweetgum	*51*	*19.3*	*70.0*
Madrona	*59*	*17.4*	*79.2*

[a]*Uncorrected for water-soluble lignin.*

resistant β-O-4 linkages would be required. The second alternative is the well-known tendency of softwood lignins to undergo intermolecular condensation reactions that may occur before the hydrolyzed lignin entities have the chance of escaping the cell wall matrix.

Of course, such physical factors as the permeability of the secondary cell wall structure could play an important role in the dissolution of organosolv lignins by limiting the accessibility of lignin bonds to the hydrolytic action or the diffusivity of hydrolyzed lignin in the solvent medium.

V. TECHNOLOGICAL POTENTIAL OF ORGANOSOLV DELIGNIFICATION IN BIOMASS UTILIZATION

As mentioned earlier, hardwoods and straws are generally delignified more readily than softwoods. These biomaterials are therefore to be considered the most promising biomaterials for application in the organosolv process.

A. Solvent Systems

Of the potential solvents the high boiling ones, such as ethylene glycol and triethylene glycol, possess the advantage of low digester pressures during the delignification. This apparent advantage is, however, more than counterbalanced by the obvious difficulties in the recovery of the solvent from the spent liquor and from the pulp fibers.

The recovery of solvent and of lignin is particularly convenient, when a mixture of water and a low-boiling solvent, such as ethanol, is used as the delignification medium. The solvent can be recovered by simple distillation and simultaneously, lignin precipitates as a solid. The principle of chemical recovery in ethol-based systems is illustrated in Figure 3.

The use of ethanol-water has the consequence of higher digester pressures (up to 30 atmospheres) than experienced in the conventional pulping processes. Any leak in the digester represents an explosion hazard. Two recent patents, one by Kleinert (1971) and another by Diebold (1978) describe designs appropriate for the organosolv pulping. The latter patent describes a series of tubular digesters containing the chip charges through which the heated pulping liquor is pumped in a countercurrent manner. After completed delignification, the alcohol remaining with the chips is driven off by direct steaming. Katzen (1976) has estimated that ethanol losses using this system may be as low as 8.35 liters/metric ton of pulp.

B. Catalysts

In selecting the catalyst for the process, it should be noted that the prevailing hydrogen ion concentration should not exceed the level appropriate to a given delignification temperature. Excessively low pH causes lowering of selectivity, reprecipitation of dissolved lignin on the fibers (Sanwal, 1978) as a consequence of condensation and/or

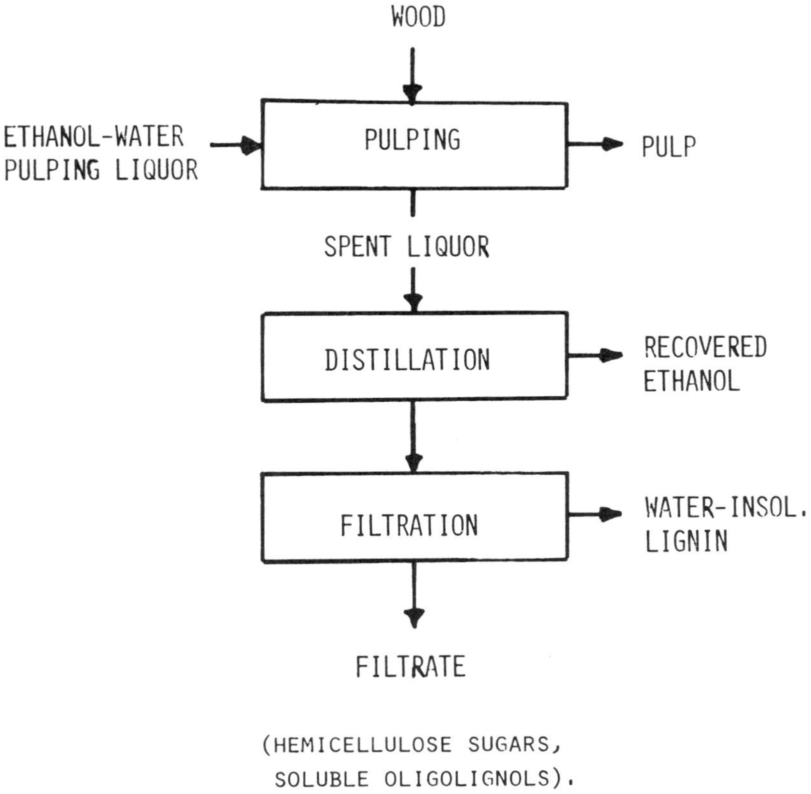

FIGURE 3. Principle of recovery in organosolv pulping.

dehydration reactions and generation of furfural from pentosans which subsequently condenses with lignin (Goyal, 1980). For mineral acids, the useful catalyst concentrations range roughly from 10^{-4} to 10^{-2} M for delignifications at 155 to 180°C. The pH of ethanol-water solutions containing oxalic acid increases during the heating period which may be advantageous (Paszner, 1980). This phenomenon is probably caused by either decomposition of oxalic acid to formic acid or by partial esterification with ethanol. The opposite phenomenon --an increase in hydrogen ion condensation by more than an order of magnitude upon heating--has been observed for aluminum salt catalysts (Sarkanen and Hoo, 1980). This is the consequence of gradual conversion of the aluminum cation to hydrated complexes. A catalytic action milder than that of mineral acids can be achieved by acid salts such as $NaHSO_4$. The relative merits of various catalysts remain to be clarified.

C. Potential of Cellulosic Fibers

Enzymatic hydrolysis to glucose, two separate applications as papermaking fibers and conversion to cellulose derivatives should all be considered as viable alternatives for the use of cellulose fibers.

1. Enzymatic Hydrolysis. The unusual reactivity of poplar fibers delignified with butanol towards enzymatic hydrolysis (Humphrey, 1979) was mentioned earlier and has been confirmed for western cottonwood fibers delignified in ethanol-water (Shafizadeh and Sarkanen, unpubl. results). In the latter case, the level of residual lignin remaining in the fibers had only a minor effect on the rate of enzymatic hydrolysis. This exceptional reactivity is probably the consequence of uncollapsed pore structure present in these fibers.

2. High Yield Paper Pulps. When the delignification in organosolv pulping is limited to residual lignin contents of approximately 10%, losses in the polysaccharide content remain minimal and the pulp yields are consequently high (Figures 1 and 2). Regardless of the high lignin content, separation to fibers is excellent. These pulps compare favorably with neutral sulfite semichemical pulps in not requiring a refining treatment after pulping.

3. Bleachable Pulps. Delignification to the level of 6% residual lignin (kappa 35) still produces pulps in higher yields than the conventional kraft process (Figures 1 and 2).

The potential of these pulps for the manufacture of fully bleached paper pulps deserves to be explored.

4. Conversion to Cellulose Derivatives. By applying sufficiently high catalyst concentrations, less than 1% of residual lignin remains in the pulp (Sanwal *et al.*, 1978) and the major part of hemicelluloses are removed through hydrolysis. Such pulps ("dissolving pulps") have potential for the manufacture of rayon and cellulose derivatives. In comparison with conventional dissolving pulps, organosolv pulps are devoid of extractive materials which in conventional pulps become chlorinated in the bleaching treatment and cause major problems in further processing.

D. Potential of Organosolv Lignins

When volatile solvents, such as ethanol, are used in admixture with water in organosolv delignification, the dissolved lignin components precipitate upon the removal of the volatile solvent by distillation of the spent liquor, as shown in Figure 3. Generally, the precipitated lignin is filterable, but occasionally tarry precipitates or colloidal suspensions are encountered. If lipophilic extractions are present in the original plant materials, these are likely to precipitate conjointly with lignin. The opportunity of obtaining lignins directly as carbohydrate-free solids is a positive feature differentiating organosolv delignification favorably from conventional pulping processes.

The solid organosolv lignins have relatively low molecular weights--$\overline{M}n$ = 800 to 1200; $\overline{M}w$ = 1700 to 2000 (S. Sarkanen *et al.*, 1980). After drying, they are directly usable as high energy fuels (∿5.7 MJ/kg). Since they are completely soluble in dilute aqueous alkali, their hydrogenation in this form to liquid hydrocarbon fuels is an attractive possibility which has not been explored as yet. Combined with the conversion cellulose to ethanol, hydrogenation of lignin, if successful, could nearly double the energy content of liquid fuels obtainable from a given quantity of biomass.

Organosolv lignins deserve to be studied also as potential chemical feedstocks, particularly as constituents for phenolic resins. Chemical modification, by controlled acidolysis or hydrogenation, may improve their suitability for such applications. The conversion of these lignins to pure phenol by hydrocracking followed by hydrodealkylation also deserves exploration.

E. Potential of Hemicellulose Sugars

Depending on the biomass species and the extent of hydrolysis, the filtrate obtained from the separation of lignin contains various amounts of pentose and hexose sugars, and acetic acid. In addition, water-soluble lignin components are present, amounting to approximately 10% of original lignin in the case of western cottonwood. These components appear to be related to β-1- and β-O-4 linked oligolignols (Sarkanen and Hoo, unpubl. results), identified by Nimz (1965, 1966) among products obtained in mild hydrolysis of beech and spruce woods.

In general, the dominant solute in the filtrates is expected to be xylose which lends itself to the following conversions: (1) chemical conversion to either xylitol or furfural, and (2) enzymatic isomerization to xylulose, fermentable to ethanol by bakers yeast (Tsao, 1980).

On the other hand, a simpler processing option for this stream consists of anaerobic fermentation to methane which could be used to generate part of processing energy.

VI. CONCLUSION

Separation of lignocellulosics to their components by organosolv delignification represents undoubtedly one of the simplest options for integrated utilization of lignocellulosic biomass and possesses realistic potential for industrial application. Besides wood and agricultural residues, application potential to species containing valuable extractive components, such as guaiule, has been demonstrated (Sarkanen and Asbell, 1979). So far, the amount of research expended in this area has been insufficient to bring the promise of the method into proper focus. Obviously, detailed investigations will be required to assess the utility of organosolv delignification in comparison with such alternative, but chemically related biomass separation processes as the Iotechsteam explosion process (Marchessault, 1978) and the autohydrolysis approach explored by Wayman (Wayman and Lora, 1979).

Finally, organosolv lignins can be of great utility in fundamental studies focused on the characterization of the chemical structure of lignins. They are readily isolated in good yields, contain most of the original lignin structures intact, yet are free of carbohydrate constituents, possess low molecular weights and are soluble in many organic solvents.

REFERENCES

Adler, E. 1977. *Wood Sci. Technol. II,* 169.

Aronovsky, S. I., and Gortner, R. A. 1936. *Ind. Eng. Chem. 28*, 1270.

Bobomolov, B. V., Groshev, A. S., Popova, G. I., and Vishnyakova, A. P. 1979. *Khimiya Dravesiny 4*, 21.

Casey, J. M. 1970. M.S. thesis, University of Washington, Seattle, Washington.

Clermont, L. P., and Bender, F. 1961. *Pulp Paper Mag. Can. 62* No. 1, T 28.

Diebold, V. B., and others. 1978. U.S. Patent 4, 100, 016.

Erikson, O., and Lindgren, B. O. 1977. *Svensk Papperstidn. 80* (2), 59.

Glassner, W. 1979. Private communication, Virginia Polytech. Inst., Blackburg, Virginia.

Goyal, G. 1980. M.S. thesis, University of Washington, Seattle, Washington.

Grondal, B. L., and Zenczak, P. 1950. U.S. Patent 2, 772, 968.

Humphrey, A. E. 1979. *Adv. Chem. Ser. 181*, 25.

Kleinert, T. N., and Tayenthal, K. 1931. *Z. angew. Chem. 44*, 788.

Kleinert, T. N. 1971. U.S. Patent 3, 585, 104.

Kleinert, T. N. 1974. *Tappi 57*, No. 8, 99.

Lee, Buyung. 1979. M.S. thesis, University of Washington, Seattle, Washington.

Marchessault, R. H., and St. Pierre, J. 1978. Proc. of World Conference on Future Sources of Organic Raw Materials, July 10-14, 1978, Toronto, Canada, in press.

Miksche, G. 1972. *Acta Chem. Scand. 26*, 289.

Nahum, L. S., and Pellegrini, P. 1970. *Svensk Papperstidn 73*, 725.

Nelson, J. 1977. *Appita 31*, No. 1, 29.

Nimz, H. 1965. *Chem. Ber. 98*, 3153.

Nimz, H. 1966. *Chem. Ber. 99*, 496, 2638.

Orth, G. O., Jr., and Orth, R. D. 1977. U.S. Patent 4, 017, 642.

Paszner, L. 1980. Private communication. University of British Columia, B.C., Canada.

Rassow, B., and Gabriel, H. 1931. *Cellulosechem. 12*, 227, 249, 290, 318.

Resalati, H. 1979. M.S. thesis, University of Washington, Seattle, Washington.

Sanwal, J. S., Meshitshuka, G., Wu, L., and Sarkanen, K. V. 1978. Catalytic Effects in Organosolv Pulping. Paper presented at the Tappi Wood Chemistry Conference, Appleton, Wisconsin, May 7-10, 1978.

Sanwal, J. S. 1978. M.S. thesis, University of Washington, Seattle, Washington.

Sarkanen, K. V., and Asbell, H. 1979. Organosolv Delignification of Plants Containing Polyisoprene. Paper presented at the Cellulose, Paper, and Textile Division, ACS/CSJ Chemical Congress, Honolulu, Hawaii, April, 1979.

Sarkanen, K. V., and Hoo, L. 1980. Paper submitted for publication in *Wood Science Technol.*

Sarkanen, S., Hall, J., Teller, D., and McCarthy, J. L. 1980. Paper submitted for publication in "Macromolecules."

Schwenzon, K. 1966. *Zellstoff Papier 15* (7), 202.

Tsao, G. T. 1980. Paper presented at the Gordon Conference, New London, New Hampshire, July.

Wayman, M., and Lora, J. H. 1978. *Tappi 61* (6), 55.

ENVIRONMENTAL CONSIDERATIONS IN WOOD FUEL UTILIZATION

William D. Kitto

Envirosphere Corporation
Bellevue, Washington

ISBN 0-12-535902-0

I. INTRODUCTION

Controlled supplies and high prices of oil and natural gas have caused public officials and energy planners to look seriously at a broad diversity of energy sources, including biomass. Typically, analysts studying the feasibility of energy sources such as biomass have focused on the economic and technical feasibility of different development scenarios. Implicit in an approach focusing on these concerns is the assumption that environmental concerns are not constraining the development of such fuels. As a result, basic environmental considerations associated with utilization of biomass fuels are not always addressed adequately.

Several comprehensive studies on the environmental impacts of regional biomass projects have been prepared. However, national assessments are less complete. Most notably, Rose *et al.* (1979) and the Department of Energy (1980) have produced studies for the projects in Minnesota and Maine, respectively. Although both of these studies analyze regional concerns, their approach facilitates extension of their findings to generic concerns. This analysis of important environmental concerns is based, in part, on these works and the reports of Hewitt and High (1979) and High and Knight (1977).

Environmental impacts related to the use of biomass as a fuel fall into two categories: (1) supply related environmental impacts and (2) impacts associated with conversion or utilization processes. Because there are many fuel supply scenarios under consideration, it is impossible to address them all in a generic analysis. As a result, it is necessary to limit environmental analysis to supply activities that hold the most promise for successful present and near term utilization. In this analysis, environmental considerations related to the use of wood fuel are analyzed. The energy conversion technology examined is direct combustion.

II. FUEL SUPPLY AND ENVIRONMENTAL IMPACTS

A. *Fuel Supply Options*

Fuel supply activities of varying intensities lead to different levels of environmental impact. An intense fuel supply scenario using fuels such as crops grown on intensively managed short rotation energy farms has considerably greater impacts associated with it than a less environmentally demanding supply option such as only using existing mill residues. These are the two extreme positions. To reflect the varying intensity of impacts, four reference supply systems are identified and subsequently analyzed to determine impacts. Four supply options used by Kitto (1979) are listed and described in order of increasing intensity: (1) existing mill residues, (2) mill residues and logging residues, (3) previously described fuels and stand improvement cuttings, and (4) all wood fuels including biomass from energy farms. These categories form the basis of this analysis.

1. Existing Mill Residues. Many forest industry plants currently produce large quantities of material that are not suitable for use in structural or fiber products. Virtually all of this material is currently used as fuel in the forest products industry, except where local supply/consumption imbalances exist. The residues consist of bark, sawdust, spent liquor and other matter, and planer shavings. Currently the U.S. uses approximately $\sim$1.7 quads (Tillman, 1980), as Figure 1 shows. Although much of this material is currently used, some quantities are not used to produce energy because of economic considerations. This residual fuel is bulk and expensive to transport. If a plant has an excess supply of these residuals it may be cost effective to incinerate them or to dispose of them in a landfill. In general, more complete utilization of this material is the least environmentally damaging regime.

From strictly an environmental viewpoint, more complete utilization of this residual material would reduce disposal activities with their attendant impacts. It could, however, lead to alterations in the market place and an associated

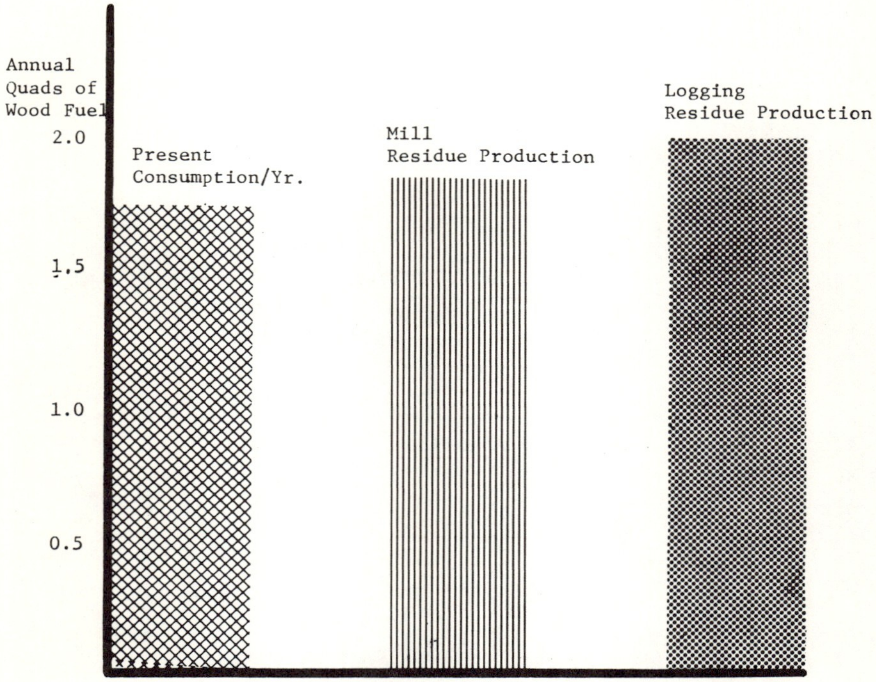

FIGURE 1. The relationship between the production and consumption of wood (fuel) residues.

increase in price for traditional forest products. The increased use of mill residues for energy production, particularly by nonforest industry firms, should be carefully analyzed to make certain that net improvements in environmental impacts do occur.

2. *Logging Residues*. In much of the United States, large quantities of biomass remain on the site following harvesting operations (see Grantham and Howard, Logging Residues as an Energy Source). The quantity of this potential fuel has been estimated in a number of studies, both at the national level (Bethel *et al.*, 1979) and in regional studies such as the ones prepared by Sampson (1979) and Howard (1979). Use of this material for fuel is constrained by the economics of collection, which is typically labor intensive and costly. Transporting this material from the remote locations where it is found to conversion facilities is also costly.

These factors offset some of the energy gains achieved by using logging residuals for energy. The economic and net energy analyses of transporting logging residues to major conversion facilities is reviewed by Bethel *et al.* (1979). In this study it was concluded that transporting wood distances greater than 25-50 miles is uneconomic. This compares with the economically acceptable transportation of coal of over a thousand miles. Thus, logging residues must be located close to the utilization facility (e.g., the pulp mill boiler) if economic operation is to occur. It is also interesting to compare these values to the estimated breakeven point for transporting wood in pickup trucks for domestic use. In one study, it was estimated to be energy efficient to haul a load of wood up to 213 miles (Burke 1980).

At the present time, logging residues, consisting of tops, branches, and other unrecovered material is typically left in the forest and subsequently burned during site preparation. Although utilization of this material for energy production is limited by high removal and transportation costs, it may become economic to remove some of this material and transport it to a conversion facility as other energy prices rise further. If such a fuel supply system becomes a reality, certain environmental impacts will occur. Logging residue associated impacts will be more significant than utilization of mill residues, but far less damaging than the most intensive supply options, such as biomass energy farms, described below.

3. *Silvicultural Residues*. As forest management practices become more intensive, silvicultural activities increase and additional residues are produced before final harvest material is produced. For example, if an intensive management regime is implemented, biomass material produced from

precommercial thinning operations and unutilized material in commercial thinning operations could be made available for use as fuel. Currently, in areas where more intensive management is carried out, these materials are left at the site or burned. Development of such a fuel supply, like mill and logging residues, is limited by economic considerations. In addition, much forest land is not presently managed intensively where thinning and other residues are produced. As such forest management becomes more common, these residues may become a significant resource of fuel. Although economic utilization of silvicultural waste material does not occur at the present time, it offers promise. More frequent and intensive management actions will increase both the quantity of this fuel and the relative ease with which it can be removed. If past trends continue, a rise in energy prices and an increase in the intensity of forest management activities will lead to greater use of this material for energy production.

4. Biomass Energy Farms. The most intensive biomass energy development would be to grow crops on a short rotation basis for the exclusive objective of producing energy. Employing standard agricultural crops and practices, rotations of one year could occur. If the objective is to use forest biomass, rotations of 5 to 10 years could be expected. In either case, energy farming represents an intensive land use and management activity where the site is totally committed to energy production. In such systems, it is anticipated that fertilizers must be applied and that certain herbicides and pesticides may be needed to ensure that growth is optimized.

Much interest and controversy has surrounded proposals to devote large tracts of land to energy production. Irrigation, fertilization, and frequent harvest have been included in many proposed energy farms. One of the most extensive studies of energy farms (Mitre, 1977) analyzed a number of aspects of such projects. In general, the Mitre report judged a number of projects to be viable. Experience and the lack of any privately funded activities of this type, however, suggest otherwise. Regardless of whether or not such proposals are economically viable, a number of environmental concerns must be analyzed and dealt with. As the following discussions indicate, environmental consequences of more intensive supply options will be significant requiring detailed analysis.

B. Environmental Impact Analysis--Supply Systems

A range of impacts are anticipated for the various supply options considered. Impacts related to air, water, land use, aesthetics, terrestrial ecology, and socioeconomic conditions are expected. Impacts in each of these areas are described below.

1. Air Quality. Air quality impacts are associated with present practices of disposing of mill residues. Such residues, if not disposed of through landfill operations, are normally burned. Most of this material is combusted using relatively new clean combustion systems. In some cases, however, material is burned in old inexpensive, but polluting, facilities. Table I shows air emissions from wood combustion compared to oil utilization.

More intense utilization of either mill or logging residues will have a positive effect on air quality. Such an effect will occur because utilization of this material for energy production will eliminate the need to dispose of this material through open burning. Impacts of such burning are

TABLE I. Emission Comparison of Wood Combustion and Residual Oil[a,b]

Pollutant	*Wood waste*[c] *Kg/h*	*Residual oil*[d] *Kg/h*
Particulates	*0.95*	*6.90*
Sulfur dioxide (SO_2)	*8.07*	*96.16*
Nitrogen oxides (NO_x)	*10.43-49.90*	*18.14*
Hydrocarbons	*19.96*	*0.32*

[a]*Reference: Tillman, 1980.*

[b]*Basis* 10^5 *GJ or ca* 10^6 *Btu/h of fuel.*

[c]*Assumes baghouse control.*

[d]*Assumes 2% sulfur in oil as fired, no stack gas cleanup system.*

Source: Boyes, 1978.

TABLE II. Emissions From Forest Burning (Prescribed and Wildfires) for the United States[a]

Pollutant	*Emissions* (10^6 tons/year)	*Percent of national total*
Particulates	*6.7*	*23.7*
Sulfur dioxide (SO_2)	*Negl.*	*Negl.*
Nitrogen oxides (NO_x)	*1.2*	*5.8*
Hydrocarbons	*2.2*	*6.9*
Carbon monoxide (CO)[b]	*7.2*	*7.2*

[a] *Reference: Hall, 1972.*

[b] *Compiled from NAPCA Publication AP-65 (U.S. Department of Health, Education, and Welfare, 1970d); comprised of 2.48 million tons from prescribed burning and 4.74 million tons from wildfires.*

significant and have been extensively studied (Geomet, 1978; Hall, 1972). Table II shows the emissions from such open burning.

Unlike air quality impacts associated with mill and logging residue supply options, emissions from using silvicultural residues for energy production will not result in a reduction of existing emissions. This is because these residues are not typically burned at present. Air quality impacts from using thinning residues, however, should not have significant air quality impacts. Burning this material under controlled conditions will not produce large quantities of harmful emissions. In addition, even if one considers fugitive dust and other emissions released during removal and transportation activities, the impacts should not be great.

Although silvicultural energy farms will likely have more impact on air quality than any of the less intense supply systems, overall supply impacts should not be significant. Activities leading to a small increase in air pollutants would include exhaust from equipment used in planting, managing, and harvesting operations. Fugitive dust released by such equipment and materials released into the atmosphere through employment of aerial fertilizer and chemical application equipment will also have an effect on air quality, but these should not be of overriding concern. In general, air quality impacts associated with even the most intense fuel supply system will not be of great environmental concern. Table III shows expected emissions from harvesting equipment used to supply a major woodburning facility (DOE, 1979).

2. *Water Quality*. Water quality impacts vary with the intensity of the supply system employed. In the case of utilization of mill residues, few, if any, additional water quality impacts are expected. In the second most intensive supply regime, logging residues, both positive and negative impacts could occur. In the most intensive supply scenarios, silvicultural residues and energy farms, negative impacts are expected. Since no incremental water quality impacts are expected from using mill residues, only impacts from the other three supply options are evaluated here.

Removal of logging residues may contribute to an increase in runoff because removing this material will eliminate many obstacles blocking runoff. Studies have shown that runoff increases as biomass is removed from the site. There is conflicting evidence whether increased runoff from cleared forest lands leads to greater erosion and increased sediment in the streams. Regardless, this effect should not be major.

TABLE III. Expected Annual Emissions of Criteria Pollutants for Harvesting Equipment at Nine Sites[a] *(tons/year)*

	Particulates	*Hydrocarbons*	*Nitrogen oxides*	*Carbon monoxide*
Small diesel powered equipment	*1.6*	*3.3*	*22.5*	*1.4*
Large diesel powered equipment	*5.0*	*5.7*	*71.1*	*4.7*
Gasoline powered equipment	*1.2*	*25.2*	*18.9*	*0.0*

[a]*Reference: DOE, 1979.*

In addition, certain benefits will occur as a result of removing the logging or thinning residues from the site. For example, removal of the logging debris for use as fuel will reduce the likelihood that this material will end up in streams.

Depending on the intensity at which biomass energy farms are managed, water quality impacts may vary from slight to significant. In addition to the impacts associated with vegetation removal described above, impacts resulting from the intrusion of the chemical used in fertilizing the site and in controlling insects and unwanted vegetation should be considered. Such analyses should be completed on a site specific basis, recognizing that the biomass farms must be actively managed.

3. *Land Use and Productivity*. Using either mill, logging, or silvicultural residues will not alter land use patterns. In each of these cases, the biomass fuel generated is not the primary product. In each of these cases, biomass production remains the secondary land use.

Land use conflicts could arise if large blocks of land are set aside as energy farms. Conflicts will likely be more intense in the future as the poplation expands and as the number of lands dedicated to special uses expands. For example, in many states land adjacent to streams must be managed to protect the stream and adjoining shoreline. Such individual variations within the larger energy farm will make uniform management difficult and more costly. Other land use designations such as flood plains, wetlands, wildlife related designations may also make it difficult to set up an expansive biomass energy farm. Such land use considerations would have to be analyzed by site specific basis, but it is reasonable to conclude that land use conflicts would be expected should any biomass plantation be seriously proposed.

Along with concerns about the availability of large acreages for energy production is the issue of whether complete utilization of the site's biomass production leads to a serious depletion of an area's productive capacity. In the case of utilization of mill residues, the potential fuel has already been removed from the site and the loss due to conversion to energy is a moot point. For the other supply options, more detailed analysis is needed.

Logging residues are generally burned on the site enabling many of the nutrients to be available to future crops. If this material is transported to an energy conversion facility, there is concern that the site will become nutrient poor. Although evidence exists showing that vegetation removal causes a deterioration in site quality, there is stronger support for the contention that removal of logging residues alone will not severely degrade a site's nutrient resources. In particular, removal of material normally left on the site as slash, is not anticipated to cause more nutrient depletion than standard timber harvesting practices. This is explained by the fact that most nutrients are fixed in a tree's foilage or small branches. This material would generally remain on the site even if damaged logs, large branches, and other material constituting slash is left on the site.

For the more intensive supply options, silvicultural residues and biomass energy farms, nutrient related problems become more important as Table IV shows. For each of these options, reduction in site quality will occur if living material, including small branches and foilages, is removed from the site. This situation would be most severe when entire trees are harvested on short rotations and used to produce energy. Where nutrient loss is a potential problem, the application of fertilizers can be used to insure that adequate growth rates are achieved. This practice occurs in agricultural activities where fertilization occurs as routinely as planting and harvesting. The future of intensive fertilization in forestry is unclear, however, as the price of petroleum derived fertilizer escalates. Nevertheless, the reduction of site quality notwithstanding application of fertilizers together with the economic and environmental constraints associated with the use of fertilizer suggest that individual proposals calling for complete use of silvicultural residues or the initiation of a biomass energy farm warrant careful scrutiny.

TABLE IV. Nutrient Drain for Stemwood Logging and Whole-Tree Harvesting Per Rotation[a]

Harvesting Method	Nutrients			
	Nitrogen	Phosphorous (pounds/acre)	Potassium	Calcium
	N	P	K	Ca
Stemwood logging	99-240	9-26	50-153	158-252
Whole-tree harvesting	263-687	30	143-390	334-409

[a]*Reference: Kimmins, 1975.*

4. *Social and Economic Effects*. Rose (1979) and the Department of Energy (1979) completed extensive socioeconomic impact analyses for biomass-fueled facilities in Minnesota and Maine, respectively. Each of these studies addressed a relative increase in employment expected if more intensive utilization of forest growth were to occur. In the former study, social and economic impacts were postulated for a facility designed to produce 1000 green tons of cord wood. This amount of wood is required to fuel a cogeneration project yielding approximately 25 megawatts. For this facility, it was estimated that 114 persons would be directly employed in harvesting operations. In this same study, indirect jobs resulting from the increased harvesting activity were estimated to be 134. This increase in employment, however, was spread over a large region because fuel had to be obtained from considerable distances to provide an adequate supply.

An increase in income to the affected region is associated with the increase in employment. The significance of this increase can only be assessed in relative terms. It will be most important in regions of low population. Other sources of income include stumpage fees paid for the biomass fuel and fees paid for materials and services rendered in maintaining the equipment needed in harvesting and management operations.

Another important socioeconomic consideration is the level of revenue to be obtained by the state and local governments. Both sales and income taxes will result in an increase in revenue to the region where any intensive management program is implemented. In addition, excise, fuel use, business and occupation, and other taxes may become sources of revenue for local governments. These revenue increases should be evaluated on a specific basis.

Socioeconomic impacts associated with more intensive use of mill residues would be negligible since present operations already provide for disposal of these residues. The additional personnel required to transport this material and dispose of it in an energy conversion facility would not be significant.

A much more significant socioeconomic impact would arise from the use of logging residues in an energy conversion facility. The process of harvesting and transporting logging residues to a central point is labor intensive. Workers would be needed at the collection site, for transportation activities, and at the conversion facility where the material is unloaded. In many areas employment levels associated with collection and transportation activities would be similar to existing employment patterns in the forest products industry. Collection of logging residues, however, would be accomplished with less skilled workers. Employment related to residue collection would tend to be seasonal closely reflecting seasonal patterns associated with traditional harvesting practices.

The net result of the factors described above is that more intensive use of biomass will increase employment and spending in a particular geographic area. This increase in economic activity will amplify seasonal employment problems and will not add diversity to a region's economic base.

Unlike more intensive use of logging residues, establishment of biomass energy farms would not necessarily have employment patterns similar to existing industries. Employment levels at biomass energy farms would be relatively low, but they would be constant year round. Personnel would be engaged in production, harvesting, and transportation activities so a range of technicians and professionals would be employed. Although employees with a range of backgrounds would be needed for this type of operation, it is not anticipated that many employees would be needed. In fact, the economic viability of biomass energy farms depends on the extent of mechanization possible. Mechanical harvesters collecting fuel over large acreages with little human support are required to make energy farms economically viable. This fact suggests that it is not the number of employees that would be employed at a biomass energy farm that is the most important socioeconomic concern. Rather, it is the large amount of land that would be set aside for energy production instead of being used for other purposes that is most significant.

5. Visual Quality. The utilization of biomass material will have an impact on the visual quality of the area from which the biomass materials are being supplied. Assessment of the effect on visual quality is necessarily subjective, but certain general conclusions can be drawn. In the case of utilizing existing mill residues, impacts will be positive. Currently, underutilized biomass material is either burned or transported to a landfill area and disposed of using conventional landfill techniques. If the latter of these disposal techniques is employed, residual material is typically stored, transported, and disposed of without regard to aesthetics. More efficient use of this material would likely lead to a reduction in some of these aesthetically unappealing activities.

The removal of this waste material generated during industrial operations generally contributes positively to the aesthetics of an area. This positive contribution is nowhere more evident than in the case of removal of logging slash. This residue is strewn about the landscape in a disorganized manner with virtually no redeeming aesthetic factors. This is significant because people using forest land generally find logging residues to be visually unappealing. Therefore, removal of this material would have a positive short-term effect on the visual quality of the harvested area.

As fuel supply options become more intense, there is a greater probability of a decrease in visual quality as a result of harvesting activities. For example, the removal of undesirable commercial species for use as fuel would decrease natural diversity making the forest less aesthetically pleasing. Other intensive forest management activities, such as precommercial thinning, also lead to unnatural appearing stands with relatively large openings between individual trees. The unnatural appearance of such stands may also reduce the aesthetic appeal at a specific area. Such changes, however, typically become less significant over time.

In vigorous stands, intensive forest management activities are not visually evident after a relatively short time. Although intensive forest management activities lead to a somewhat unnatural appearing forest stand, failure to adequately remove residues generated during management activities leads to far more significant aesthetic impacts. For example, if precommercial thinning residues are allowed to remain in the forest they become dried out and discolored presenting both a fire hazard and an aesthetically undesirable element in the landscape. Because production of forest management residues could occur with or without the understanding that residues would be used for energy production, two conclusions should be draw. First, if activities associated with intensive forest management are going to occur primarily to increase supplies of biomass fuel, then it can be concluded that biomass fuel supply requirements have had a deleterious effect on the environment. On the other hand, if intensive management activities will occur regardless of fuel supply needs, it can be assumed that positive impacts will result from the use of residues for energy production.

It is anticipated that biomass energy farms would be even age stands with virtually no diversity in constituent plant species. Such stands tend to be monotonous and aesthetically unappealing. The degree to which these stands are unnatural appearing varies with time. Such concerns are also subjective and vary according to the bias of the viewer.

In addition to a stand's appearance, a number of roads where equipment quarters will probably be needed to efficiently harvest material grown on a biomass energy farm. These roads will further compound the unnatural appearance of the energy farm. As a result of these site modifications, it may be appropriate to consider the unique attributes of the biomass energy farm as an aesthetic attraction in itself rather than as it compares to a natural forest. In this light, aesthetic impacts are not as negative.

6. *Wildlife.* The removal of dead and dying vegetation or commercially undesirable species for use as fuel may make economic sense, but it could represent a loss of habitat for certain species. Conversely, activities which increase the diversity of a forest canopy often contribute to beneficial impacts on widllife. Therefore, impacts on wildlife may be of either a positive or negative nature, depending on the specific type of supply option being considered.

In determining wildlife impacts associated with fuel production schemes, it is important to review the impacts on wildlife of other related activities. For example, it has been shown that various timber harvesting schemes influence wildlife to various degrees. In general, an even aged management scheme leads to less diversity and lower habitat quality for most species. On the other hand, several studies have demonstrated that the edge effect associated with the clearcut harvesting practices is beneficial for some species. The edge effect refers to that portion of a harvested area adjacent to an area that has not been harvested. Here, wildlife are able to take advantage of the attributes of both communities that have experienced disturbance and those which have not. Typically in such areas, large mammals do particularly well because of the abundance of browse and availability of shelter.

Analyzing the impact of individual fuel supply options on wildlife can be addressed in light of the analytical framework provided above. In general, activities that lead to increased diversity enhance the habitat. In the case of more intensive production scenarios, habitat modification can be an important issue.

Removal of dead, diseased, or undesirable species for energy production may be a logical forest management activity from a commodity production standpoint, but it does not necessarily lead to improvement of wildlife habitat. Dead trees or snags are important in the maintenance of stable populations of certain birds. This fact is recognized by several land management agencies that have established snag management programs. In the particular case of biomass energy facilities, this concern over loss of wildlife habitat has been the focal point of much of the environmental opposition of Consumers Power Company's proposed biomass electric-generating facility in Hersy, Michigan. For other facilities, where "undesirable" species are planned to be used for energy production, similar opposition may develop.

Biomass energy farms will consist of intensely managed vegetation harvested after a relatively short rotation. Vegetation communities that develop under such intensely managed conditions tend to consist of few species spread over a narrow range of age classes. Consequently, biomass energy farms represent a less than ideal setting for wildlife. This intensive management of a limited number of species is similar to timber management activities in many respects, but the impact on wildlife is more pronounced because the vegetation is being managed over very short rotations.

Although the abundance of a diversity of wildlife species will be reduced in an intensely managed short rotation energy plantation, certain species will prosper, particularly those species requiring young vegetation for forage or shelter. In many regions, species that prosper under such conditions are the most desirable big game species. Such factors should be analyzed on a site specific basis recognizing the underlying condition that even though individual species may be better off, species diversity and the overall quality of wildlife habitat will be reduced.

In addressing the impacts of biomass energy farms on wildlife, it is necessary to spell out exactly what type of harvesting activities would be carried out. Studies (Ressler, 1972; Pengelly, 1972) have shown that a variety of harvesting schemes have different impacts. Moreover, the extent to which large areas are managed on an even age basis has a very significant impact on wildlife. The larger the area being managed on a similar harvesting schedule, the lower the quality of the wildlife habitat. Similar issues exist for agricultural residues. Such factors make it essential to analyze any project area to determine if the sacrifice in quality that would occur from establishing a large tract managed in the same fashion is warranted. Such an analysis would certainly consider any endangered or threatened species that would be located in the proposed project area. Such concerns can be of paramount importance in project planning.

III. ENERGY CONVERSION PROCESSES

Environmental concerns related to the conversion of biomass to energy are distinct from those associated with the various supply options. Regardless of how the biomass material was supplied, impacts that occur are a direct result of the conversion process employed. A number of conversion systems of varying feasibility can be identified. The conversion process considered here is direct combustion for raising industrial process steam and/or electricity generation. Gasification, although likely to be a proven conversion process in the near future, is not considered here (for a review of gasification technology, see Levelton *et al.*, 1978).

A. *Direct Combustion Environmental Impacts*

Direct combustion of wood fuel to produce heat that is used to generate steam for either industrial processing, direct heating, or electricity generation is an established practice. A number of combustion techniques can be employed to convert biomass to energy. The attributes, advantages, and disadvantages of the most important combustion systems are summarized elsewhere (Bethel *et al.*, 1979; Junge, 1975). For a meaningful analysis of environmental concerns, however, it is useful to look at impacts affecting specific resources. In this analysis, air quality, water quality, and solid waste disposal impacts are addressed. Other impacts such as aesthetic, socioeconomic, or land use are left for more detailed site specific analysis.

1. Air Quality. Air quality impacts from wood burning facilities are a result of either gaseous or particulate emissions. Gaseous emissions from wood reflect the chemical make-up of wood as well as combustion conditions. The chemical content of wood is revealed in the ultimate analysis of wood fuel shown in Table V. The most significant difference between wood and other fossil fuels is the relatively little sulfur present in wood. As a result, SO_2 emissions are low and insignificant when compared to coal. Oxides of nitrogen, NO_x, are potentially more troublesome, and studies reported by Tillman (1980) reveal that wood NO_x emissions are quite variable. In spite of this potential variability, the conclusion reached by Schweiger (1980) that NO_x discharges can be controlled within regulatory limits with proper boiler design seems justified. Consequently, it is reasonable to conclude that gaseous emissions are not of major concern for wood fuels.

Unlike gaseous emissions, particulate emissions from wood burning systems are quite high. The quality of material released varies greatly from operation to operation and is influenced by many factors. Schweiger (1980) identified the following factors as being important:

(1) Species of tree and logging region;
(2) Type and amount of fuel burned in combination with wood waste (if any);
(3) Distribution of fuel-particle sizes;
(4) Ash content of fuel;
(5) Firing method;
(6) Fuel distribution in the furnace;
(7) Air distribution over and under the grate;
(8) Grate heat-release rate;
(9) Furnace heat-release rate, residence time, and upward gas velocity;

TABLE V. Proximate and Ultimate Analysis of Softwoods and Hardwoods (wt %)[a]

Analysis	Softwoods	Hardwoods
Proximate		
Volatiles	40.6	52.4
Fixed carbon	12.4	12.9
Ash	1.0	2.5
Moisture	46.0	32.2
Total	100.0	100.0
Ultimate (dry basis)		
H	6.1	6.2
C	53.0	51.0
O	38.8	39.7
N	0.2	0.4
S	0.0	0.0
Ash	1.7	2.7
Total	100.0	100.0

[a]*Reference: Tillman, 1980.*

(10) Furnace configuration;
(11) Design of fly-carbon reinjection system; and
(12) Amount of fly carbon reinjected.

Obviously, the influence of each of the factors listed above varies for each system making it difficult to accurately quantify particulate releases at a generic level. A range of particulate emissions between 0.5 and 8 grains per standard cubic foot was identified (Schweiger, 1980) as being typical for wood burning facilities. Other information (Boyes, 1979) suggests that with available control technology, more limited emission levels can be achieved.

Although wood burning systems have a high potential to produce particulate emissions, pollution control equipment is available to limit emissions to acceptable regulatory levels. Mechanical collectors, wet scrubbers, electrostatic precipitators, and fabric filters are the most widely employed particulate control equipment.

Cyclone and multicyclone collectors have been used extensively to control particulate emissions. These systems, however, are frequently not sufficient to meet air quality regulations. Consequently, they are often used in series with scrubbers, electrostatic precipitators, or fabric filters. In these systems, the cyclones typically trap the majority of larger pieces leaving the hard to contain particles for the more sophisticated control equipment.

Wet scrubbers are often used in conjunction with cyclone collectors. These devices are generally effective in capturing the small-sized particles, but require a wastewater treatment system and are quite expensive. Wet scrubbers also can follow load charges without sacrificing collection efficiency (Schweiger, 1980). This is an important factor in the pulp and paper industry.

Electrostatic precipitators also are used in controlling particulate emissions. Such systems require energy to operate and are not ideally suited for wood fuel systems. If, however, wood is burned in conjunction with other fuel such as coal, precipitators can be evaluated in a more favorable light. The use of electrostatic filters in conjunction with other control devices is also effective in limiting emissions to acceptable levels.

A fabric filter, or baghouse, collection system is viewed by some (Boyes, 1979) as the most effective control strategy. There is resistance, however, within the forest industry and among design engineers to using this type of control for wood combustion systems. This resistance stems from the fact that baghouse equipment has a reputation for being unsafe and highly susceptible to fire. Primarily because of this fact, fabric filters are not widely employed in spite of their ability to limit emissions.

Studies (Hall *et al.*, 1975) have quantified the effectiveness of various control equipment. Results of these studies are shown in Table VI. Not shown in this table is the fact that a high cost is associated with the most effective particulate control equipment. Consequently, problems related to particulate control are primarily of a regulatory or economic nature. Stated differently, the problem is: Are the economic costs of employing a certain control strategy worth the benefits? Answers to these questions are normally determined through regulatory compliance activity.

TABLE VI. *Particulate Emission Control Efficiency*[a]

Collector equipment	*Collection efficiency*
Mechanical collection	90.0
Wet scrubber	98.0
Dry scrubber	98.0
Electrostatic precipitator	99.9
Fabric filter	99.9

[a]*Reference: Ekono, 1980.*

2. *Water Quality*. Wastewater treatment, as well as intake and discharge needs, is the primary activity associated with wood energy facilities that affect water quality.

Perhaps the primary activity affecting water quality is the use of water to transport boiler ash from a steamplant. Such a transport system requires a wastewater treatment facility. Wastewater systems are typically designed to meet regulatory standards which insure that discharges do not have a significant impact on the aquatic environment. As a result, wastewater treatment, like air quality impact mitigation, often becomes a regulatory and economic issue rather than a technical one.

Wood burning steam producing facilities also require makeup water for steam production and cooling purposes. In general, such needs are not great because of the limited size of wood burning facilities. As noted earlier, and explained in Bethel *et al.* (1979) the bulky nature of wood fuel effectively limits the size of wood burning facilities to 50 megawatts. Water needs for a typical 40 MW wood fired full condensing power plant would be approximately 1000 gpm. Intake and discharge restrictions would likely be placed on a wood fired facility even though the water needs are relatively low. Such a facility would also trigger regulatory review as described previously.

While water needs are relatively low, restrictions on water use can be important. Generally, such concerns are most significant if the facility draws or discharges water into a small water body. In these circumstances, small withdrawals or discharges can represent a significant alteration of the hydrologic regime. Of particular importance are thermal discharges during periods of low flow. The problem is further compounded by the fact that wood energy facilities are located largely in response to resource availability. Consequently, the availability and suitability of a good water supply may be a secondary concern.

3. Solid Waste Disposal. Disposal of ash produced during the combustion process causes solid waste problems in addition to the water quality concerns described earlier. Fortunately, the components of wood ash are relatively nontoxic and do not pose a major health threat. Moreover, they are not classified as hazardous wastes for regulatory purposes.

The nontoxic nature of wood ash is revealed in Table VII. This table was developed based on Brown's (1973) work analyzing wood fly ash, boiler ash, clinker, and slag produced at a hog fuel fired mill in Eugene, Oregon. The extremely small quantities of hazardous material in these wastes attest to the fact that wood waste can be disposed of in an environmentally acceptable manner. Furthermore, some investigators (Hall *et al.*, 1975; Howard, 1970) have suggested that this material can actually be applied as fertilizer, yielding a net benefit. The results of such applications, although not harmful, have not demonstrated that such use of this material justifies wider use.

4. Other Impacts. A range of social, biological, and physical impacts could result from development of specific wood energy conversion facilities. Such impacts, however, need to be addressed on a site specific basis. Consequently, these potentially important concerns are not addressed here, but are instead left for detailed studies.

IV. ENVIRONMENTAL REGULATION AND WOOD FUELS

On January 1, 1970, President Nixon began the new decade by signing into law the National Environmental Policy Act. This gesture marked the beginning of the era of environmental concern and regulatory constraint.

Since that time, numerous laws and regulations were adopted to protect environmental quality. In addition, many local, state, and federal agencies were established to promulgate and enforce these regulations. The resulting regulatory requirements have added considerable cost and delay to the startup of any new facility that has a direct or indirect effect on the environment. Biomass energy facilities are no exception and are subject to numerous environmental regulations.

TABLE VII. *Spectrographic Analysis of Hogged Fuel Ash*[a]

Components	Concentration (ppm)
Silicon (Si)	19.6
Aluminum (Al)	3.6
Calcium (Ca)	2.9
Sodium (Na)	2.1
Magnesium (Mg)	0.8
Potassium (K)	0.3
Titanium (Ti)	0.1
Manganese (Mn)	0.016
Zirconium (Zr)	0.006
Lead (Pb)	0.003
Barium (Ba)	0.010
Strontium (Sr)	0.002
Boron (B)	0.003
Chromium (Cr)	<0.001
Vanadium (v)	<0.001
Copper (Cu)	<0.001
Nickel (Ni)	<0.001
Mercury	Nil
Radioactivity	Nil

[a]Reference: Brown, 1973.

Perhaps the most complete listing of the environmental laws and regulations that must be considered for any federally related action was developed by the Council on Environmental Quality (CEQ) (CEQ, 1979). From this source, a comprehensive list showing agencies with jurisdiction over biomass energy related environmental concerns can be developed. These agencies, and their responsibilities, are best reviewed by topic area.

A. Air Quality

The Clean Air Act and its 1977 Amendments have led to one of the most confusing set of environmental regulations in existence at the present time. The process established under this legislation requires the involvement of federal, state, and local agencies in a wide range of overlapping activities. The compounded and complicated review by these agencies and the lack of a clear authority or well defined standard which is applicable to a particular source may be the cause of more delay, expense, and confusion than the regulations themselves. Sets of standards have been developed which protect air quality by (1) dictating the level of emissions permissible from new polluting sources or (2) establishing criteria for pollutant levels in the ambient air for individual geographic areas. This situation often means that separate regulatory agencies review the same facilities. Further compounding the confusion often resulting from overlapping jurisdictions is the fact that air quality regulations are presently the subject of considerable debate and controversy. Recent court decisions, most notably Alabama Power *v.* EPA, have greatly altered portions of the EPA's air quality program leaving much uncertainty in its wake. Recognizing that the present program is still subject to change, a review of its requirements is necessary.

The Clean Air Act, along with the 1977 Amendments, affects the siting and licensing of new biomass energy facilities and the conversion of existing fossil fueled facilities to ones using biomass fuel in three important ways. First, the Prevention of Significant Deterioration (PSD) regulations regulate facilities located in areas which meet the national ambient air quality standards (NAAQS). Second, facilities in nonattainment areas (pollutants levels above NAAQS) fall under the emission offset policy. Third, new source performance standards (NSPS), are developed for each type of emitting facility. These sets of standards, as they relate to biomass facilities, are discussed below.

1. Prevention of Significant Deterioration. The PSD regulations are intended to protect air quality in areas where existing air quality is presently acceptable. In general, the regulations permit certain facilities to be constructed, but limit the total level of emissions within a geographic area. As described by Patterson (1979) the PSD regulations establish the maximum incremental changes in ambient levels of pollutants on a "first come-first serve" basis. To administer this program, EPA has designated all regions other than nonattainment areas as either Class I, II, or III. Class I areas, such as National Parks and designated wilderness areas, are governed by the most restrictive regulations virtually prohibiting any type of development. Other jurisdictions, such as Indian Reservations, may petition for Class I designation. Moreover, Class I regulations can control activities that occur outside of but adjacent to Class I areas. As discussed later, this is especially important in regard to visibility standards for Class I areas. Class II areas allow moderate industrial growth. Most industrial growth is permitted in Class III areas. A new or "major" modified source located in either a Class I, II, or III area must go through a preconstruction review and permit process.

The definition of the term "major" source has been the source of much controversy and litigation. The confusion has resulted because the law defines any source with a "potential to emit" more than 250 tons per year of any pollutant as a major source. In addition, certain types of industries singled out in the law as being major if they have the potential to emit more than 100 tons per year of a pollutant.

Adding further controversy to the debate over what constitutes a major source is the discussion of the definition of "potential to emit." Prior to the Alabama Power Company *et al.* v. EPA case, EPA has interpreted "potential to emit" as emission levels without considering the use of any control equipment. Analyzing wood boiler emissions under this interpretation meant that even fairly small wood boilers would have to go through the PSD preconstruction review process because of particulate emissions. As required by the Alabama Power court decision, EPA has changed its interpretation of the term "potential to emit," so that emission levels will be based on the level of controlled emissions. Because particulate control technology is quite efficient in wood boilers, it is expected that virtually all emission levels can be kept below the 250 tons per year for all pollutants.

Although this new interpretation will reduce the number of new facilities likely to go through the PSD preconstruction review process, the level of preconstruction review for modified sources may actually increase. The increase could occur because EPA has proposed relatively low *de minimis* standards.

These would mean that modifications of a major source resulting in increased emissions that are well below the threshold for new sources would have to go through a complete PSD preconstruction review process.

To illustrate the potential impact of these standards on the development of biomass fueled facilities, consider the following example. An existing, moderate sized wood processing facility is being expanded and management is considering installation of a new cogeneration system. The expanded facility will be only slightly larger than the old plant, but will require larger boilers because of the added steam demands. In such a situation, it is highly likely that the proposed modification would trigger a more strenuous review process even if relatively efficient control technology is used. The rigorous review would occur because the *de minimis* thresholds are so low. Conversely, a similar new facility probably would not have to go through the PSD review even if pollution control equipment is not the best available. This latter situation occurs because it is relatively easy to keep emission levels from biomass fueled boilers below the 250 tons/year threshold level.

A specific source subject to the PSD regulations must do the following (Patterson, 1978):

(1) Demonstrate compliance with NAAQS's by performing air quality evalutions;
(2) Demonstrate compliance with PSD increments by performing air quality devaluations and modeling;
(3) Perform assessments of the direct and indirect effects of the source on visibility, soils, and vegetation;
(4) Determine whether the source is applying "Best Available Control Technology (BACT); and
(5) Hold a public hearing on the proposed PSD application.

Perhaps more important from a scheduling standpoint is the potential need for an air quality monitoring program. Such a program typically requires a complete year of monitoring data and additional time for using modeling and other analysis techniques. Such a one year study has obvious economic effects on organizations where time is a critical factor.

Under EPA's proposed regulations, the *de minimis* standards are important because they facilitate a best available control technology (BACT) review for all sources above the standards. Such a review is often more stringent than the applicable new source performance standards and can lead to the installation of costly pollution control equipment that the applicant might not otherwise use. For example, EPA may attempt to require a control strategy using baghouses even though an applicant may demonstrate that they are too costly and susceptible to fire. The requirement that certain relatively small modified wood boilers will be subject to a BACT determination could discourage the construction of such facilities.

Related to this concern is the proposed regulations' provision that all pollutants, not just those which qualify the facility as a "major emitting facility," require a BACT determination. Consequently, it is possible that through the standard permitting process, control of pollutants other than particulates may be required.

2. *Nonattainment Emission Offset Policy*. Nonattainment areas are designated in each state's implementation plan for each of the different pollutants. These areas are typically located in urban or industrialized regions. It is generally more difficult to comply with the air quality regulations in such areas. The procedure for achieving approval is to reduce an existing source of pollution by an amount greater than or equal to the emissions of the planned new source. Meeting the requirements for locating a facility in a nonattainment area is a site-specific task. Consequently, such activities can only be fully addressed on a project-by-project basis. Nevertheless, it is significant to recognize that emission offsets are needed before a project can be constructed in a nonattainment area.

In a nonattainment area, it is also mandatory that the proposed source employ the Lowest Achievable Emission Rate (LAER). LAER is determined by the most stringent existing standards for similar sources or by demonstrated results from actual test cases. It is also determined without regard to cost or economic feasibility. To construct in a nonattainment area, an applicant must also demonstrate that all his other sources are in compliance with all air quality regulations or on a compliance schedule.

3. *New Source Performance Standards*. The Clean Air Act requires EPA to develop New Source Performance Standards (NSPS) for all major emission sources. These standards are to be developed in the early 1980s. Standards for biomass facilities have been given a relatively low priority, however, and are not scheduled for completion near the time all standards will be final. In the interim, the other requirements of the Clean Air Act will dominate the regulatory process. This lack of NSPS standards is not as significant a concern as one might think, because other aspects of the review process will require effective and sophisticated technology to be employed.

It is currently the EPA's policy that if the biomass combustion facility is designed to burn coal, or can be modified easily to do so, or intends at any time to mix coal with the biomass (except at startup), the proposed unit must meet the NSPS for fossil fuel fired boilers. These boilers fall into the select group of sources for which the 100 tons per year pollutant emission level is used to categorize the source as a major source. (The PSD provision would also be applied to such boilers if this source emitted above 100 tons per year of any pollutant.) NSPS for fossil fuel fired boilers are quite detailed and extensive, and they require an in-depth analysis which is not appropriate here.

B. *Water Quality*

Although the regulation relating to water quality will not impose as many restrictions on the operation of wood energy facilities as air quality regulations, they are a factor that must be considered. The primary way in which water quality regulations apply to biomass energy facilities is through the National Pollution Discharge Elimination System (NPDES). These regulations were promulgated under the Clean Water Act as amended in 1977. The regulations are most important in terms of their effect on siting new facilities.

Requirements imposed in conjunction with the NPDES process depend on effluent limitations and water quality standards. Effluent limitations are developed from major sources of pollution. In the case of biomass energy facilities, however, there has been relatively little experience with either combustion or gasification facilities. As a result, sophisticated effluent standards for biomass energy facilities are yet to be promulgated. Nevertheless, biomass energy facilities must comply with specified standards as they are promulgated. In particular, EPA has identified a list of pollutants that are regulated. This list (44 FR 4450) classifies pollutants as conventional, nonconventional, and toxic pollutants.

Conventional pollutants identified by EPA include biological oxygen demand, total suspended solids, pH, fecal coliform, and oil and grease. Since activities other than the actual conversion of biomass are likely to occur at biomass energy facilities, some pollutants will be discharged. In addition, any discharge, including that from storm sewers, requires an NPDES permit. As a result, biomass energy facilities will be required to limit a range of pollutants. The Clean Water Act amendments also require that facilities employ the best conventional control technology (BCT) for pollutants by 1984. This requirement will insure that discharges will be controlled where necessary. Nonconventional discharges, which includes thermal pollution, must achieve the best available technology (BAT) by 1984. These standards are more stringent than the BCT requirements. Requirements for BAT can be waived if it is demonstrated that thermal limitations are not needed to protect indiginous populations of shellfish, fish, and wildlife in the affected water body. Because of this requirement and the requirements described under water quality standards, biomass energy facilities will have to employ some control technology on their discharges. In particular, since biomass facilities will often be located adjacent to small streams, cooling towers will likely be needed on even small facilities.

1. *Water Quality Standards*. Along with regulations directly limiting pollution discharges, standards have been developed for specific water bodies which dictate that certain ambient quality levels be maintained. For example, a given water body can be designated as one of extraordinary quality and very high standards are set for it. Such standards are typically developed by individual states under the Clean Water Act. These regulations have the effect of limiting the likelihood of a biomass conversion facility being located adjacent to a water body of high quality. Furthermore, the regulations are particuarly restrictive in the case of small streams or lakes. This is the case because the smaller the water body, the more likely even a small level of pollutants will affect the existing quality.

2. *Other Related Water Quality Regulations*. Two other federal water quality related permits related to water requirements of a given plant are the permits issued under Section 404 of the Federal Water Pollution Control Act and under Section 10 of the River and Harbor Act. These permits are administered by the Corps of Engineers and are required if a structure is located in a navigable waterway. Through these permitting processes, which are normally handled jointly, the Corps of Engineers fulfills its responsibilities to keep navigable waterways free of obstacles. It is generally not difficult to design facilities that will be granted these permits. This permit process, however, is also important because the Corps must fulfill the requirements of NEPA before issuing any permits. Consequently, it informs other agencies of the proposed project and formally seeks the input and advice of these agencies before issuing any permits. Through this program, federal and state agencies, as well as the public, have the opportunity to call to the Corps' attention environmental problems associated with a given project. Consequently, concerns such as a proposed site's value as a historical site that otherwise may be overlooked can be considered in project planning.

C. *Resource Conservation and Recovery Act (RCRA)*

A third set of regulations potentially impacting biomass conversion facilities are those related to the Resource Conservation and Recovery Act. These regulations, however, are not fully developed and do not pose as significant concern as either the Clean Air or the Clean Water Act requirements. The relatively small quantities and benign qualities of biomass conversion facility wastes limit the potential regulatory obstacles imposed by the RCRA.

The RCRA defines pollutants in two categories. First, hazardous wastes are defined and rather restrictive disposal parameters established for their disposal and control. Second, the series of regulations apply to all wastes determined not to be hazardous. This latter set of regulations would apply to biomass energy facilities including such things as ash and other process byproducts. Regulations applying to these non-hazardous wastes are not yet finalized, but will not be nearly as restrictive as those applying hazardous wastes. Requirements that will be most important to biomass facilities relate to the disposal of materials in wetlands or flood plains. These restrictions which are being considered by EPA could limit the number of disposal sites under consideration and possibly lead to an increase in the cost of disposal through the added transportation costs required.

D. *Land Use*

Any potential biomass conversion facility must comply with local land use regulations. Building permits and zoning variances would likely be needed for any conversion facility in an urban area. Other land use considerations, as the impact on wetlands, flood plains, or shorelines, could bring in some special regulatory requirements. Such requirements are site-specific and can be analyzed only in regard to a specific project.

V. TRADEOFFS FROM UTILIZING BIOMASS FUELS

A. *Offset Use of Traditional Fuels*

Although biomass fuels have a limited energy producing potential, in certain circumstances use of this material may offset the need to burn other fuels. Obviously, any reduction in the amount of oil or natural gas consumed has favorable ramifications as far as the balance of payments is concerned.

Moreover, for certain small facilities using biomass instead of coal may make economic sense if the biomass fuel is available locally and the coal is not. Also, installation of relatively sophisticated pollution control technology may be required if coal is used.

Generally, mandatory pollution control expenditures are proportionately higher for small coal burning facilities, than for smaller biomass facilities. In particular, alleviation of the requirements to control SO_2 emissions could have significant economic implications. As noted earlier, the small volume of sulfur present in biomass fuels along with the size limitations of these facilities virtually eliminates the need to be concerned with SO_2 emissions.

B. Reduce Open Burning

On site burning agriculture or forestry residues following harvest is a standard practice in much of the United States. Removal of this material enables more efficient planting of the next crop. In actively managed forest lands, it also reduces the fire hazard associated with leaving large quantities of dry material on the site. These residue removal objectives could be met if forestry and agricultural residues are removed from the site and used to produce energy.

Although off site use of logging and agricultural residues may be appealing on the grounds that it utilizes an otherwise wasted resource, it generally fails most economic tests. There is a high cost associated with gathering and transporting this material to a central collection and distribution point. This costly labor intensive process is especially expensive from an energy and economic standpoint if fuel gathering takes place independently of harvesting activities.

The cost disadvantages of harvest site residue removal would be somewhat offset by the improvement in air quality that would result from elimination of slash burning activities. As shown previously, logging residues burned in the open emit far greater quantities of pollutants than the same material burned under controlled conditions. This absolute emission reduction is most keenly felt in the rural areas where open burning takes place. Such burning, however, has also been known to affect urbanized areas with higher population densities. Table VI, showing collection efficiencies of certain equipment, shows the impact of burning in the woods or in a modern boiler.

The significance of this pollution source is supported by the fact that a number of states currently have regulatory programs covering open burning activities. A reduction in burning will reduce the need for concern about pollution resulting from forestry or agricultural activities. These benefits, whether economically achievable or not, can be significant in a given location and thus must be analyzed when specific biomass energy development scenarios are proposed.

As noted in the discussion of regulatory concerns, the question of visibility impairment in wilderness areas could also provide the impetus for using logging residues for energy production. The tradeoff of the high costs for collecting this material weighed against the reduction or elimination of visibility impairment in many Class I Air Quality Areas will be made as the federal visibility standards are implemented. Greater utilization of logging residues for energy production could be the result of the regulatory effort to maintain high visibility standards in wilderness areas.

C. Completing Demands for Biomass Fuel

If a large amount of biomass material is destined for use as fuel, there could be a restructuring of the market for other related products. For example, if a number of facilities are constructed that will require hog fuel or logging residues as their primary energy source, there could be increased demand for chips or other material currently used for producing pulp, paper, particle board, and other products. If the demand for this material as an energy source were to become strong, prices of other forest products would likely rise. Perhaps of greater concern were the unusual circumstances that could arise such as would occur under the fuel adjustment clause that pertains to utilities. Under this clause, it might be possible for a utility using biomass fuel to pay very high prices for biomass material traditionally used for nonenergy related purposes. The high cost would not deter purchase of such material because under the fuel adjustment clause, the high price could be passed on to the consumer. This example illustrates one of the ways increased use of biomass for energy could affect nonenergy related uses.

VI. CONCLUSIONS

Utilization of wood for energy will cause environmental impacts relating to both fuel supply and energy conversion activities. Fuel supply impacts depend on the wood supply system employed and on specific conditions at the site from which the fuel is acquired. In general, impacts increase as more intensive supply options are employed.

A. *Fuel Supply Impacts*

In the least intensive supply option, utilization of existing mill residues, impacts are negligible. These residues are presently almost fully utilized, thus reducing their potential for providing additional energy in the future. The next supply option, logging residues, features positive as well as negative impacts. Reduction in the need for open burning of logging residues improves the site's appearance and reduces air emission because material burned for energy is burned under controlled conditions where emissions can be better controlled. Negative impacts from using logging residues result from potential nutrient removal and the added emissions and other impacts associated with transporting the fuel to the conversion site.

Increased use of silvicultural residues would have many effects similar to those of increased use of logging residues. In both cases, presently unused material would be removed from the forest and burned in relatively controlled conditions. Removal of this material, although generally uneconomic at the present time, would increase fuel supplies at relatively small environmental costs.

Unlike the other supply systems, impacts of biomass energy farms can be solely attributed to utilization of biomass for energy because no benefits associated with production of forest products are realized. Nutrient removal, establishment of intensively managed monocultures, and the removal of large areas from other productive uses are the significant impacts of biomass energy farms. These impacts, along with the fact that energy farms have not been demonstrated to be economically efficient, make them a less appealing option regardless of the energy conversion system employed.

B. *Energy Conversion Impacts*

The most significant impacts that will occur at a wood energy conversion facility will affect air quality. Specifically, particulate emissions from wood fueled boilers can be high. Fortunately, particulate emission control technology exists and is effective. The cost of such control equipment is high, however, and considerable debate focuses on what level of control is needed given the cost of employing a particular control technology. This debate is prompted by regulatory requirements that require the organization building a facility to employ the best available control technology. Because such technology is constantly improving, it is difficult to quantify emissions at a generic level. Nevertheless, it is reasonable to conclude that although potential particulate emissions are high, control technology can limit emissions to acceptable levels.

Water related impacts occur as a result of steam generation, ash transport and general cleaning activities. For steam production, make-up water is needed. In major facilities, large quantities of water are needed while in smaller ones, the quantity used is minimal. Similarly, both the nature and extent of water discharges vary with plant size and design. The resulting impacts also vary. Other impacts that result from ash transport or cleaning activities are normally regulated and contained to acceptable levels. Treatment programs are also established, if needed. The lack of toxic material found in other fuels further minimizes the impact to water resources.

The relatively benign nature of the material remaining after wood is burned makes solid waste (ash) disposal problems minimal. Wood ash contains very little toxic material. Consequently, impacts of disposing of this material are minor unless there is a lack of care in selecting suitable waste disposal sites. Other site-specific impacts related to aesthetic, noise, land use, or other areas of environmental concern could occur at a specific biomass energy conversion facility. These types of impacts are important and must be evaluated for specific projects. These are, however, beyond the scope of a generic analysis.

C. *Regulatory Concerns*

At least as important as the actual environmental impacts associated with biomass energy facilities are the regulatory requirements that affect energy facility siting and licensing. Regulations related to air and water quality, waste disposal, fuel use, facility siting, and numerous other environmental/energy concerns must be considered before individual projects proceed. To project planners, these regulations can impose economic or time constraints that have a major influence on determining a project's feasibility.

Air quality regulations are the most significant ones affecting biomass direct combustion systems. Regulations at both the state and federal levels apply to biomass facilities. Some of these regulations are presently undergoing extensive revision as a result of recent court decisions. Because of these current revisions in the Prevention of Significant Deterioration (PSD) considerable uncertainty and delay has been added to the project planning process. Added costs for detailed studies are another result to the present air quality program.

Water related regulatory concerns are less significant than those for air quality. Thermal discharges are regulated, but it is generally possible to reduce such discharges to acceptable limits with proven technology. Project costs are, however, affected by compliance with these regulations.

Other sets of regulations, such as those related to facility siting waste disposal, fuel use, or a range of environmental concerns, could affect biomass energy facilities. The effect of such regulations cannot be fully assessed in the absence of a specific project. These regulations may, however, have a significant bearing on a project's feasibility and need to be carefully analyzed on a case-by-case basis.

REFERENCES

Bethel, James, and others. 1979. Energy from Wood. A Technology Assessment. College of Forest Resources, University of Washington, Seattle, Washington.

Burke, Edwin J. 1980. How Far Should You Go for Wood. Western Wildlands Vol. 6, No. 1, Missoula, Montana.

Ekono. 1980. Personal Communcations, July-August.

Geomet. 1978. Impact of Forestry Burning Upon Air Quality--A State of the Knowledge Characterization in Washington and Oregon, Draft final Report, EPA Region X, Seattle, Washington, GEOMET Report No. EF-664.

Hall, E. H., and others. 1976. Comparison of Fossil and Wood Fuels. Battelle Columbus Laboratories (for the Envkronmental Protection Agency), Columbus, Ohio.

Hall, J. Alfred. 1972. Forest Fuels, Prescribed Fire, and Air Quality. Pacific Northwest Forest and Range Experiment Station, Forest Service, U.S. Department of Agriculture, Portland, Oregon.

Hewett, Charles E., and High, Colin J. 1979. Environmental Aspects of Wood Energy Conversion. Resource Policy Center Paper DSD #145, Thayer School of Engineering, Darmouth College, Hanover, New Hampshire.

High, Colin J., and Knight, Susan E. 1977. Environmental Impact of Harvesting Noncommercial Wood for Energy: Research Problems, Resource Policy Center Paper DSD #101, Thayer School of Engineering, Darmouth College, Hanover, New Hampshire.

Howard, E. J. 1970. A Survey of the Utilization of Bark as Fertilizer and Soil Conditioner, *Pulp and Paper 71*:23-24, 53-56.

Howard, James O. 1979. Wood Energy in the Pacific Northwest—An Overview. General Technical Report PNW-94, U.S. Department of Agriculture, Pacific Northwest Forest and Range Experiment Station.

Junge, D. C. 1975. Boilers Fired with Wood and Bark Residues. Research Bulletin 17, Forest Research Laboratory, Oregon State University, Corvallis, Oregon.

Kitto, William D. 1979. A Technical and Environmental Appraisal of Wood Energy: Western Wildlands, Vol. 5, No. 2. Montana Forest and Conservation Experiment Station, Missoula, Montana.

Mitre Corp. 1977. Silvicultural Biomass Farms, Vol. I-V, Mitrek Div., McLean, Virginia.

Patterson, W. D. 1979. Federal/State Licensing Overview. Edison Electric Institute, Envirosphere Company Second Annual Conference on Environmental Licensing and Regulatory Requirements Affecting the Electric Utility Industry, October 21-24, Washington, D.C.

Pengelly, W. L. 1972. Clear-cutting: Detrimental Aspects for Wildlife Resources, *Journal Soil and Water Conservation*, 255-258, November-December.

Resler, R. A. 1972. Clear-cutting: Beneficial Aspects for Wildlife Resources, *Journal Soil and Water Conservation*, 250-254, November-December.

Rose, D. W., and Olson, K. 1979. Social Economics and Environmental Impacts of a 25 MW Wood-Fueled Power Plant. *Journal of Environmental Management 9*, 131-143.

Sampson, G. R. 1979. Energy Potential from Central and Southern Rocky Mountain Timber. Research Note RM-36R. Presented at the Colorado State University Natural Resource Days, March 10, 1978, Ft. Collins.

Schweiger, B. 1980. Power from Wood, *Power 124*(2), S-1-S-32.

Tillman, D. A. 1980. Fuels From Waste, *in* "Kirk-Othmer Encyclopedia of Chemical Technology", Wiley Interscience, New York.

Tillman, D. A. 1978. "Wood as an Energy Resource," Academic Press, New York.

WOOD FUEL PREPARATION

W. Ramsey Smith

College of Forest Resources
University of Washington
Seattle, Washington

ISBN 0-12-535902-0

I. INTRODUCTION

Wood residues used for fuel originate from a large number of sources and therefore when accumulated contain large variations in material type, particle size and moisture content. The greater this variation, the greater the difficulty in obtaining the maximum potential efficiencies and minimum particulate emissions allowable in wood fired boilers. In many of these installations these problems are handled by increasing design capacities for the furnace, fuel handling and transport processes, excess air equipment and air pollution control equipment. Of course, these increases create much higher capital expenditures and operational expenses, promoting the desire to provide the facility with wood fuel as uniform as possible. However, the greater the uniformity the greater the cost of preparation involved. The benefits gained by providing more uniform fuel therefore must be weighed against the cost of preparation.

So that a better understanding of the causes and effects of these variables may be obtained, it is the purpose of this chapter to:

(1) examine the major sources of wood fuels and discuss their inerent variability with respect to particle size, moisture content and ash content;

(2) examine the influences of particle size and moisture contents on combustion efficiencies;

(3) discuss methods presently available for fuel preparation to enhance handling and burning efficiencies; and

(4) discuss the general advantages and disadvantages in fuel preparation techniques.

II. WOOD FUEL SOURCES AND VARIABILITY

Wood fuels are presently obtained from forest and mill residues, mill residues being the largest source due to their perspective location with respect to furnace location. Forest residues are consumed to a limited degree because of high extraction, conversion, and transportation costs. This, however, is changing with the dramatic increase in costs for fossil fuels. Not only are the overall costs becoming more competitive but it has also provided the driving force to increase residue extraction technology. This source will become increasingly important since it consists of the greatest quantity of viable wood fuel.

A. Forest Residues

Forest residues may arbitrarily be classified in two major groups, standing and grounded as indicated by Zerbe (1977). Standing residue consists of trees that are diseased, insect infested, damaged by fire, dead and standing in stagnated stands, taken out in precommercial thinnings, and urban removals. Ground wood residue is that left in the woods after removal of merchantable stock, i.e., tops and limbs, stumps and cull logs, plus others such as trees that are dead and down, wind blown, or resulting from road and interstate construction. Not all of this material will be used as fuel, or should be, since any which may be converted into acceptable chips or fiber for the pulp and paper industry or composite board industry should be sold as such.

The portion of material that is used for fuel is broken down by hogging. This can range in particle size from 4 inches to fine material, $\leq$ .25 inch, depending on the type hog used. Its moisture content will vary depending on species and wood type, i.e., bark, sapwood or heartwood, and can range from 23% MC_G[1] to over 77% MC_G. Johnson (1975) listed the range in moisture content of hogged bark being 25-75% MC_G at the mill site. This range is due to storage condition of logs prior to debarking and the debarking process used.

The greatest variability for this type of fuel will occur from differing material types extending from rotten cull logs to leaves and branches. Each has different chemical constituents, bark contents, moisture contents and hence heating values.

Mill Residues

Sawmills produced the majority, 78%, of all mill residues produced in 1970, plywood mills followed with 15% of the total and all other wood product industries at 7% (Howlett *et al.*, 1977). These totaled 3806 million cubic feet, of which 80% were softwoods. The large variation in particle sizes, moisture content, and ash content for these residues is inherent because of the different processes from which they originate. This variation may be better understood by examining each mill type more closely.

[1]*Moisture contents in this chapter will be expressed on a wet weight basis, i.e.,*

$$MC_G = \frac{\text{Weight of water}}{\text{Total original weight}} \times 100$$

1. Sawmill Residues. Figure 1 is a generalized schematic of a sawmill showing those steps which affect the physical characteristics of wood fuels, derived from work done by Tillman (1979). The type of mill and processes used will directly affect the characteristics of the residue. Log storage probably has the greatest effect in the sawmill system, followed by debarking and then the other processes combined.

The type of log storage a particular mill uses will depend on mill location, species, resources available and individual preferences. The two major types are dry and wet storage. The dry storage can be further broken down into paved and unpaved yards and the wet into fresh and salt water storage.

A paved log yard is the best with respect to using resulting wood residues for fuel. It keeps to a minimum contamination of bark with rocks, dirt and sand, plus it aids in reducing the quantity of this contaminated material from entering in the yard debris which can be collected and burned. Even though it is a dry yard, the logs may need to be sprinkled to help prevent decay, end checks and splits. This will, of course, add some moisture to the bark but will also tend to wash some dirt and sand from the logs.

Unpaved yards, on the other hand, are the worst for bark contamination. Logs placed on the ground can contain embedded dirt and rocks plus become caked with mud during wet weather. The logs above can become coated with dust raised by the heavy log transport equipment during dry weather. Therefore, depending on the storage type, bark ash contents can range from 1% to 20% by weight.

Pond storage will keep the logs cleaner but at the expense of increasing moisture content. Leman (1975) shows that Douglas-fir and Hemlock bark increase rapidly in moisture content the first five days of storage then decrease to a slower, but constant rate for at least the following 40 days. Bark moisture content, therefore, may be increased to as high as 80% MC_G depending on storage time. He also points out that log bundles (required by some marine districts) versus flat rafting of logs tend to: (1) increase absorption of moisture uptake from the greater hydrostatic pressure on lower logs, and (2) prevent the logs from losing dirt and sand which is usually removed with the greater water to log surface contact when logs are free floating. Fresh water storage is better than salt water due to potential salt water contaminations. Salt contained in the bark and other wood material burned in the furnace will corrode internal parts if condensation occurs. This means that high stack temperatures must be obtained, affecting boiler efficiencies.

The debarking operation is the greatest single wood fuel producing process in the sawmill. It can affect the potential fuel's moisture content, particle size, heat content and ash

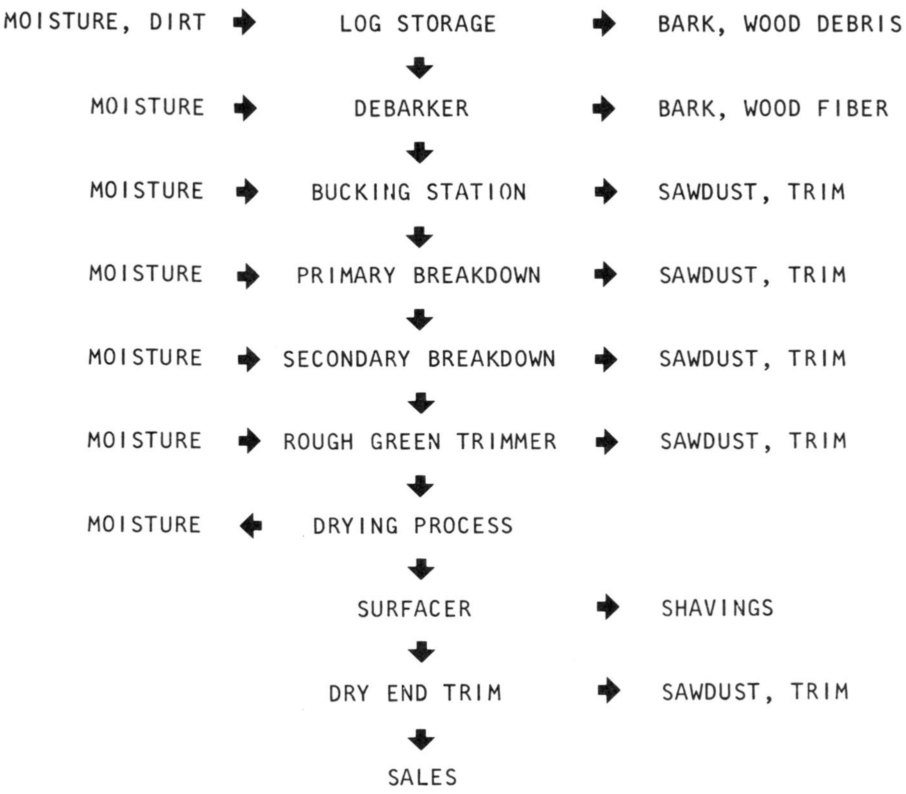

FIGURE 1. Sawmill processing inputs which affect fuel characteristics and residues produced at each machine center which can be used for fuel.

content, the amount of which depending on type of debarker used. The most common debarkers are cambio, mechanical ring, rosserhead, hydraulic, drum and a "pole shaver" or "buzz barker." The type used depends on the log mix and type of industry. Table I contains estimated influences of debarker types on wood fiber content versus bark content, moisture content and ash content of the resulting residue.

The particle size produced by each is dependent on log species and quality. In general, the particle size produced might be ranked: pole-shaver ≃ rosserhead < hydraulic < drum < cambio < mechanical ring. The quantity of fines generated would follow in the same order. The pole-shaver, rosserhead and hydraulic would also tend to produce a more

TABLE I. Influence of Debarker Type on Residue Fuel Quality

		Wood fiber content (%)	Change in moisture ($\%MC_G$)	Ash content reduction (%)
a	Cambio	Nil[a]	-1	Nil
b	Mechanical ring	1.8[b]	-1	Nil
c	Rosserhead	10-50	-1	Nil
d	Hydraulic	1-2	+25 ± 10[c]	-2[c]
e	Drum	10.3[b]	-1	Nil
f	Pole-shaver	50.8[b]	-1	Nil

[a] Williston, 1976.
[b] Weldon, 1966.
[c] Tillman, 1979.

uniform particle size than the remaining three due to their operational characteristics.

The log then proceeds through a combination of processes including a bucking saw, headsaw, pony saws, resaws, edgers and trim saws for conversion into lumber. During this journey approximately 26% of the log is converted into coarse residues, 13% into sawdust and 49% into rough lumber. The remaining 12% was the bark removed in the previous step (Corder *et al.*, 1972). The coarse residues, consisting primarily of slabs, edgings and lumber trim, generally should go into pulp chips, a much more valuable product than hog fuel, even though some does find its way to the fuel pile. That which does is reduced in the hog to the particle size distribution previously discussed.

Sawdust, however, can have an appreciable increase in moisture content. The sawing processes involved may use water for saw cooling at a rate of 2-4 gal/min (Williston, 1979). The majority of the water is usually carried away with the sawdust and on into the fuel pile. The moisture content increase due to this source is dependent on water input rate with respect to sawdust volume generated and density of the species being cut. To get an idea of the resulting increase in moisture content, assume the following: a double arbor over-under rotary gang edger cutting a cant 12 inches deep and 24 inches wide into 2 × 12's at a feed rate of 80 feet per minute. Using carbide teeth with a 0.18 in kerf, 19.2 ft^3 of sawdust will be produced per minute of operation. The increase in moisture content, assuming a 4 gal/min water rate, with respect to species density is shown in Figure 2. Since the density of most commercial species lie in the range from around 45 lbs/ft^3 to 20 lbs/ft^3, a moisture content increase of 3.7% MC_G to 7.9% MC_G can occur. In practice, depending on the particular sawing process, this range could realistically extend from no moisture added upwards to 20% MC_G.

Once the lumber is manufactured it may be sold either green or dried, the choice depending on mill facilities, and product and/or the lumber market. Softwood dimension lumber may or may not be kiln dried but is usually surfaced. In general 40% and 50% of all surfaced material is green and the remainder has been initially dried. The majority dried is brought down to a moisture content below 16% MC_G to meet minimum dry dimension lumber specifications, and some down to less than or equal to 13% MC_G. Other softwood lumber contained in shop grades and practically all hardwood lumber not used in pallet stock, is brought down to a level between 6% and 11% MC_G depending on end use. Therefore, any subsequent cutting operations will produce dry residue. In the sawmill industry this consists of the dry end trimming operation which produces a minor amount of sawdust, as compared to the rest of the mill, and dry planer shavings. All material designated for the fuel pile such as bark or edgings and trim not sold as pulp chips, are sent to the hog for size reduction.

2. *Plywood Mill Residues*. The plywood mill residues may be generally categorized in three groups: veneer log residues, green-end residues and dry end residues. Table II lists these categories and types of residues found in each with their perspective contributions to residues as a whole and to those used for fuel.

Almost two-thirds of plywood residues used for fuel come from the veneer log residues as can be seen in Table II. Approximately one-fourth of these are trim ends or "lily pads" produced where the veneer bolts at the plant site are sized to the correct lathe length. Plants using water storage for their logs will saturate these ends, since primarily end grain is exposed resulting in high moisture contents, up to 80% MC_G. If salt water storage is used, then problems of salt contamination are more prevalent, as might be expected.

Other log storage and debarking techniques which were discussed for the sawmill applies as well to veneer bolts. The mechanical ring and rosserhead type debarkers are most commonly used.

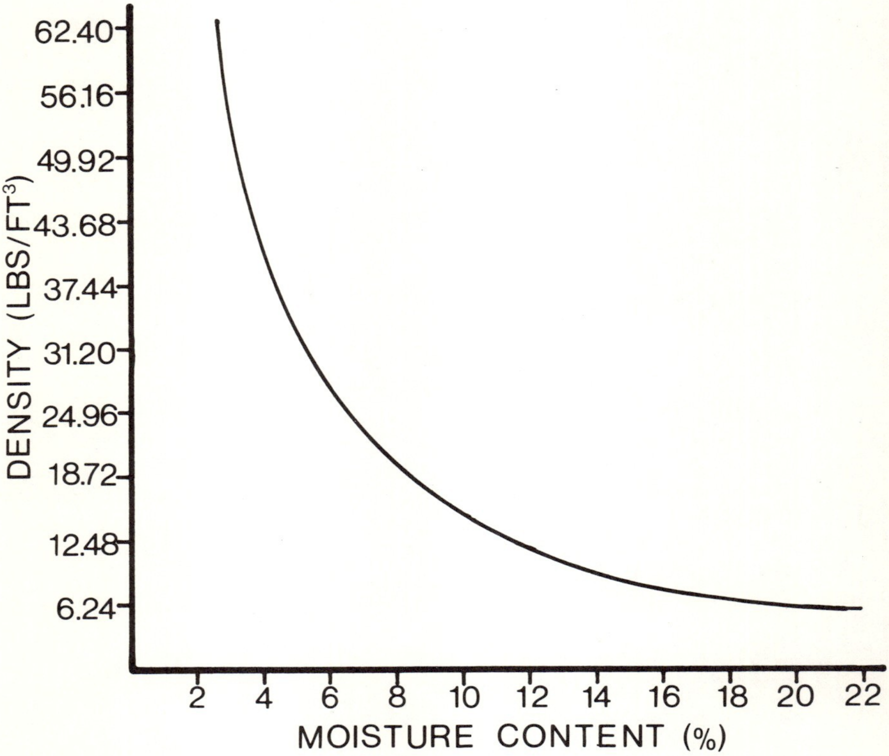

FIGURE 2. Moisture content increase with respect to species density for the addition of 4 gallons of water per minute to a gang edger cutting a 2 × 12 cant at 80 feet per minute.

Green-end residues are essentially either used or sold for purposes other than fuel, i.e., green veneer for pulp and composite board manufacture and cores for pulp chips and studs. The dry-end residues unsuitable for pulp, are those used primarily for fuel. This material has been dried to low moisture contents prior to their disposal, 2% to 7% MC_G. All of the dry end residues other than sanderdust go through a hog producing particle sizes similar to those discussed

TABLE II. Plywood Mill Residue Categories[a]

Veneer log residues	*Contribution to total residues*	*Contribution to residues used for fuel*
Veneer log residues		
Log trim or "lily pads"	*7.86*	*16.30*
Bark	*22.76*	*47.22*
Subtotal	*30.62*	*63.52*
Green-end residues		
Round-up, spur trim, and veneer trim	*43.15*	*Nil*
Cores	*8.65*	*Nil*
Subtotal	*51.80*	*Nil*
Dry-end trim		
Reclip-veneer reject, veneer breakage, jointer shavings, veneer reject at glue spreader, and plywood panel trim	*14.37*	*29.81*
Sander Dust	*3.21*	*6.67*
Subtotal	*17.58*	*36.48*
TOTAL	*100.00*	*100.00*

[a]*Data from Corder, et al., 1972.*

previously. Sanderdust, resulting from the product finishing step, is collected directly from that process having a range in particle size from 8×10^{-5} inches to 3×10^{-2} inches (Junge, 1975).

3. *Other Mill Residues.* The remaining wood products industries which may contribute to wood residue fuel supply consist primarily of other composite board manufacturers such as particleboard, fiberboard and flakeboard and secondary wood product manufacturers such as the furniture industry,

door manufacturers and container product producers.[1] The residues resulting from these sources are low in moisture content ranging from 2% to 11% MC_G since they use dried lumber as their raw material. Particle sizes, however, still will vary from sanderdust as small as 8×10^{-5} inches to larger hogged material reaching 4 inches in size.

III. EFFECTS OF WOOD FUEL VARIABLES IN WOOD FIRED BOILER SYSTEMS

The variation of particle size, moisture content and ash content as just discussed can be large within a given mill. When these individual mills are located in an industrial complex and all of their residues are combined to feed a central boiler installation, this fuel variability is greatly increased and boiler efficiency is affected. The overall effects of fuel variability may be better understood through a closer examination of each.

A. Particle Size

The wood fuel particle size can vary from 8×10^{-5} inches to 4 inches after the hogging operation. This is an overall size variation of 5×10^{4} times which is like mixing toothpicks with logs over a mile and a half long. This in turn, greatly affects the rate of combustion and heat content per unit volume which directly affects the design and specifications of the installation.

Wood burns primarily in the gaseous state since 70% to 86% of wood and bark are composed of volatile material (Mingle and Boubel, 1968). The remaining 14% to 30% consists of fixed carbon and ash. The rate of combustion then, is directly proportional to the time it takes for the required heat to reach and volatilize this material which is dependent on the exposed surface area per unit volume of the particle. This ratio increases with decreasing particle size. Larger particles not only have a smaller ratio but also tend to insulate their interiors with the formation of char or that nonvolatile portion of the material which additionally reduces the rate of combustion.

The percentage of the nonvolatile portion depends on species and wood type, i.e., wood or bark. Bark, with a few

[1] *The pulp and paper industry will not be considered here since it is unique in the type and use of residues for fuel.*

exceptions, generally contains 10% more fixed carbon than wood. This fixed carbon burns in a solid state at a reduced rate and therefore requires a longer residence time in the furnace.

Particle size combined with moisture content and fuel compaction also influence the fuel's bulk density, i.e., weight per unit volume, which directly affects heat content per unit volume. Bulk density is used to control the quantity of fuel fed to the furnace per unit time in response to fluctuations in steam demand. As the demand for steam or heat increases, the fuel feed rate and excess air volume increases. Since particle size regulates the rate of combustion it also directly affects the response time required to meet this increased demand. Therefore, the larger the material the longer the response time and the greater the volume of excess air used.

The volume of excess air affects combustion efficiency in a number of ways. The greater amount of outside air put through the system, the greater the requirements in horsepower to supply this air, and the greater the thermal energy lost in heating up this air to combustion temperature. Also, since excess air requirements are less for larger pieces than for smaller, because of their differences in rate of combustion, a change in particle size will change the level of air required.

Very small particles or fines (<1/4 in) can perform a double disservice. Their increased rate of combustion requires an increase in excess air which in turn increases the total air velocity through the furnace. This tends to carry unburned portions of the fines out the stack as flyash, decreasing combustion efficiency and increasing pollution control problems. As this flyash particle size decreases, cyclone collection efficiency also decreases, as discussed by Junge(1975). He shows that a single large cyclone will start becoming less effective as flyash particle size is reduced below 40 micrometers and the efficiency of multiple small cyclones will be drastically reduced for particle size below 8 micrometers.

Particle size also affects the transportation, storage, and handling safety. As particale size decreases there is an increasing tendency to cause "dusting", i.e., particles being carried in the wind. This problem is increased as moisture content decreases which will be discussed more fully in the following section.

B. Effects of Moisture Content in Wood Fired Boiler Systems

Moisture content of wood fuels is probably the single most important controllable variable affecting their combustion. It can directly influence: (1) net heating value, (2) rate of combustion, (3) particulate emissions, (4) boiler efficiency, (5) degree of control over the combustion process, (6) furnace design, and (7) fuel handling and transport.

Moisture contained in any fuel lowers the available heat content of that fuel in two ways. First, that portion of the fuel by weight which is moisture does not contribute to the heat value. This means that one pound of wood fuel which has a higher heating value of 8000 Btu per pound when oven dry has only 4000 Btu per pound input to the furnace potentially available at 50% MC_G. Second, all moisture contained in a fuel must first be evaporated before the fuel can go through a combustion process. Since evaporation is an endothermic process, heat is required and consumed which otherwise could be used to produce process steam or heat. Therefore, the greater the moisture content the greater the reduction in fuel heat content as shown in Table III for various higher heating values. These values are for losses attributable to moisture content; no other factors are considered.

The rate at which fuel combusts is also greatly affected by the amount of moisture it contains. Again, the moisture must be initially removed from a given area of the fuel particle before the combustion process can begin, in essence delaying the process. Tillman (1978) using data obtained from Koehler (1924) shows this effect with respect to ignition time and temperature. Koehler determined the time required for ignition of air dried and oven dried samples of tamarak, longleaf pine and hemlock at various temperatures ranging from 392°F to 752°F. His data show that it may take from 2 to 6 times longer for air dried fuel to ignite than oven dry fuel, the time differential decreasing as temperature is increased. However, the effect is still there and can increase appreciably with an increase in moisture content.

Junge (1975) discusses the relative effect moisture content and particle size has on the rate of combustion. He indicated that bark with a typical 45% MC_G and a surface to volume ratio of 1 gave a relative combustion rate of 2 whereas sawdust at 35% MC_G and a ratio of 6 had a relative combustion rate of 17; and 8.5 factor increase. He also indicates that reject mat furnish at 6% MC_G and a ratio of 8 has a 130 relative rate of combustion, an approximate 7.6 factor over sawdust and a factor of 65 over bark. This information shows that not only are moisture content and particle size very influential in combustion characteristics of the fuel, but

TABLE III. Net Heating Values for Various Higher Heating Values as Affected by Moisture Content

MC_G (%)	Higher heating value (Btu/lb)			
	7000	8000	9000	10,000
0	7000.0	8000.0	9000.0	10000.0
10	6199.8	7099.8	7999.8	8899.8
20	5399.6	6199.6	6999.6	7799.6
30	4599.4	5299.4	5999.4	6699.4
40	3799.2	4399.2	4999.2	5599.2
50	2999.0	3499.0	3999.0	4499.0
60	2198.8	2598.8	2998.8	3398.8
70	1398.6	1698.6	1998.6	2298.6

fuel originating from various sources also are very influential. Since each mill source and machine center in that mill can have different moisture contents and surface to volume ratio, very drastic differences can then be encountered from one type to another.

An additional result of the endothermic reaction resulting from evaporation of moisture is its reduction of flame and hence furnace temperature which when combined with the reduced rate of combustion can affect the quantity of particulate emissions. Particle size quantity of excess air and moisture content greatly affect the particle residence time. Heavier particles will stay in the combustion zone longer than lighter particles, lose their moisture and combust, giving up their heat. Lighter particles tend to be picked up within the air current and burn in suspension. If the lighter particles are laden with moisture their combustion rate may be slowed enough such that they may be carried out through the heat transfer zone without having completed combustion, increasing the load on the pollution control devices. The smaller and lighter the particle, the less the residence time and the greater the chance for its removal prior to complete combustion. Also since water vapor requires approximately 5700 times the volume than it does as a liquid, air velocity is increased as moisture is evaporated, again reducing particle residence time. Johnson (1975) reported on a study by Parkesh and Murray done at the University of British Columbia in 1971 which quantified this effect. These studies show that as moisture content increases, average combustion rate decreases for a given combustion zone temperature. As the combustion zone temperature decreases particulate matter or flyash increases exponentially. This increase in particulates not only increases pollution control requirements but also decreases boiler efficiency since potential fuel is lost.

Boiler efficiency also suffers from other moisture content effects. Again, the greater amount of moisture present, the greater amount of heat consumed for evaporation and the less amount of heat available for steam production. For example, a boiler rated to produce 550,000 pounds of steam per hour with ovendried fuel can only produce 400,000 pounds per hour with fuel at 55% MC_G. This is a reduction in heat of 27.3%. Also, when too much moisture is present, the fuel cannot produce enough heat to sustain its combustion and a furnace blackout occurs. The fuel moisture content therefore must be considered for correct furnace design. For a given steam output the furnace size must increase with an increase in moisture content, and the rate of feed must also increase, increasing the size of fuel handling and feeding equipment. The sum total is an increase in the cost of the entire system than if ovendry or low moisture content fuel were readily available.

The amount of moisture present, combined with the variability of particle size also greatly affects the degree of control which can be obtained over the process. The greater the variability of these two factors implies a greater difficulty in providing uniform heat for a constant steam output. Without uniform heat capabilities the ability of a process to adequately respond to changes in steam demand is greatly impaired. As size variation occurs, changes in combustion rate occurs, fluctuating heat output. This variation in heat output greatly increases with changes in moisture content. Also, as combustion rate changes, the required oxygen level changes hence changing the excess air requirement. Depending on the situation, the air requirement may be insufficient to complete the combustion when dry material is present. If a sudden increase in moisture occurs then the air introduced in the furnace may be too excessive increasing the percent excess air and the chance of fines passing through. This results in unnecessary power consumption, plus unnecessarily lowering the combustion zone temperature from the extra amount of air requiring heat up, and again can increase the quantity of particulate matter requiring collection.

Water is roughly one half to two-thirds heavier than wood, therefore higher moisture content fuel also requires more energy to transport. It does however provide a benefit in that moisture has an adhesive quality so that smaller particles or fines will better adhere to larger particles. This means during transport in open containers or conveyors there is less of a problem of "dusting," i.e. wind picking up and carrying off the fines as dust. In a closed area such as in storage, covered conveyors, etc., dusting can be extremely hazardous. In concentrated form, backfires can result traveling to the fuel pile or an explosion may result.

Effects of Ash Content

The ash content of wood fuels has been found to range from 0.1% to over 20% by weight, depending on previous handling processes and storage techniques. The quantity of this nonorganic noncombustible portion of the fuel primarily affects wear and tear on boiler components, maintenance of these components, size of air pollution control equipment, and ash removal techniques.

Abrasive ash such as sand and grit greatly increase wear and tear on feeding and handling equipment. Once in the furnace ash can plug passages on grates and can coat heat exchange sections, reducing heat transfer efficiencies. If the ash fusion temperature is reached, it will produce molten slag which can adhere to grate surfaces. Sand and other abrasive material can also erode boiler tubes and sea salt can corrode heat exchangers and breachings if condensation occurs. All of this results in higher maintenance costs due to the greater frequency which grates and tubes must be cleaned or blown off, plus increases replacement costs with time.

Ash is also carried in the gas stream through the heat exchange section and into the pollution collection system. It can range from 40% to 100% of the particulate emissions depending on the burning quality of the boiler. The remaining portion is unburned fuel particulates. Again, depending on ash type, mechanical collectors can be eroded or corroded with condensation of salt.

Ash handling and removal systems are dependent on the amount and type of ash the fuel contains. As an example, assume a 100,000 pound per hour boiler consuming 46,000 pounds of wood fuel per hour at 50% MC_G. Based on the above extremes, 552 pounds to 55,200 pounds of ash must be removed per day, posing quite a disposal problem.

IV. WOOD FUEL PREPARATION SYSTEMS

Any step or process used to prepare fuel for burning requires energy inputs and capital expenditures. On one extreme certain types of wood burning facilities are designed to burn only small sized, dry material in suspension, therefore require the maximum preparation if small dry residues are not available. The other extreme consists of older dutch oven furnaces which can burn wet material with large size variations. Most installations, however, fall in between these two, therefore, the extent and type of preparation, if any, should be justified.

A. *Cleaning and Particle Size Classification*

One of the best ways of obtaining clean fuel is by keeping the logs and bark from becoming embedded with dirt and rocks, for example using paved log yards which are lightly sprinkled. Unfortunately for the boiler operation, the decision on the type of log storage and handling facilities used are based on other factors. Therefore, for most installations, fuel cleaning is done during particle size classification. There are, however, several techniques used primarily for cleaning, but not very widely. These are techniques using air and water segregating by differences in density. Air is blown through a screen which lifts the lighter bark and carries it along while the denser rock and dirt fall through. Water is used in much the same way by placing the material in a flume, skimming the floating bark and wood particles from the top and letting the rocks and dirt sink to the bottom. This latter technique has the obvious disadvantage of adding moisture to the fuel.

Particle size classification should begin at the respective residue generating source. Bark, wet sawdust, dry sawdust, dry trim ends, and dry sanderdust should ideally be kept separate and fed separately into the boiler to better control fuel input variability. If these are the only types of residue used, then furnaces designed just for this type of material can be used, greatly reducing the problems previously discussed. However, when they do have to be burned with other types, extra storage space and handling operations are required which can increase costs beyond the apparent benefits gained. Therefore, other methods such as shaker screens and disc screens are used.

Shaker screens with an approximate one inch opening were the first type used, letting the smaller diameter dirt and rocks fall through. They do not provide the capacity required, therefore, are presently rarely used for fuels. Disc screens are the most common, due to their higher capacities, lower energy requirements and ability to withstand the abrasive nature of wood fuel components. They work on the principle of discs rotating on a series of shafts spaced at a distance equal to the desired maximum particle size. The turning discs carry the larger pieces to a conveyor which transports them to the hog, as the smaller material passes through to a conveyor below. The tumbling action of the process helps shake dirt and rocks loose from the bark.

B. Size Reduction

Reduction of wood fuel particles by grinding or "hogging" to some maximum size is normally performed in either a knife-type hog or hammer hog. The knife type, better known as chippers, cut the wood into chips. They are prone to dulling, especially from abrasive material existing in most bark. Because of this, they are only used on debarked material such as end trims and slabs, which go to the pulp chip pile unless they have been dried.

The hammer type hogs are used primarily for wood fuel, especially bark. These machines, used in the industry for many years, normally produce particle sizes from 2 to 4 inches maximum, but are also made to reduce material much finer, averaging 1/4 inch if desired. A secondary grinder is used if material finer than 1/4 inch is desired. This type hog reduces the size of wood particles by hammering them through a series of anvil points spaced according to the desired maximum particle size desired. The smaller this size the greater the horsepower requirements. Since this machine requires a greater force to reduce the material than the knife-type hogs, horsepower requirements are also greater.

C. Moisture Content Reduction

Moisture content reduction can be the most important controllable variable, as discussed previously. The degree of control, however, is a question of both energy and cost economics. The energy economics will be discussed here and the cost economics left to be determined by the specific installation since costs will vary with vendor, installation idiosyncracies and date of purchase.

Moisture can be removed from wood residue in a number of ways. Air drying has been examined with respect to bulk storage techniques (White, 1978) but was found that appreciable reduction in moisture content occurred only with mixed shortleaf and pine bark. Sawdust and hardwood bark moisture contents were not reduced appreciably over a six month period.

If wood fuel is very wet, a mechanical press can be used to squeeze out the moisture to approximately 55% MC_G. This method is somewhat slow and can have high maintenance costs, but has been proven a viable method when very high moisture contents are present.

Hot hogs are also used, a process which combines the grinding and drying operation. Hot gases, up to 1200°F are fed into a modified hammer-type hog simultaneously with wet fuel. The gases and fuel mix well during the grinding providing a good surface to air contact. The resulting particles are then conveyed in the hot gases to a low-efficiency cyclone, drying as it goes. Enroute to this primary cyclone, the stream passes through a classifier and all overs are routed back through the hog. The primary cyclone separates the fines and dirt from the larger pieces which are subsequently sent to dry storage. The gases then carry the fines and light dirt fraction through a high-efficiency cyclone where they are deposited and the gas emitted in the atmosphere or routed back through the system. The fines can then be used to fuel the heat source for the hot hog or be disposed of, as the mill desires. The main problem with this system is the chance of fire if the right mixture of dry material, temperature, and oxygen occur in the hog. This is diminished by recirculating a portion of the combustion air which has a lower oxygen content back through the system.

Another type of conveying-drying system is the hot conveyor. This system consists of placing hogged material on a perforated vibrating conveyor through which hot gases pass and mix with the material. The entire unit is located under a stationary hood which collects the gases and directs them through a high efficiency cyclone before venting. The cyclone removes the fines and dirt entrained in the air and again either sends them to the burner or the waste pile. This operation can handle gases up to 600°F from an auxiliary burner or stack gases. Since the gases are in contact with the material a shorter length of time, less drying will be done than in the hot hog process.

Rotary dryers are also used to remove moisture from large quantities of hog fuel. These dryers operate at inlet temperatures from 1000°F to 1600°F, "flash drying" the material as it enters and passes through. Theoretically, the moisture on the particle will keep the surface temperature down to 212°F, or below, however, in practice the surface can dry out, become overheated creating a blue haze. The blue haze results from low grade oxidation occurring on the particle surface emitting volatile hydrocarbons into the discharge gas.

This gas has to then either be passed back through the burner, since it is combustible, or through high efficiency collectors.

Each of these processes require some type of heat source which must be obtained from fuel which could otherwise be used to provide steam or heat, the primary objective. Therefore, the only way these predrying processes can be feasible is if they can remove the moisture more efficiently than the boiler, efficient enough to carry any extra capital and operating costs it involves. The extra capital costs involved are those incurred for the installation of a dryer and extra handling and transfer processes required greater than the cost reductions obtained in the boiler installation from its being provided a more uniform fuel. The specific costs involved will be site specific as noted above, but the overall efficiencies can be examined.

This examination will be made by determining the energy efficiencies of (1) initially drying the fuel in a rotary dryer having an auxillary burner utilizing the wood fines and the other wood fuel as needed, and (2) the same system using stack gases in lieu of the auxillary burner. The overall efficiencies will then be compared with burning the wet fuel directly in the boiler. For this comparison assume a boiler producing 100,000 pph steam at a pressure of 300 psia and superheating the steam to 700°F. At these conditions the steam has an enthalpy of 1368.9 Btu/lb.

Since energy efficiencies of each of these processes will be explored, it would be advantageous to point out how the different variables affect boiler efficiency. With this understanding the results for this example may be better explained and applied to other boiler operations.

The efficiency of the boiler depends primarily on the fuel moisture content, ambient temperature, stack gas temperature, percent excess air used, higher heating value of the fuel and miscellaneous heat losses due to radiation, etc. The relationships of these variables are shown in Figure 3 with ambient temperature and miscellaneous losses held constant at 60°F and 3%, respectively. If the curve representing 0% excess air, 350°F stack gas temperature and 8000 Btu per pound higher heating value is taken as a base, then by doubling each variable, their respective influences on boiler efficiency can be determined. As can be seen in Figure 3, the higher heating value has the greatest influence. This is because the more heat available per unit weight of material, the less of an effect the losses in the boiler will have. Moisture content shown to be second to higher heating value for the influence of boiler efficiency is the largest heat loss factor. Therefore, for a given pound of wood fuel at 55% MC_G, the heat loss attributable to moisture would be the same whether the higher heating value of 8000 Btu per pound or 10,000 Btu per pound.

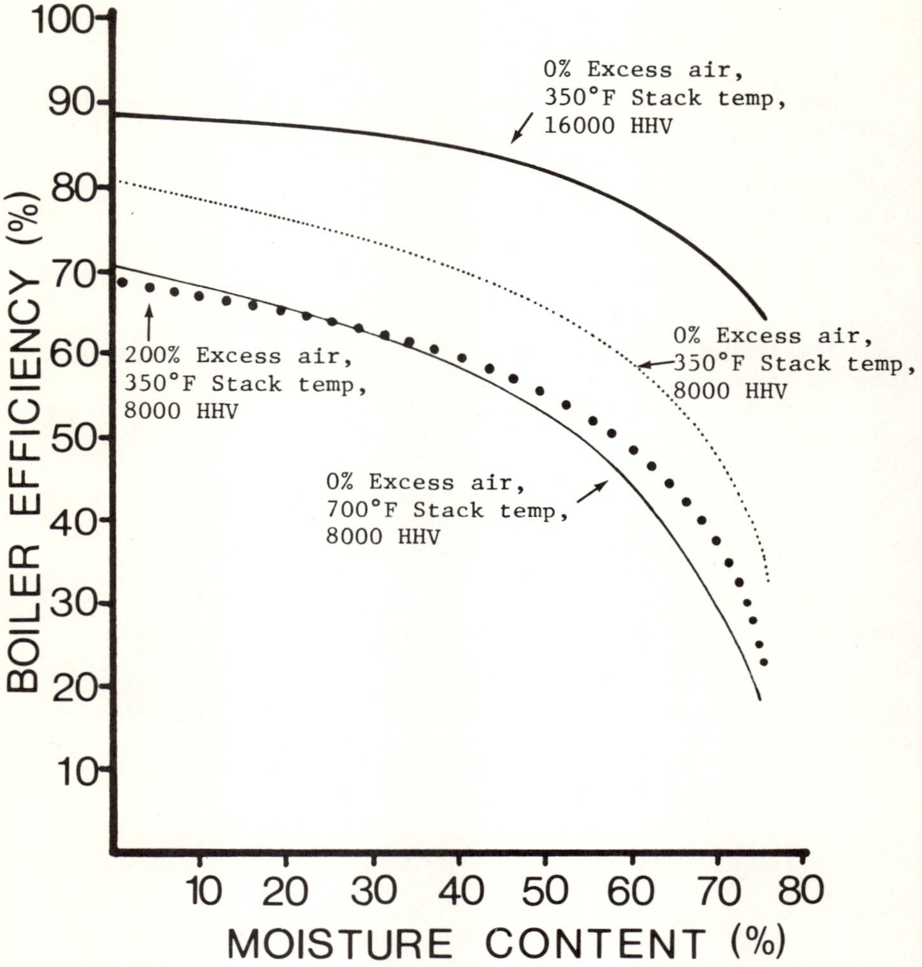

FIGURE 3. The effects of moisture content, stack gas temperature, excess air, and higher heating values on boiler efficiency.

However, on an efficiency basis, the 8000 Btu per pound value will be lower since it will have a greater percentage of its available heat, 3600 Btu's absorbed by this loss than the higher heat content material at 7200 Btu. At moisture contents below approximately 24% MC_G excess air is more influential on boiler efficiency than stack temperature, however, this reverses for moisture contents above this point.

The comparison base for this examination is burning the wet fuel directly in the boiler. If 50% excess air and 8000 Btu per ovendry pound higher heating value are assumed, the boiler efficiencies and quantities of fuel required are given in Table IV. So depending on fuel moisture content, the furnace may require anywhere from as much as 107.28 tons of wet fuel per hour for 75% MC_G to as little as 11.03 tons of ovendry fuel per hour. The 75% figure is not realistic since it actually can not burn continuously without supplemental fuel. Wood fuel must be at or below approximately 68% MC_G before it will sustain combustion. At 75% MC_G either mechanical pressing to 55% MC_G or some type of predrying is required if a self-sufficient wood fuel operation is desired.

Wood fuel dried prior to being placed in the furnace has additional considerations before an accurate analysis can be made. Initially, the optimum amount of moisture which should be removed in the dryer has to be determined. This will depend on the initial moisture content of the fuel, the efficiency obtained in the dryer plus the efficiency obtained in the furnace at the new moisture level. The overall efficiency of the dryer, based on the available heat in the fuel entering

TABLE IV. Quantity of Wood Fuel Required at Various Moisture Contents for 100,000 pph Steam Boiler

Initial MC_G (%)	*Boiler efficiency (%)*	*Available heat per lb input Btu/lb*	*Available heat for stem production (Btu/lb input)*	*Quantity wood fuel required (tons/hr)*
75	31.9	2000	638	107.28
65	49.3	2800	1380	49.58
55	59.0	3600	2124	32.22
45	65.2	4400	2869	23.86
0	77.6	8000	6208	11.03

the dryer, will decrease as more moisture is removed. This is because heat is required to remove that moisture. Wood fuel used to supply heat to the dryer can not supply heat to the boiler, therefore with respect to boiler output or efficiency it is a negative value. However, as long as it is removing moisture more efficiently than in the boiler there is an overall net gain which may make it profitable. For example, Figure 4 shows the relationship between dryer efficiency and boiler efficiency for wood fuel initially at 75% MC_G. As more moisture is lost in the dryer more heat is consumed by that particle and hence it is "costing" the operation in thermal energy to remove that moisture. However, at the same time the moisture is being lost in the dryer, boiler efficiency is improved due to a drier fuel entering the furnace. The optimum amount of moisture which should be removed therefore is at that point where the rate in efficiency loss in the dryer becomes greater than the gain in boiler efficiency. Another way this may be determined is by adding the efficiencies of each process and finding the moisture content loss which has the greatest value. This has been done for the efficiencies shown in Figure 4 and is plotted above them. For this example, then 15% MC_G should be removed in the dryer and the material fed into the furnace at 60% MC_G. Table V provides these values for initial moisture contents of 75% MC_G, 65% MC_G, and 55% MC_G.

The actual efficiency of the dryer is necessarily a major consideration. Raddin (1975) and Dokken (1975) reported dryer efficiencies ranging from 1400 to 2400 Btu consumed per pound of water evaporated, the more efficient being rotary drum dryers drying material with high initial moisture contents. Therefore, for the example with the dryer using an auxillary heater, the wood fines and supplementary wood required to provide the heat required are assumed to have a 0%-4% MC_G, originating from material already passed through the dryer. The auxillary burner is assumed to have a 60°F ambient temperature, a 212°F outlet temperature, 8000 Btu per ovendry pound higher heating value fuel input, and 25% excess air. This provides a burner efficiency of 87.2%. It was additionally assumed that 1400 Btu were consumed per pound of water evaporated for material with an initial 75% MC_G, 1500 Btu per pound for 65% MC_G and 1600 Btu per pound for 55% MC_G material. No material was examined having an initial moisture content below 55% since these minimum values were very close to or below those for direct burning.

The stack gas installation uses gases from the boiler which have passed through a high efficiency cyclone to remove hot combustible materials to minimize fire hazards. The quantity of these gases which may be obtained, and their temperature, depend on furnace conditions. These conditions can be

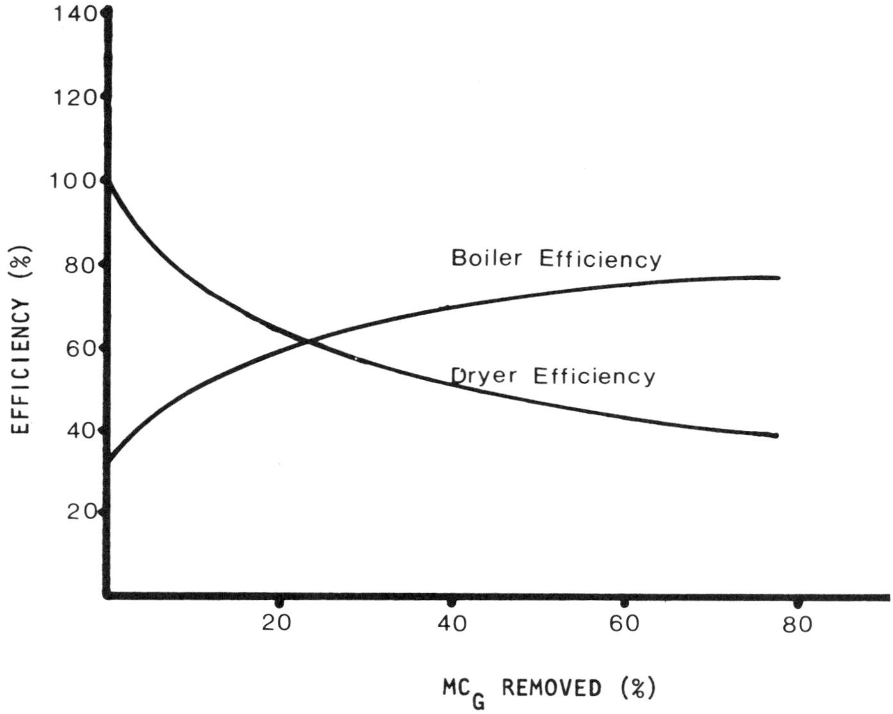

FIGURE 4. Drying efficiencies for wood fuel from 75% moisture content.

changed, however the resulting change in overall boiler efficiencies must also be considered. For example, the gases would be better for drying with higher percentage excess air and higher temperatures, however, each of these occur at the expense of boiler efficiency. The gain, of course, is the increase in boiler efficiencies due to lower moisture content, which is the more influential variable as shown previously.

The quantity of gas available in this assumed boiler installation has been determined for each pound of material input at various moisture contents. This information is contained in Table VI. The available thermal energy for any given temperature drop can be calculated using these values by knowing the heat capacity of this flue gas. If a 50% drying efficiency is assumed, then the new boiler efficiencies for the process using stack gases can be calculated for various moisture contents. Since the moisture content variable is more influential than either the stack gas temperature or the

TABLE V. Operational Efficiencies for Drying Wood Fuel with Auxillary Dryer Prior to Combustion

MC_G reduction in dryer (%)	Lbs H_2O evap. per lb input (lbs)	Dryer Btu consumption per lb H_2O (Btu)	Boiler eff. at lower MC_G level (%)	Dryer efficiency (%)	Fuel quantity required for boiler (tons/hr)
75-65	.286	1400	49.3	77.1	90.08
75-60	.375	1400	57.8	69.9	84.71
75-55	.444	1400	59.0	64.3	90.18
65-60	.125	1500	57.8	92.3	45.81
65-55	.222	1500	59.0	86.4	47.98
65-50	.300	1500	62.4	81.6	48.03
55-45	.182	1600	65.2	90.7	32.14
55-40	.250	1600	70.5	87.3	30.91
55-35	.242	1600	68.5	87.6	31.67

TABLE VI. Quantity of Stack Gases Per Pound Fuel Input Available for Drying Wood Fuel at Various Fuel Moisture Contents and Excess Air Percentages

	Weight of stack gas (lbs)				
	Percent of excess air				
MC_G (%)	0	25	50	100	150
75	2.29	2.68	3.06	3.83	4.60
65	2.81	3.35	3.89	4.97	6.05
55	3.33	4.03	4.72	6.11	7.50
45	3.84	4.69	5.54	7.23	8.93
0	6.17	7.71	9.26	12.34	15.43

percentage excess air used, fuel at 0% MC_G going into the boiler should promote the largest.

The values for various moisture content reductions and resulting changes in boiler efficiencies, stack gas temperatures and percentage of excess air are given in Table VII. The stack gas temperatures were kept as high as possible and the excess air percentages varied to obtain the heat quantities required since at 0% MC the latter has a greater influence on the boiler efficiencies.

The results of each method of moisture removal can now be directly compared. Table VIII is a summary of all results. This shows that using stack gas to dry incoming fuel with high initial moisture contents of 75% and 65% can reduce wood fuel inputs by 42.34 and 7.14 green tons per hour, respectively. If it were not possible to locate the dryer next to the boiler, thereby not being able to utilize the stack gases, the stand alone dryer is still more efficient by 22.57 and 3.77 green tons per hour, respectively. At 55% moisture content, the dryer with an auxillary heater is only slightly better than using stack gas or burning directly. Forty-five percent moisture content material and below should be burned directly in all cases.

Given these results for the operation outlined, predrying the fuel prior to feeding the boiler may or may not be economically feasible depending on various other factors. If for example the fuel has an initial 65% MC_G, then predrying by

TABLE 7. Boiler Efficiencies for Moisture Removal with Stack Gases Assuming a Dryer Efficiency of 50%

Inital MC_G (%)	Final MC_G (%)	H_2O evap. per lb input (lb)	Btu required @ 50% eff. (Btu/lb)	Stack temp (°F)	Excess air (%)	Boiler eff. (%)	Wood required (tons/hr)
75	0	.75	1766	700	150	52.7	64.94
65	0	.65	1530	700	108	57.6	42.44
55	0	.55	1295	700	77	61.3	31.02
45	0	.45	1060	700	45	65.0	24.38

TABLE VIII. *Summary of Drying Method Results*

Initial MC_G (%)	Wet fuel feeding boiler		Dryer using auxillary heat source		Dryer using stack gas	
	Boiler efficiency (%)	Quantity wood required (tons/hr)	Boiler efficiency (%)	Quantity wood required (tons/hr)	Boiler efficiency (%)	Quantity wood required (tons/hr)
75	31.9	107.28	57.8	84.71	52.7	64.94
65	49.3	49.58	57.8	45.81	57.6	42.44
55	59.0	32.22	70.5	30.91	61.3	21.02
45	65.2	23.86	77.6	25.12	63.8	24.38

utilizing the stack gas would require that the 7.14 green tons per hour saved from burning wet fuel would compensate for any increase in capital and operating expenses. This amounts to 2.5 ovendry tons per hour or 21,891 ovendry tons per year.

The above computations have made assumptions which may be different in various situations. The greatest for the stack gas operation was a 50% dryer efficiency. If this is increased to 90% the results would be somewhat better as shown in Table IX. With these efficiencies our 65% MC_G example would save 9.28 green tons per hour or 3.25 ovendry tons per hour. This amounts to 28,452 ovendry tons per year, a 30% increase. Therefore, the parameters surrounding the installation must be covertly known before an honest decision can be made.

V. SUMMARY AND CONCLUSIONS

The major sources of wood fuels in the future may be forest residues, however, presently are the sawmill and plywood industries. Each of these industries produces wood residue with large variations in particle size, moisture content, and ash content. A range of 8 to 10^{-5} inches up to large pieces of lumber are produced ranging from 2% to 80% MC_G and containing from 0.1% to over 20% in ash content. These variations occur from the various types of mills having a large range of machine centers producing these residues. The combustion of these fuels is greatly influenced by these variations affecting boiler efficiency. The fuel can, however, be made more uniform through preparation techniques, increasing boiler efficiencies but at a cost. The extent of preparation, therefore must be examined with respect to the monetary potential through efficiencies gained throughout the system.

The most common type of preparing wood fuels for better uniformity is thoroughly mixing all materials from all sources. In this way large variations still exist between pieces but are more uniform per unit change. This does work to some extent, however, it is extremely difficult, if not impossible to get a thorough mixing. This results in sudden changes in particle size and/or moisture content.

Particle size variation is controlled primarily through hogging to a 4 inch minus. Some combustion processes require particle sizes as small as 1/4 inch minus, requiring higher horsepower. The extent of size reduction depends on the characteristics of the residue and hence combustions process used.

Moisture content, the most influential variable presents the greatest problems in boiler efficiencies. The moisture

TABLE IX. Boiler Efficiencies for Moisture Removal with Stack Gases Assuming a Dryer Efficiency of 90%.

Initial MC_G (%)	H_2O evap. per lb[a] input (lb)	Btu required @ 90% eff. (Btu/lb)	Stack temp. (°F)	Excess air (%)	Boiler efficiency (%)	Wood required (tons/hr)
75	.75	981	700	95	59.2	57.80
65	.65	850	700	83	60.6	40.30
55	.55	719	700	70	62.1	30.62
45	.45	589	700	57	63.6	24.46

[a] Theoretical.

content can be reduced prior to the furnace by drying the fuel in rotary dryers with auxiliary heaters or using heat from stack gases. Using only stack gases can increase boiler efficiency. Dryer efficiency here is a function of stack temperature, hence the total gains of the system may be modest. Further these increases have to be sufficient enough to carry any increase in capital outlay and operational expenses for the dryer operation.

Fuel preparation has many advantages and disadvantages which have to be considered for each operation before the economics can be determined. This depends on the fuel types and sources and dictates the type of combustion unit which should be used.

REFERENCES

Corder, S. E. and others. 1972. Wood and Bark Residue in Oregon. Trends in Their Use. Res. Pap., School of Forestry, Ore. St. Univ., No. 11.

Dokken, M. 1975. The Technology of Drying Wood Particles. Report, East. For. Prod. Lb., CANADA No. OP-X-109E.

Howlett, K. and others. 1977. Silvicultural Biomass Farms, Volume VI: Forest and Mill Residues as Potential Sources of Biomass. Mitre-TR7347, ERDA Contract Sponsor, Project Number 2170, May.

Johnson, R. E. 1945. Some Aspects of Wood Waste Preparation for Use as Fuel, *in* "Wood and Bark Residues for Energy," (compiled by S. E. Corder). Proceedings of a conference held May 31, 1974, School of Forestry, Ore. St. Univ., Corvallis, Oregon.

Junge, D. C. 1975. Boilers Fired with Wood and Bark Residues. Res. Bull. 17, Nov., For. Res. Lab., Ore. St. Univ. Corvallis, Oregon.

Koehler, A. 1924. "The Properties and Uses of Wood." McGraw-Hill, New York.

Leman, M. J. 1975. Special Environmental Problems Originated by Burning Bark from Saltwater Borne Logs, *in* "Wood and Bark Residues for Energy" (compiled by S. E. Corder). Proceedings of a conference held May 31, 1974. School of Forestry, Ore. St. Univ., Corvallis, Oregon.

Mingle, J. G. and Boubel, R. W. 1968. Proximate Fuel Analysis of Some Western Wood and Bark. *Wood Science 1*(1):29.

Raddin, H. A. 1975. Drying Hardwood Flakes and Fibers: Review of Techniques. FPRS No. MS-75-557.

Tillman, D. A. 1978. "Wood as an Energy Resource." Academic Press, New York.

Tillman, D. A. 1979. The Sawmill as a Fuel Generator: Sources and Causes of Variability. Col. For. Res., Univ. of Wash., Seattle, WA.

Weldon, D. 1966. The Volume and Location of Southern Yellow Pine Bark in East Texas. Texas For. Serv., Circ. No. 101.

White, M. S. and DeLuca, T. A. 1978. Bulk Storage Effects on the Fuel Potential of Sawmill Residues. *For. Prod. J. 28* (11):24.

Williston, E. M. 1976. "Lumber Manufacturing: The Design and Operation of Sawmills and Planer Mills." Miller Freeman, San Francisco.

Williston, E. M. 1979. Personal communication.

Zerke, J. I. 1977. Wood in the Energy Crisis. *Forest Farmer 37*(2):12, Nov.-Dec.

Index

Proceed——of the International Symp
on Nitrite in Meat Products

The symposium was sponsored by:

Commodity Board for Livestock and Meat, Rijswijk
Ministry of Public Health and Environmental Hygiene, Leidschendam
Organisation for Nutrition and Food Research TNO, The Hague

Organized under auspices of:
The Research Group for Meat and Meat Products TNO, Zeist

Proceedings of the International Symposium on Nitrite in Meat Products

held at the Central Institute for Nutrition and Food Research TNO, Zeist, the Netherlands, September 10-14, 1973

Editors: B. Krol and B. J. Tinbergen

pudoc

Wageningen
Centre for Agricultural Publishing and Documentation
1974

ISBN 90 220 0463 5

Cover design: Pudoc, Wageningen

Printed in the Netherlands by Krips Repro, Meppel

Contents

Preface

It has long been known that small amounts of saltpetre are indispensable for the formation of the characteristic meat colour. About 1890, it was found that nitrite was in fact the active compound responsible for the colour, the mechanism of it being discussed by Haldane in 1901 in his classic publication.

Several decades passed before another characteristic of nitrite was discovered: a potent inhibitor of micro-organisms, among them pathogens, in many meat products. In particular the inhibition of toxin formation by *Clostridium botulinum* was established. More recently the role of nitrite in cured meat flavour was recognized.

For a few years, attention has been focused on the human health hazard arising from possible formation of carcinogenic *N*-nitroso compounds from nitrite and secondary amines in meat. Debate started in many countries on the pros and cons of the use of nitrite in meat products. Research was intensified in various Institutes all over the world.

We had the impression that it might be useful to bring experts together in an international meeting on nitrite in meat products. The reactions received from those currently engaged in the relevant problem indicated that such a conference was welcomed indeed by everybody.

Accordingly, the symposium took place at Zeist in September 1973 and was attended by experts from chemical, microbiological, toxicological and technological disciplines and some public health authorities.

All the participants felt that the papers and discussions had contributed to a better understanding of several problems. Their prompt publication together with resolutions of the meeting should assist others involved in the problem of nitrite in meat products.

We are grateful to the staff of Pudoc, Wageningen, for their competence in preparing the proceedings for publication so quickly.

B. Krol
B. J. Tinbergen

Word of welcome

H. de Boer

Research Group for Meat and Meat Products TNO, Zeist

I am glad to welcome you here on behalf of the Research Group for Meat and Meat Products TNO, under whose auspices this Symposium has been organized. For those not familiar with the structure of meat research in the Netherlands I may explain, that this Research Group includes the main governmental institutes and university departments for research on meat production, meat technology and meat hygiene.

Because of the significance of international co-operation of research workers, who work on the different aspects of this complex problem, the Dutch authorities decided to sponsor the symposium. We greatly acknowledge the financial support by the Commodity Board for Livestock and Meat, by the Ministry of Public Health and Environmental Hygiene, and by the Organization for Nutrition and Food Research TNO.

The organization of this Symposium is largely due to the work of the staff of the Central Institute for Nutrition and Food Research TNO, who are our hosts, and to the efforts of several staff members of the Netherlands Centre of Meat Technology.

This symposium brings together scientists working on the different aspects of the problem. I will not take too much time by extensive words of welcome but merely express how much we appreciate your response and the presentation of your latest research, thus providing material for a better understanding on the whole problem. We are grateful that Dr Aunan, Dr Rubin, Professor Kotter and Professor Mossel have accepted the invitations to act as chairmen in the different sessions. As you know, the problem of nitrite in meat products has various facets, each of which represents a specific field with its specific starting points. Even though it will be difficult after each session to synthesize each of the separate fields – analysis, microbiology, technology and toxicology – it will be more difficult to integrate these aspects into general conclusions.

I think we cannot expect this symposium to provide direct implications about the use of nitrite in meat curing. Our purpose may be to use this unique confrontation of knowledge and views from different facets for a better mutual understanding, allowing collaboration and leading to new approaches in some common framework.

I hope your work of the next days will be productive and provide prospects for the future.

Analytical session

Reporters: J. H. Houben, W. J. Olsman, P. S. van Roon

Proc. int. Symp. Nitrite Meat Prod., Zeist, 1973. Pudoc, Wageningen.

Formation of curing pigments by chemical, biochemical or enzymatic reactions

K. Möhler

Abteilung Technologie der Nahrung – Bromatologie – Technische Universität München, 805 Weihenstephan bei Freising, W-Germany
Present address: Techn. Univ. München, Abt. Technologie der Nahrung, 805 Freising/Vöttung, W-Germany

Abstract

Chemical, biochemical and enzymatic reactions for the formation of curing pigment nitric oxide myoglobin (or nitric oxide hemoglobin) are discussed. All systems are derived from metmyoglobin which is the first product formed in the chain of reaction. It is reduced chemically by SH groups, biochemically by the co-enzymes NADH and FMN or FAD, or enzymatically by a NADH-dependent dehydrogenase system under co-function of ferrocytochrome c. Under similar or identical conditions nitrite is reduced to nitric oxide which is bonded to the myoglobin (hemoglobin).

Curing pigment – properties and nomenclature

Curing pigment is the red agent, which is formed from the muscle pigment myoglobin by addition of nitrite under various conditions. This pigment retains its colour after heating. The collective term of curing pigment includes the corresponding reaction products of the blood pigment hemoglobin, because blood is occasionally a component of sausages. Blood is also left behind in small, varying amounts in the muscle tissue after slaughter. A detailed illustration of the physical properties of hemoglobin (Hb) can be found in Weissbluth (1967). Calculations and theoretical conceptions as to the type of bonding of the oxygen onto the Hb-iron and the charge distributions in the oxymyoglobin (O_2-Mb) originate from Bayer and Schretzmann (1967). Such data are as yet not available for NO-myoglybin (NO-Mb) but a lot of basic data can be derived from the two publications mentioned. A publication of Smith and Williams (1970) is an important reference for the relation between the reactions of Hb and Mb and their spectra.

It is important for all further considerations, that all Mb-Bonds are of a complex nature. Protohaem IX is the active centre in the ferromyoglobin = deoxymyoglobin = Mb. The 5th coordination position is occupied by the tertiary imidazol-nitrogen of the so called proximal histidine. The divalent iron is penta-coordinated in Mb, i.e. the bonding of O_2 results without having to substitute a ligand. Similarly NO, with the characteristics of a radical, can be bonded. Metmyoglobin (MetMb) – iron is always hexa-coordinated, the 6th coordination position is occupied by a water molecule if no negative ions such as the nitrite ions are present. Through heating above the coagulation temperature of protein, the globin is denatured. There is no proof, however, that through this the bonding conditions of the protoheme are changed. The Tarladgis hypothesis (1962): in cooked, cured meats two coordination positions of the protoheme are occupied by NO, could not as yet be confirmed. The solubility of curing pigment in acetone (and other organic solvents) is probably caused by the exchange of the imidazol-N-bond on the 5th coordination position with acetone.

The result of structural examinations definitely show that the name nitrosomyoglobin, and nitrosohemoglobin, is incorrect since curing pigment has no connection with nitroso bonding. The name nitrosyl myoglobin is also incorrect as nitrosyl bonds are considered as mixtures of anhydrides of nitrous acid and other acids (e.g. nitrosyl chloride). The only correct name can be nitrogen monoxide myoglobin (nitric oxide myoglobin = NO-Mb). The same is true for hemoglobin.

Fundamentals of NO-myoglobin formation

According to present opinion the first reaction of nitrite[1] with muscle pigment is an oxidation of Fe^{2+} in Mb to Fe^{3+} in MetMb. Simultaneously, nitrate is produced in an autocatalytic reaction. In model experiments with minced beef muscle, the amount of nitrate formed varies greatly (Möhler, 1967; Walters et al., 1968). However, from the statistical evaluation of more test material the following reaction equation can be formed (Möhler, 1967):

$$4\ MbO_2 + 4\ NO_2^- + 2\ H_2O \rightarrow 4\ MetMbOH + 4\ NO_3^- + O_2 \qquad (I)$$

Met-Mb and nitrate are formed in equimolecular amounts. Deviations can be caused by different factors, e.g. by varying the concentration of oxygen. This could be established with hemoglobin (Möhler & Baumann, 1971). In the next section it is implied that nitrate and MetMb are formed in equivalent amounts since the amount of MetMb has not yet been exactly determined. It is unknown whether or not MetMb appears always as an intermediate step in the formation of NO-Mb.

In meat products which are commonly manufactured with nitrite, usually a higher content of nitrate is found then would be expected with Equation I. This is due to the secondary oxidation (Möhler, 1967) in which the dismutation of nitrous acid could also play a role, as could the oxidation of NO, formed through nitrosothioles, according to Mirna & Hofmann (1969).

There are no specifications about whether other decomposition products of nitrite in commonly used concentrations have an influence on the formation or stability of curing pigments. In a balance of nitrite turnover during the formation of curing pigment, nitrite is constantly lost. In the study of these reactions NO and NO_2 were found, which were freed as nitrose gases. Another gaseous component is N_2O (Möhler & Ebert, 1971; Walters & Casselden, 1973). There is also a possibility that small amounts of ethyl nitrite are formed (Walters & Casselden, 1973).

1. When nitrate is used for curing in practice, the reduction to nitrite through micro-organisms in necessary.

The formation of curing pigments by chemical reactions

Application of gaseous NO

If oxygen is completely excluded, NO-Mb is formed in Mb-solutions by adding gaseous NO as occurred with NO-Hb in Hb-solution. After excess NO has been removed with N_2, the pigment solution is relatively stable to oxidation. According to Sancier and co-workers (1962) NO-MetHb is formed when NO is introduced to MetHb-solutions, whose spectrum coincides largely with that of NO-Hb but shows an overall higher extinction. Chien (1969) was able to show that the so-called NO-MetHb was actually NO-Hb; he always obtained the same end-product whether gaseous NO was added to Hb or MetHb. He therefore assumed that MetHb is reduced by NO whereby NO becomes NO_2^- and where Hb reacts with excess NO, NO-Hb is the result. From crystal suspensions of Mb or MetMb Dickinson & Chien (1971) obtained, by supplying sufficient amounts of gaseous NO, in both cases NO-Mb. Compared to that with Hb, this reaction was somewhat slower. If O_2 is excluded no reoxidation of NO can follow. Hence no special mechanism is required for the reduction of Fe^{3+} to Fe^{2+} if enough NO is available.

Application of nitrite

The formation of NO-Mb in vitro usually results from a reaction of Mb-solutions with nitrite, without exclusion of air oxygen and in the presence of a strong reducing agent such as sodium hydrosulphite ($Na_2S_2O_4$) e.g. according to Walsh & Rose (1956). The excess nitrite and hydrosulphite can be removed by dialysis. Comparable experiments with muscle mince are not available and would not be of use for practical purposes.

NO-Mb is formed in muscle mince by adding nitrite without any special reducing agents and by heating. In these two model tests, as in manufacture of frankfurters, luncheon meats etc., only chemical procedures will occur (Möhler, 1967). Denaturation caused by heating eliminates enzyme and co-enzyme activity. However conformational changes on proteins do occur in connection with the activation of SH groups, formation of S-S groups and H_2S.

The following mechanism-theory for the formation of NO-Mb begins with MetMb-nitrite. Through intermolecular reaction it is transformed to NO-Mb in one step, whereby the reactions of the SH groups act as electron donors. Since the SH groups of the myoglobins are probably not sufficient, other muscle proteins can act as reducing agents. However the effect seems to be connected with reactions such as $2\ SH \rightarrow S\text{-}S + 2\ H^+$ since, for instance, cystein or glutathion additives have no effect.[1]

One must consider however, that absolute proof of activity or inactivity of –SH groups is impossible under these conditions. Inhibiting the reaction by the usual SH

1. The positive findings of Reith & Szakály (1967) are due to a strong pH-reduction through use of cysteine hydrochloride. In my experiments these effects did not occur when the initial pH value was maintained. See text.

blockers has up-to-now been unsuccessful (exception: mercury acetate in high concentration according to Möhler (1967). A positive effect of SH groups can therefore only be noticed on a higher yield of NO-Mb which is normally about 70%.

Notable increases of the yield of NO-Mb can be achieved by addition of reducing agents such as ascorbic acid, erythorbic acid (synonym: isoascorbic acid) or thioglycollic acid and thioacetamide. This finding is not opposed to simultaneous intramolecular reduction of Fe^{3+} and NO_2^-. Steric conditions apparently play a role, since the formation of NO-Mb ceases when a large amount of water is added (10 parts H_2O to 1 part muscle meat). NO-Hb formation in blood is heavily diminished under similar conditions (Möhler & Baumann, 1971).

A transfer mechanism of pure chemical nature was discussed by Mirna & Hofman (1969). With SH groups nitrous acid forms nitrosothioles. These products, perhaps unstable, produce the curing pigment by splitting of NO and forming disulphide. The condition is, however, that at least temporarily there are sufficient hydrogen ions in the reaction so that nitrous acid can be formed from nitrite. If the pH value is near 3, red nitrosothiol can easily be formed from cystein and nitrite. As stated above, cystein has no effect on the formation of curing pigment in meat products at the usual pH value of 6. The question is now whether or not this transfer mechanism is possible under usual conditions of meat processing.

Fox (1966, 1968) assumed that a temporary bond exists between ascorbic acid and nitrite for the transfer of NO. Since additives such as ascorbic acid, glucono-δ-lacton etc. promote the formation of curing pigment but are not essential for the formation of NO-Mb, the reaction possibilities in reference to this subject will not be discussed.

Formation of curing pigments by biochemical reactions

Non-enzymatic reactions

Non-enzymatic formation of NO-Mb from MetMb was described by Koizumi & Brown (1971). NADH in presence of FMN or FAD was used as reducing agent. NADH by itself is not active. If oxygen is completely excluded, MetMb is reduced to Mb as shown in Fig. 1. Mb reduced nitrite to NO and is itself oxidized to MetMb. NO is immediately bonded to excess Mb, while MetMb returns to the circulatory system. Two points are especially important for the process of reaction. The system NADH/FMN or FAD cannot reduce nitrite to NO. The authors deduced from this that the co-enzyme mixture only affects the reduction of MetMb to Mb. Secondly, nitrite is reduced to NO by Mb under anaerobic conditions. This reaction was described for Hb by Brooks as early as 1937. The proof of a similar reaction with Mb relies upon spectrophotometric measurements of Koizumi & Brown (1971). It was admitted however, that under these test conditions NO-Mb and O_2-Mb are difficult to separate. A methylene blue-diaphorase system can be used instead of the co-enzyme mixture, whereby the enzymatically reduced methylene blue causes the reduction of MetMb to Mb. The reduction of nitrite to NO is impossible.

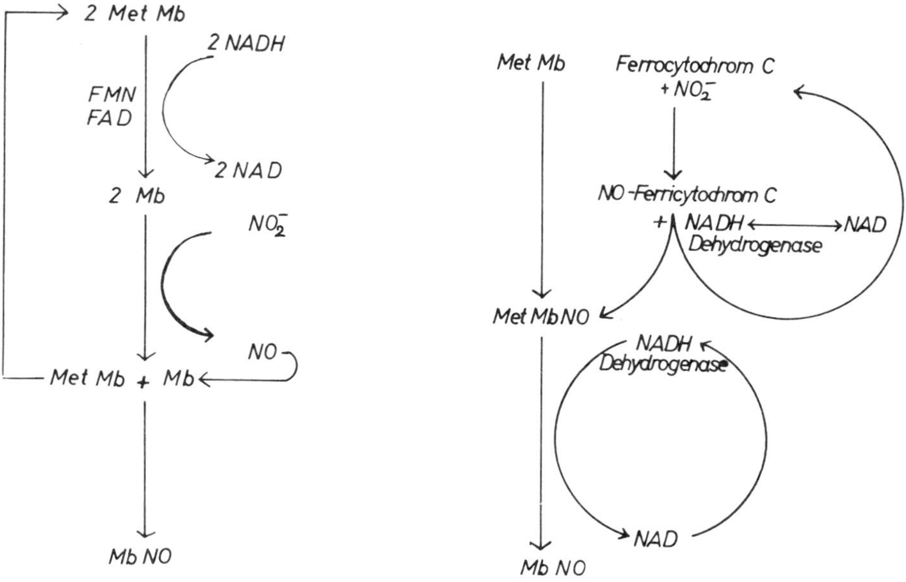

Fig. 1. Non-enzymatic and enzymatic formation of NO-Mb.
Left: formation of nitrosylmyoglobin, according to C. Koizumi and W. D. Brown.
Right: formation of nitrosylmyoglobin, according to C. L. Walters and coworkers.

Enzymatic formation of NO-Mb

The enzymatic biochemical interpretation of the formation of NO-Mb is closely connected with the name of Clifford Walters (1968). Principally his explanation states that nitrite takes the place of oxygen in a part of the respiratory chain. Then the oxidations caused hereby are repaired by the NADH-dehydrogenase system. Fig. 1 shows the somewhat abbreviated reaction-mechanism. The starting point here is also MetMb which should be produced when nitrite is added under anaerobic conditions. In addition, ferrocytochrome c has been oxidized by nitrite to ferricytochrome, and NO, as product of reduction, has been bound to cytochrome. Nitrosylferricytochrome c is now reduced by NADH with the aid of the dehydrogenase system. Ferrocytochrome c can return into circulation. Since nitrosylferrocytochrome is instable, NO separates and is transferred to MetMb. Thus, hypothetic NO-metmyoglobin developes. If the NADH-dehydrogenase system then reduces the MetMb, the desired curing pigment is produced. The initial reaction course of Mb is dependent on anaerobic conditions as they exist, according to Walters, e.g. in the interior of a muscle. If the hitherto discussed non-enzymatic system of Koizumi & Brown is compared with the enzymatic system of Walters and co-workers, two reduction steps can be observed in both systems. The first reduction step in each system is connected to NADH, whereby in one case the co-enzymes FMN or FAD are active, in the other the dehydrogenase system. Contrary to the reaction according to Koizumi & Brown that according to Walters requires NADH also for the second reduction step. An analysis of both systems shows that two reduction equivalents are necessary for the formation of one NO-Mb.

References

Bayer, E. & P. Schretzmann, 1967. Reversible Oxygenierung von Metallkomplexen. Structure and Bonding, Vol. 2, 181–250, Berlin, Heidelberg, New York, Springer Verlag.

Brooks, J., 1937. Proc. R. Soc. London, B 123: 368.

Chien, J. C. W., 1969. Reactions of nitric oxide with methemoglobin. J. Am. Chem. Soc. 91: 2166.

Dickinson, L. Ch. & J. C. W. Chien, 1971. An electron paramagnetic resonance study of nitrosylmyoglobin. J. Am. chem. Soc. 93: 5036–5040.

Fox, J. B., 1966. The chemistry of meat pigments. Agr. Fd Chem. 14: 207–210. see also:

Fox, J. B. & S. A. Ackerman, 1968. Formation of nitric oxide myoglobin: mechanisms of the reaction with various reductants. J. Fd Sci. 33: 364–370.

Koizumi, C. & D. Brown, 1971. Formation of nitric oxide myoglobin by nicotin amide adenine dinucleotides and flavins. J. Fd Sci. 36: 1105–1109.

Mirna, A. & K. Hofmann, 1969. Über den Verbleib von Nitrit in Fleischwaren. I. Umsetzung von Nitrit mit Sulfhydrilverbindungen. Fleischwirtschaft 49: 1361–1366.

Möhler, Kl., 1967. Bilanz der Bildung des Pökelfarbstoffs im Muskelfleisch. Habil.- Schrift, Techn. Univ. München, vergl.

Möhler, Kl., 1970. I. Mitteilung, Oxydation von Muskelpigment und Nitrit. Z. Lebensmittel-u. -Forsch. 142: 169–179.

Möhler, Kl., 1971. Bilanz der Bildung des Pökelfarbstoffs im Muskelfleisch. II. Mitteilung, Hitzekatalysierte Bildung des Stickoxidkomplexes. Z. Lebensmittelunters. u. -Forsch. 147: 123–127.

Möhler, Kl. & M. Baumann, 1971. Bildung von Pökelfarbstoff im Blut durch die Einwirkung von Natriumnitrit auf Hämoglobin. Z. Lebensmittel unters. u.-Forsch. 147: 258–266.

Möhler, Kl. & H. Ebert, 1971. Reaktionsprodukte von Nitrit in Fleischerzeugnissen. Nachweis und Bestimmung von Distickstoffoxid. Z. Lebensmittel unters. u. -Forsch. 147: 251–258.

Reith, J. F. & M. Szakály, 1967. Formation and stability of nitric oxide myoglobin. I. Studies with model systems. J. Fd Sci. 32: 188–193.

Sancier, K. M., G. Freeman & J. S. Mills, 1962. Electron spin resonance of nitric oxide hemoglobin complexes in solution. Science 137: 752–754.

Smith, D. W. & R. J. P. Williams, 1970. The spectra of ferric haems and haemoproteins. Structure and bonding, Vol. 7, Berlin, Heidelberg, New York, Springer Verlag, p. 1–45.

Tarladgis, B. G., 1962. Interpretation of the spectra of meat pigments II – Cured meats. J. Sci. Fd Agric. 13: 485-491.

Walsh, K. A. & D. Rose, 1956. Factors affecting the oxidation of nitric oxide myoglobin. Agr. Fd Chem. 4: 352–355.

Walters, C. L., A. McM. Taylor, R. J. Casselden & N. Ray, 1968. Investigation of specific reducing systems in relation to meat curing. Res. Rep. 139, British Fd Manuf. Ind. Res. Ass., Leatherhead.

Walters, C. L. & R. J. Casselden, 1973. The gaseous products of nitrite incubation with skeletal muscle. Z. Lebensmittelunters. u. -Forsch. 150: 335.

Weissbluth, M., 1967. The physics of hemoglobin. Structure and bonding Vol. 2, Berlin, Heidelberg, New York, Springer Verlag, p. 1–125.

Discussion

Assessment of the reaction equation for pigment formation

The equation $4\,MbO_2 + 4NO_2^- + 2H_2O \rightarrow 4MetMbOH + 4NO_3^- + O_2$ is based on experiments, in which nitrite was mixed with raw meat and which lasted for 1 – 2 h.

Between 100 and 1000 mg/kg nitrite, the amount of nitric oxide myoglobin formed is independent of the nitrite concentration, according to Möhler.

Action of reducing agents

Ascorbic acid improves the cured meat colour formation, but does not give a 100% yield of the pigment. Pigment formation also occurs at temperatures below 70°C, but at a slower rate. Reducing agents like thyoglycollic acid and thioacetamide enhance the colour formation when added to meat products; cysteine and glutathion, however, have no such effect. The reason for this discrepancy is unknown.

Nitrosation by nitric oxide myoglobin

According to Professor Roughton, University of Cambridge, the dissociation constant of nitric oxide hemoglobin is so small that it cannot be measured. Thus, it is unlikely that nitric oxide hemoglobin could nitrosate amines but this point is under study by Dr R. Bonnett of Queen Mary College, University of London.

Non-bacterial reduction of nitrate

Up to now no indication of a non-bacterial reduction of nitrate to nitrite in raw meat has been found. In vegetables however this can occur. The nitrite formed here is immediately converted to lower oxidation stages of nitrogen.

Proc. int. Symp. Nitrite Meat Prod., Zeist, 1973, Pudoc, Wageningen

Determination of free and bound nitrite

A. Mirna

Federal Institute for Meat Research, Department for Chemistry and Physics, Kulmbach-Blaich, W-Germany

Abstract

Procedures for the determination of free and bound nitrite are of interest for investigations into the different ways of nitrite reaction and its various reaction products. Compounds reacting directly with Griess reagent are referred to as free nitrite. Heavy metal ions such as Hg^{2+}, Cu^{2+} or Ag^{+} cause the cleavage of the nitrosyl ion(NO^{+}) from nitroso compounds, especially nitroso thiols. The NO^{+} formed is related to the amount of bound nitrite.

Extraction conditions (solvent, temperature, pH) have a distinct influence on the determination of the content of free and bound nitrite in meat products.

Introduction

The effect of curing additives and technological procedures on the curing process in meat products is mainly judged by the changes in the content of nitrite, nitrate and nitric oxide myoglobin (NO-Mb). In general the reliability of analytical methods is checked by recovery tests. Whether a compound found by analytical procedures was present in a product originally or was caused by artefacts during the procedures cannot always be proved adequately.

Apart from interests of mainly technological nature in this connection the question is raised, to what extent the presence of free and bound nitrite may bring about health risks.

Investigations on frankfurter and dry sausages were carried out to study analytical and compositional problems.

Methods

Free nitrite

The term 'free nitrite' is used for all compounds reacting directly with Griess reagent. For colorimetric measurement of the azo dye formed, a clear solution is required; therefore the influence of the solvent and that of the deproteination procedure on the nitrite content must be known. In most methods described in the literature, extraction with water of 100 °C is proposed. Acidic samples are adjusted to neutral or slightly alkaline pH values to prevent any further side-reactions of nitrite as far as possible. For extraction 80% acetone was used (acetone-method); by means of this extract NO-Mb was determined as well (Mirna & Schütz, 1972).

Unfortunately, at present, non-destructive procedures for the determination of the nitrite content are unknown. The indirect way to oxidize nitrite to nitrate followed by electrometric measurement with ion-sensitive membrane electrode systems does not seem very promising at present (Gerhardt & Haller, 1973).

Bound nitrite

The term 'bound nitrite' refers to the content of nitrite detectable with Griess reagent after cleavage of nitroso compounds with Hg^{2+}, minus the free nitrite. This method was developed by Saville (1958) for the determination of small amounts of nitrosated sulphydryl groups. Acidic solutions of nitrosothiols are comparatively stable, in neutral and especially in alkaline media degradation occurs rather fast resulting in the formation of NO^+ (Mirna & Hofmann, 1969 and Saville, 1958). Experiments were carried out to investigate the stability of various nitroso compounds against cleavage with Hg^{2+} in aqueous acetone, following deproteinization with Carrez II solution (300 g $ZnSO_4$/1000 ml).

The amount of NO bound on structural proteins is determined after removal of components soluble in 80% acetone (Mirna, 1970). The method of Mirna, yields slightly lower values than the procedure as modified by Olsman & Krol (1972).

Results

Model experiments on bound nitrite

Various nitrosated compounds were checked for cleavage of the nitrosyl group by Hg^{2+} in aqueous acetone. According to these results the procedure is not specific for nitrosothiols exclusively. Among the nitrosation products of amino acids, the histidine derivative shows no reaction with Griess reagent, but that of arginine reacts strongly positive; *N*-nitrosoproline and *N*-nitrosohydroxyproline give only a faint reaction. The cleavage of the NO groups from *N*-dimethyl and *N*-diethyl-nitrosamine as well as from *N*-nitrosopiperidine, *N*-nitrosopyrrolidine and *N*-nitrososarcosine-ethylester is lower than that of *N*-nitrosomethylurea. The main product of the reaction between nitrite and creatinine (Archer et al., 1971) is creatinine-5-oxime (at 25°C) and l-methylhydantoin-5-oxime (at 0°C), respectively; both compounds do not split with Hg^{2+} and have no formation of NO^+.

The effect of Hg^{2+} on some *N*-nitroso compounds in an acetate buffer of pH 5.2 is a decreasing stability with time, in the order *N*-dimethylnitrosamine, *N*-nitrososarcosine-ethylester and *N*-nitrosomethylurea. After 10 days 16%, 22% and 32%, respectively of the NO was split off. Similar results were gained in a phosphate buffer solution of pH 5.5, although with somewhat lower conversion rates (Mirna, 1970).

Conditions of extraction

Cured meat products undergo very small change of colour after cooking in

Table 1. Influence of extraction solvents, temperature and addition of nitrite on the content of NO-Mb and nitrite both expressed in mg NO/kg in frankfurter and dry sausages. Nitrite added for recovery test: 22.0.

Extraction solvent	Temperature (°C)	Addition of nitrite	NO-Mb	Nitrite	Recovery (%)
Frankfurter					
Water	20	–	.	21.4	
	20	+	.	41.9	93
	65	–	.	21.1	
	65	+	.	41.8	94
80% Acetone	4	–	0.9	26.8	
	4	+	1.1	43.3	75
	20	–	1.3	27.9	
	20	+	1.4	46.6	85
	40	–	1.2	24.2	
	40	+	1.0	43.6	88
	50	–	2.2	23.5	
	50	+	2.6	41.8	83
	65	–	2.7	21.3	
	65	+	3.7	38.5	78
Dry sausage					
Water	20	–	.	14.8	
	20	+	.	37.0	101
	65	–	.	22.9	
	65	+	.	45.1	101
80% Acetone	4	–	2.0	15.3	
	4	+	2.3	36.9	98
	20	–	3.7	17.5	
	20	+	3.9	40.2	103
	40	–	2.8	21.3	
	40	+	3.0	41.5	92
	50	–	3.5	25.7	
	50	+	3.9	46.3	89
	65	–	6.0	34.4	
	65	+	5.8	53.3	86

water. The same treatment in slightly alkaline solution leads to a remarkable fading or even to the destruction of NO-Mb, NO bound in a NO-Mb complex becomes partly free and is therefore determined as free nitrite.

Extraction with water or 80% acetone, respectively, at various temperatures influences the nitrite values found in frankfurters as well as in dry sausages (see Table 1).

As a reference the nitrite values of frankfurters determined in water extracts

Table 2. Influence of pH on the content of NO-Mb and nitrite both expressed in mg NO/kg in frankfurter sausages and dry sausages.

Product	Extraction solvent	Temperature (°C)	Addition of saturated borax solution	NO-Mb	Nitrite
Frankfurter	water	100	+	.	10.6
	80% acetone	20	–	1280	15.8
	80% acetone	20	+	1230	16.3
Dry sausage	Water	100	+	.	2.1
	80% acetone	20	–	5030	1.7
	80% acetone	20	+	4330	2.9

were used; no influence of the temperature in the range of 20°C to 65°C on the nitrite content was observed. The recovery rates amounted to 93% and 94%, respectively. The nitrite content found with the acetone method was higher at lower temperatures than that found by extraction with water; at 65°C there was no difference between the two extraction procedures. The rates for the recovery values decreased in general with increasing temperature, the maximum being in the range of 20 to 40°C. On the other hand NO-Mb values showed an increase with rising temperature. Subsequent addition of nitrite was followed by only a small increase of the NO-Mb values.

Extraction of dry sausages with water at 20°C resulted in a lower nitrite content than with the treatment at 65°C. This was apparently caused by insufficient extraction of coarse particles in the sample. In recovery tests the added nitrite was found quantitatively to be independent of the temperature during the extraction. The nitrite values, determined by the acetone method, increased as the temperature was elevated whereas recovery rates decreased. In the range from 4°C to 20°C the recovery on the average was quantitative. A more pronounced increase of the NO-Mb values with higher temperature was observed with dry sausages than with frankfurters; a tendency toward higher NO-Mb values in the experiments with subsequent addition of nitrite could be observed.

The influence of alkaline media on the extraction results in a reduction of the NO-Mb content, but in an inconsistent increase of nitrite in both types of products (Table 2).

Curing pH

A sausage mixture (beef) was cured at various pH values; after protein precipitation with Carrez II a distinctly lower nitrite content was found than with the samples which had been previously treated with Hg^{2+}. The amount of nitrite depended not only on the pH, but also on storage time of the sausage mixture. To eliminate the influence of NO-Mb, which is always present in cured meat, an analogous experiment with myofibrils, not containing Mb, from bovine muscle was carried out. (Fig. 1). Results for sausage mixture and myofibrils were similar. For myofibrils the difference between free + bound and free nitrite on the first day was

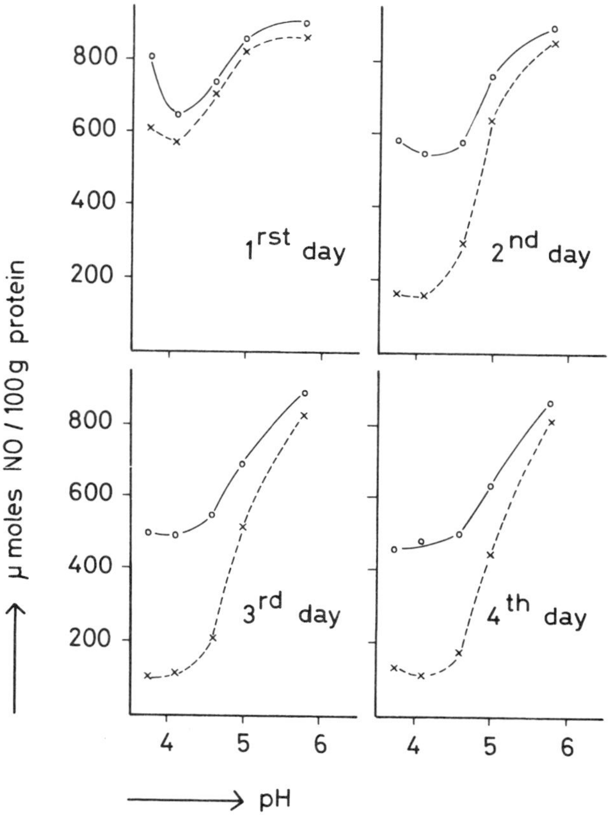

Figure 1: Influence of pH and time on the content of free and bound nitrite in myofibrils from bovine muscle. pH adjusted with acetic acid – sodium acetate – sodium hydroxide. Storage at 4°C.

———— : Hg^{2+} Carrez II (free and bound nitrite)
- - - - - - - - -: Carrez II (free nitrite)

very small; for sausage mixture this difference was more pronounced.

At pH values above 5.8 a rather small amount of bound nitrite was found. Nitrosated compounds probably are only present in those cured meat products, e.g. dry sausages, which normally have a lower pH. The results of investigations on frankfurter sausages (pH $>$ 6) showed nearly no differences in content of free and free plus bound nitrite. Any difference probably was the result of stability with time of some nitroso compound.

Discussion

Many investigations have been made on the influence of micro-organisms, food additives, temperature, pH and redox potential values and the length of storage on the decrease in nitrite content of meat products. In the interests of public health it is more important to know about different reaction products of nitrite than about the amount of nitrite added or remaining after processing.

Thorough work on the influence of pH on the content of NO-Mb and nitrite in sausage emulsions has been carried out by Ando et al. (1961, 1963). According to these investigations the amount of NO-Mb formed (colour formation value), the colour stability (colour retention value), as well as the content of sulphydryl groups decrease with increasing pH value, whereas the amount of nitrite increases. To avoid interferences in the nitrite determination an oxidation of reducing substances with potassium ferricyanide followed by the removal of excess reagent with lead acetate was proposed by Nagata & Ando (1967). Möhler (1970, 1971) has also shown that there is a statistical relationship between cured colour development, pH value and the amount of nitrite consumed. Among the reaction products of myoglobin the formation of a salt-like compound of metmyoglobin with nitrite is discussed. Because this kind of reaction product has a rather low stability in water or in deproteinizing agents, nitrite is quantitatively released (Ando et al., 1961). Investigations on complexes of iron nitrosyl compounds have shown that in stronger alkaline solution a cleavage of nitrous acid occurs (Swinehardt, 1967). The different trends in the NO-Mb and nitrite values of the tested groups of meat products lead to the conclusion, that depending on the composition of the sausage mixture and processing techniques, various reaction products are formed from nitrite.

In the literature so far few publications are known dealing with different binding forms of nitrite in meat products. Olsman & Krol (1972) observed that in heated fine ground meat products the amount of bound nitrite increased with lower pH values; during storage practically no further changes were observed. The addition of Fe^{2+} caused the amount of bound nitrite to increase during an 80-days storage period. This increase could be attributed only partly to the formation of nitrosothiols. Probably the iron adsorbed on the protein has the ability to bind NO^+ as an iron nitrosyl complex. To what extent iron not bound in myoglobin can react with nitrite in a similar way needs further investigations.

Analogous experiments with dry sausage mixtures have shown that in the first five days of ageing about 15% of the added nitrite is protein bound. The addition of glucono-δ-lacton or sodium ascorbate caused a further decrease of the nitrite content (Mirna, 1970).

The previously discussed possibility of formation of iron nitrosyl complex is interesting for other reasons. Such compounds may react in alkaline solution with secondary amines to form *N*-nitrosamines, which might cause artefacts during purification for isolation of *N*-nitrosamines (Loach et al., 1973 and Maltz et al., 1971). On the other hand, in the presence of ketones or compounds with 'acidic' carbon-bound hydrogen, oximes of the corresponding organic compounds are formed (Swinehardt, 1967).

The determination of bound nitrite described in this method gives only a rough characterization of the binding forms, because nitroso compounds of various structures are covered. The procedures of deproteinization applied in weak alkaline media lead to reactions which may systematically affect the values of bound nitrite as well as of free nitrite. In order to reduce the extent of such changes as far as possible, it is proposed to extract the sample with 80% acetone at 20°C in the dark.

In many cases the nitrate content is calculated from the nitrite values before and after reduction with cadmium; the errors as mentioned above are therefore also of influence on the reliability of nitrate determination.

References

Ando, N., Y. Kako & Y. Nagata, 1961. Studies on the color of meat products. I. Effects of orthophosphates on the color of cooked sausages. Bull. Meat Meat Prod. 1: 1–8.

Ando, N., Y. Kako, Y. Nagata, T. Ohashi, Y. Hirakata, N. Suematsu & E. Katamoto, 1963. Studies on the color of meat and meat products. II. Effects of polyphosphates on the color of cooked sausages. Bull. Meat Meat Prod. 2: 1–6.

Archer, M. C., St. D. Clark, J. E. Thilly & St. R. Tannenbaum, 1971. Environmental nitroso compounds: reaction of nitrite with creatinine. Science 174: 1341–1342.

Fox, J. B. & S. A. Ackerman, 1968. Formation of nitric oxide myoglobin: mechanisms of the reaction with various reductants. J. Fd Sci. 33: 364–370.

Gerhardt, V. & S. Haller, 1973. Nitrat- und Nitritgehalsbestimmungen in GdL-haltigen Rohwürsten. Fleischwirtschaft 53: 548–554.

Loach, K. W. & M. Carvalho, 1973. A modified Simon-Lewin-reaction: N-nitrosamines as intermediates in the reaction of secondary amines with nitroprusside and acetaldehyde. Analyt. Letters 6: 25–29.

Maltz, H., M. A. Grant & M. A. C. Navaroli, 1971. Reaction of nitroprusside with amines. J. org. Chem. 36: 363–364.

Mirna, A. & K. Hofmann, 1969. Über den Verbleib von Nitrit in Fleischwaren. I. Umsetzung von Nitrit mit Sulfhydryl-Verbindungen. Fleischwirtschaft 49: 1361–1366.

Mirna, A., 1970. Über die Umsetzung von Nitrit in Fleischwaren und dessen Verteilung in verschiedenen Fraktionen. Proc. 16th Eur. Meet. Meat Res. Workers 1: 681–691, Varna, Bulgaria.

Mirna, A. & G. Schütz, 1972. Verfahren zur gleichzeitigen Bestimmung des Pökelfarbstoffes sowie von Nitrit und Nitrat in Fleischerzeugnissen. Fleischwirtschaft 52: 1337–1338.

Möhler, K., 1970. Bilanz der Bildung des Pökelfarbstoffes im Muskelfleisch. 1. Mitt. Oxydation von Muskelpigment und Nitrit. Z. Lebensmitteluntersuch. u.-Forsch. 142: 169–179.

Möhler, K., 1971. Bilanz der Bildung des Pökelfarbstoffs im Muskelfleisch. II. Mitt. Hitzekatalysierte Bildung des Stickoxidkomplexes. Z. Lebensmitteluntersuch. u.-Forsch. 147: 123–127.

Nagata, Y. & N. Ando, 1967. A colorimetric method for the determination of nitrite coexisting with various food additives. Shokuhin Eiseigaku Zasshi 8: 532–539.

Olsman, W. J. & B. Krol, 1972. Depletion of nitrite in heated meat products during storage. Proc. 18th Meet Meat Res. Workers 2: 409–415, Guelph, Ontario, Canada.

Saville, B., 1958. A scheme for the colorimetric determination of microgram amounts of thiols. Analyst 83: 670–672.

Swinehardt, J. H., 1967. The nitroprusside ion. Co-ord. chem. Rev. 2: 385–402.

Discussion

Black Roussin salt as a curing agent

In the model system to study the formation of free and bound nitrite and the development of colour, black Roussin salt was used instead of nitrite. From the content of free and bound nitrite, it could be concluded that the seven NO groups in the black Roussin salt are not bound in the same way. The amount of 5.8% free nitrite corresponds well to one NO group (theoretical value 5.2%). The bound nitrite amounts to 30.3% (theoretical value 30.9) based on the formula $NH_4[Fe_4S_3(NO)_6NO_2].H_2O$.

Studies on colour formation showed, 15 to 30 minutes after grinding in mixtures with 82 mg/kg nitrite curing salt, the usual greyish colour, whereas with black Roussin salt at levels of 82 and 8 mg/kg, respectively a distinct red colour appeared. This pigment was only formed on the surface and remained stable for about 30 to 40 hours at 4°C.

It was not possible to extract the red pigment formed with black Roussin salt with 80% acetone. This fact indicates that the 'curing effect' of black Roussin salt is not caused by the reaction of nitric oxide with myoglobin.

Properties of black Roussin salt

The black Roussin salt is a complex coordination compound of Fe, NO and S^{2-}, the exact structure of which has not been completely elucidated. The toxicology of the salt is unknown too. Its use is certainly not allowed in countries with a positive list of food additives. It was remarked that there was a contradiction between the stabilities at acid pH of nitrosothiols and nitrite esters. There is no explanation for this at present.

Conditions of extraction for nitrite determination

The extraction liquid for the nitrite determination (80% acetone or water) is less important than the temperature of extraction. However the pH should be in the mildly acid region. One should be careful with the use of $HgCl_2$-containing agents for deproteinization, because bound nitrite can be released as free nitrite by mercuric ions. The results of the Griess-method for the determination of nitrite depends on the extraction procedure used. There is a strong need for accurate artefact-free methods for the determination of free and bound nitrite.

Proc. int. Symp. Nitrite Meat Prod., Zeist, 1973. Pudoc, Wageningen

Low-molecular meat fractions active in nitrite reduction

B. J. Tinbergen

Central Institute for Nutrition and Food Research TNO, Utrechtseweg 48, Zeist, the Netherlands
Dept.: Netherlands Centre for Meat Technology

Abstract

Buffered mixtures of sodium nitrite and low-molecular water-soluble fractions from minced beef muscle were anaerobically heated and stored at 18 °C. Subsequently, the free nitrite content was examined at regular intervals. The formation of nitrosomyoglobin was studied in vitro to screen the low-molecular fractions for their ability to reduce nitrite in the presence of ferric metmyoglobin.

The fractions strongly reduced the nitrite content in the heated mixtures during storage at 18 °C. The activity was pH- dependent.

The observed ability of the low-molecular fraction to reduce nitrite to NO to form nitrosomyoglobin under anaerobic conditions was found to be proportional to the concentration of the fraction.

It was found that an amino acid or a lower peptide – probably with an SH group – as well as a non-amino acid could be involved in the nitrite reduction.

Introduction

During heating and storage of a cured meat product the content of nitrite continuously decreases. The mechanism responsible for the nitrite depletion has only been partly elucidated. At the beginning of the heating process some nitrite is oxidized to nitrate (Möhler, 1970) and another part is used for the formation of the characteristic colour of the heated product. For the latter, nitrite has to be reduced to nitric oxide before it enters into the colour formation reaction with the ferrous muscle pigment myoglobin.

In an unheated minced muscle incubated with sodium nitrite, the remaining enzymatic reduction activity should be held responsible for a part of the nitrite depletion (Taylor & Walters, 1967). However during storage of the heated product this activity can be ruled out, in which case endogenous chemical reductants are considered active in nitrite reduction (Fox & Ackerman, 1968). The quantity of nitrite required for the formation of both nitrate and nitrosomyoglobin, however, is far less than that actually disappearing during heating and storage of the meat product.

Results obtained in a quantitative study of the effects of a number of factors on nitrite depletion in heated meat products during storage (Olsman & Krol, 1972; Olsman, 1973) and the recent findings of Ando, Nagata & Okayama (1971) that substantial colour forming activity could be ascribed to the dialysable part of the sarcoplasm fraction brought us to the work described below.

Nitrite depletion during storage at 18 °C was studied in anaerobically heated model systems with the water-soluble low-molecular fractions from minced beef muscle. The formation of nitrosomyoglobin was studied in vitro to screen the fractions for their ability to reduce nitrite in the presence of ferric metmyoglobin.

Experimental

Water-soluble fraction

Two kg beef muscle, trimmed free of external fat and connective tissue, was ground twice and further homogenized in a high-speed laboratory mixer under addition of 8 000 ml determineralized water at 0 – 2 °C. The slurry was centrifuged at 15 000 X *g* and the supernatant volume reduced to 500 ml in a rotating vacuum evaporator at 10°C. The concentrate was dialysed against 10 000 ml demineralized water in cellulose tubing for 48h at 2 °C. The dialysate was concentrated by evaporation in vacuo at 10°C to a volume of 400 ml. The concentrated water-soluble low-molecular fraction was stored at –20 °C in 25 ml portions until further use. The fraction thus obtained is denoted below as D.

Fractionation by ion exchange

With a method for the quantitative separation of free amino acids and low-molecular peptides (e.g. glutathione) from organic non-amino acids, sugars etc., described by Salminen & Koivistoinen (1969), 30 ml of fraction D was adjusted to pH = 3.6 and passed through a column with Dowex 50 W X 4 (H^+ form, 50 – 100 mesh). The amino acids were sorbed on the column. Sufficient washing with demineralized water removed neutral compounds. The eluate and washings were combined and concentrated by evaporation in vacuo to the original volume of the dialysate, i.e. 30 ml. The obtained fraction is indicated below with D-A.

Amino acids were eluted from the resin with 10% ammonia and the effluent was evaporated to dryness. 5 ml 0.1 M NaOH was added and the evaporation repeated. The dry residue was dissolved in water, the pH adjusted to 5.5 and the volume brought to 30 ml. The amino acid fraction is denoted below as A. A recombined fraction was prepared by mixing equal volumes of D-A and A, followed by evaporation to half the volume in vacuo. The recombined fraction is denoted below as R.

All fractions D, D-A, A and R were tested for both nitrite depletion and colour forming ability in model systems described below.

Fractionation with ultrafilters

In order to determine whether the nitrite reducing activity of D was linked to a specific molecular weight range, the dialysate was passed through Amicon Diaflo[0] ultrafilters type UM2 and UM05, allowing the passage of solutes with molecular weights below 1000 and 500 daltons[1], respectively. The fractions with mol. wt. below 1000 and 500 daltons and the solute retained by the UM2 filter and redissolved in distilled water (mol. wt. above 1000), were tested for their nitrite reducing ability.

Determination of free nitrite

The determination of free nitrite was carried out with Griess reagent according to the ISO method (1971). Extinction was measured at 527 nm.

Determination of SH content

The content of SH groups in the samples under study was determined by direct amperometric titration with 0.001 M $AgNO_3$ in a Radiometer autoburet assembly, as described earlier (Tinbergen, 1970).

Nitrite depletion in heated model systems

In 4 ml screw-capped glass vials 1 ml nitrite-phosphate buffer (0.3 M NaH_2PO_4, 9% NaCl and 0.03% $NaNO_2$) was mixed with 2 ml sample by passing through oxygen-free nitrogen via an injection needle. In all depletion experiments in heated model systems the buffer pH was varied between the values 5.5; 6.0; and 6.5. The vials were closed with a rubber stopper. The stopper had a cross-groove allowing the needle to remain in position during closing, thus avoiding air enclosure. The needle was then pulled out and the stopper held in position by tightly screwing the cap. The vials were heated for 20 min at 95 °C in a waterbath and cooled to room temperature. They were stored at 18 °C until further analysis. Control samples contained 1 ml nitrite-phosphate buffer + 2 ml H_2O.

Formation of cured meat colour in a model system

In a Thunberg tube, 2 ml of the sample under study (pH adjusted to 5.5), 0.1 ml antibiotics solution (40 000 iu Penicillin-G + 10 000 u Polymixin B sulphate per ml H_2O) and 0.2 ml horse-heart metmyoglobin (25 mg per ml H_2O) were gently mixed. The tube was closed with a hollow sidearm containing 0.5 ml $NaNO_2$ (560 mg/l) in H_2O. The tube was evacuated after which the contents were mixed and placed in an Optica CF4R spectrophotometer. The rate of nitrosomyoglobin (NO-Mb) formation was followed by the increase in absorption at 578 nm, corresponding to the maximum of the α-peak of NO-Mb.

Nitrite depletion in a comminuted beef product with added dialysate

A comminuted beef product was prepared with a composition of 1.5% fat, 17% protein and 73% water. In a similar product a part of the added water was replaced

1. 1 dalton = $1.660\,41 \times 10^{-27}$ kg.

by the dialysate D in such a way as to double the concentration of the low-molecular water-soluble meat fraction. Both products contained 0.02% $NaNO_2$ and were heated for 70 min at 100°C. Nitrite depletion as a function of the time of storage was studied.

Results and discussion

The nitrite depletion during heating and 24h storage at 18°C under oxygen-free conditions by the dialysate D and D fractions obtained by ion exchange and ultrafiltration is given in Table 1.

It can be seen that a water-soluble dialysable fraction D of beef muscle strongly reduces the $NaNO_2$ content in a heated anaerobic system, the activity being pH-dependent. When amino acids and lower peptides are removed from D the depletion is markedly diminished. However, as can be seen in Fig. 1, during further storage of the vials without amino acids (D-A), the nitrite content slowly decreases suggesting the possible reducing activity of a non-amino acid compound still present in the D-A fraction and not absorbed on the Dowex column.

From the data in Table 1 for the A fractions and the A lines in Fig. 1 it can be seen that A has a strong reducing activity although less than that of the recombined fractions R. Compounds active in nitrite depletion can be expected to have molecular weights not exceeding 500 daltons.

The ability of D to reduce nitrite to nitric oxide to form NO-Mb under anaerobic conditions is shown in Fig. 2. The NO-Mb formation is plotted as a function of time for the fraction D and dilutions with water 1:1, 1:4 and 1:8, respectively. The reaction rates for the corresponding dilutions can be compared by estimating the slopes of the linear parts of the curves. The resulting ratios are 1:0.56:0.24:0.1, respectively. This means that the reaction rate is proportional to the dialysate concentration. According to Fox & Ackerman (1968) this would make the involvement of a sulphydryl compound (e.g. cystein) unlikely, for they found the NO-Mb formation to be proportional to the square root of the concentration of cystein. However, when the SH-alkylating reagents vinylpyridine (VP) or N-ethylmaleimide (NEM) are added to a mixture of dialysate D,$NaNO_2$ and metMb, the formation of

Table 1. Percentage of free nitrite content remaining in buffered mixtures of nitrite and dialysate heated at 95°C for 20 min under oxygen-free conditions after 24 h storage at 18°C.

pH	Control	D	D-A	A	R	Mol.wt.>1000	Mol.wt.<1000	Mol.wt.<500
5.5	100	32	76	30	33	99	37	43
6.0	100	71	93	66	59			
6.5	100	89	98	79	81			

control : 1 ml nitrite-phosphate buffer + 2 ml H_2O
D : water-soluble low-molecular fraction of minced beef muscle
D-A : fraction D without amino acids and lower peptides
A : amino acid fraction of D
R : recombined fraction of D-A and A.

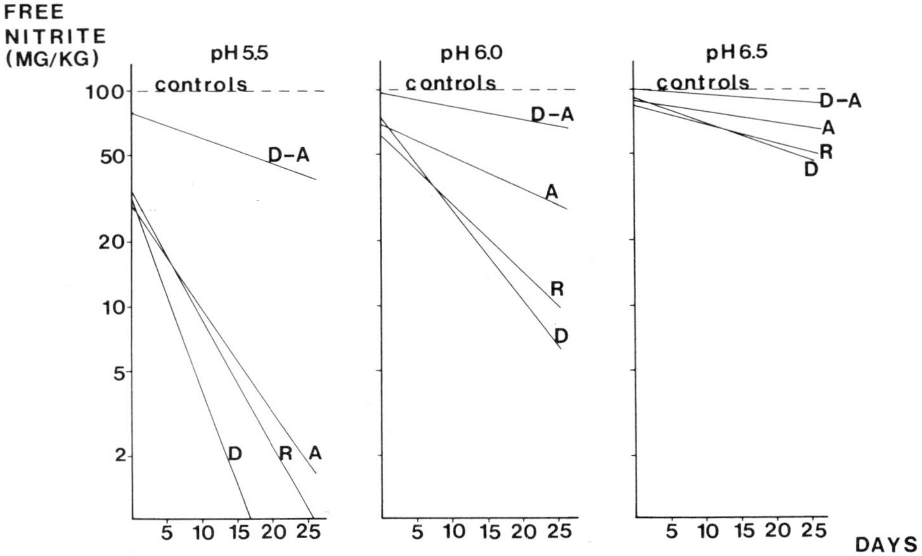

Fig. 1. Nitrite loss at different pH-values in solutions, stored at 18°C after cooking.

NO-Mb is partly inhibited, as can be seen in Fig. 3. It justifies the assumption that the SH/SS redox system is involved in the nitrite reduction. Cystein for example is known to be very reactive towards nitrite compared with other amino acids (Mirna & Hofmann, 1969; Olsman & Krol, 1972).

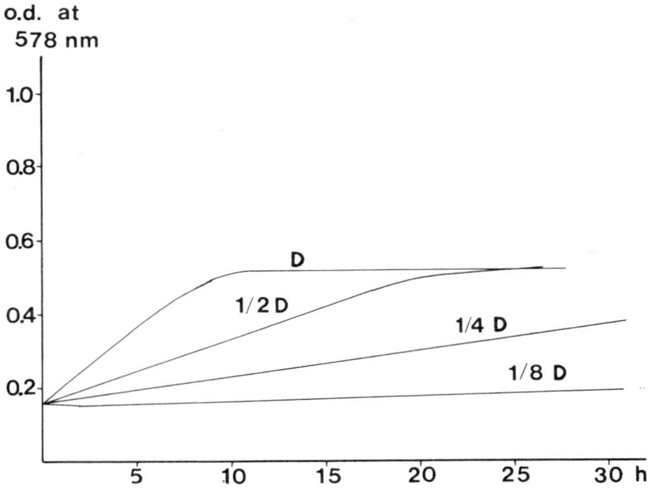

Fig. 2. Formation of NO-Mb under anaerobic conditions by different concentrations of dialysate D.

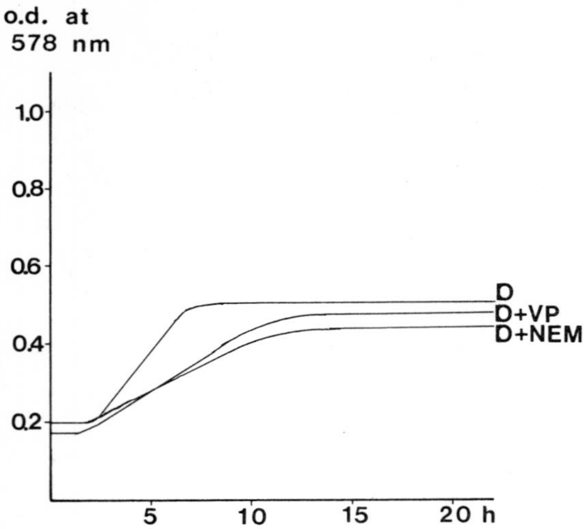

Fig. 3. The effect of SH-alkylation on the formation of NO-Mb under anaerobic conditions. D= dialysate; VP= vinylpyridine; NEM= N-ethylmaleimide.

The SH content of the dialysate, as determined by amperometric titration with $AgNO_3$ (Tinbergen, 1970), was found to be 2.3 – 2.5 mM SH. Reduction with 4 ml 5 mM $Na_2S_2O_5$ (pH adjusted to 7.5) raised the SH content to 4 –4.5 mM SH. This indicated the presence of 2 mM SS. If the dialysate is considered to have a

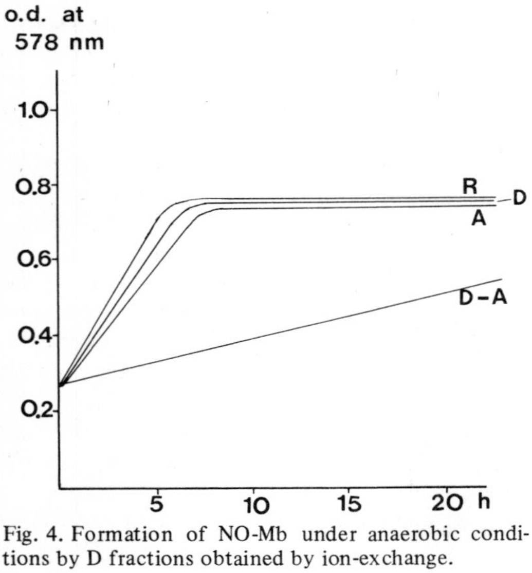

Fig. 4. Formation of NO-Mb under anaerobic conditions by D fractions obtained by ion-exchange.

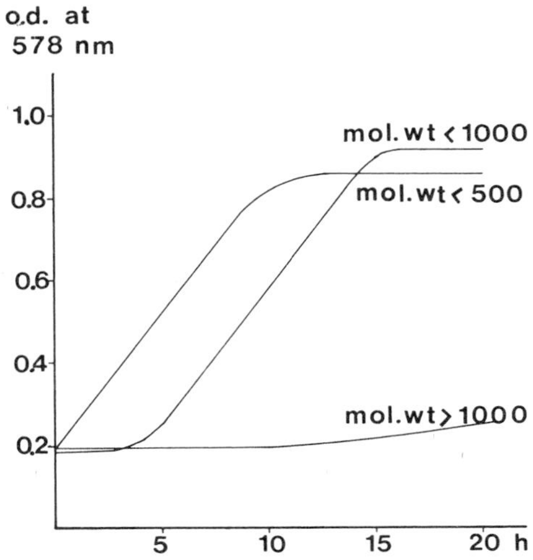

Fig. 5. Formation of NO-Mb under anaerobic conditions by D fractions obtained by ultrafiltration.

concentration 5 times as high as the water-soluble fraction of beef muscle, this content of 4 – 5 μmol SH + SS per ml dialysate is in fairly good agreement with the reported glutathione content in muscle tissue (Olsman & Krol, 1972). Only a part of the SH + SS content of D was detected in the fraction with a molecular weight below 500. No explanation for the observed reduction can as yet be given.

Removal of amino acids from D by ion-exchange resins strongly reduces the rate of NO-Mb formation as can be seen in Fig. 4. Recombination of the D-A and A fractions restores the original nitrite reducing activity. The residual activity of D-A demonstrates the presence of a non-amino acid compound involved in the NO-Mb formation in vitro.

As can be seen from Fig. 5 the colour forming ability of D is associated with a fraction with a molecular weight below 500. This is in good agreement with the findings on nitrite depletion in heated vials, reported above. The addition of dialysate to a comminuted meat product – thereby doubling the concentration of low-molecular water-soluble fraction – resulted in a significant increase of the reaction rate constant for the nitrite depletion. However, the increase was only 20%. Lowering the pH of the product from 6.24 to 5.75 did not change the effect. Olsman & Krol (1972) found that the addition of glutathione increased the reaction rate, although not to a significant extent.

The conclusion can be drawn that the nitrite reducing activity of the dialysate can be ascribed to both an amino-acid or lower peptide – probably with an SH group – and a non-amino acid compound.

Further investigations on the nature of the compounds involved should be carried out.

References

Ando, N., Y. Nagata & T. Okayama, 1971. Effects of the low-molecular fraction of sarcoplasm from porcine skeletal muscle on the behaviour of nitrite and the formation of cooked cured meat colour in rapid curing process. Proc. 17th Eur. Meet. Meat Res. Workers, Bristol, England 1971, p. 227–233.

Fox, J. B. Jr & S. A. Ackerman, 1968. Formation of Nitric Oxide Myoglobin: Mechanisms of the reaction with various reductants. J. Fd Science 33: 364–370.

ISO/TC34/SC6/WG2, N183: Fourth draft proposal. Determination of the nitrite content of meat products. (Febr., 1971).

Mirna, A. & K. Hofmann, 1969. Über den Verbleib von Nitrit in Fleischwaren I. Umsetzung von Nitrit mit Sulfhydrylverbindungen. Fleischwirtschaft 49: 1361–1366.

Möhler, K., 1970. Bilanz der Bildung des Pökelfarbstoffs im Muskelfleisch. Z. Lebensmittel unters. u.-Forsch. 142: 169–179.

Olsman, W. J. & B. Krol, 1972. Depletion of Nitrite in heated meat products during storage. Proc. 18th Meet. Meat Res. Workers, Guelph, Ontario, Canada, 1972, p. 409–415.

Olsman, W. J., 1973. About the mechanism of the nitrite loss during storage of cooked meat products. This symposium, Session III.

Salminen, K. & P. Koivistoinen, 1969. A scheme for the simultaneous determination of free amino acids, organic non-amino acids, sugars, nitrogen and pectic substances in plant materials. Acta chem. scand. 23: 999–1006.

Taylor, A. McM. & C. L. Walters, 1967. Biochemical properties of pork muscle in relation to curing. Part II. J. Fd Science 32: 261–268.

Tinbergen, B. J., 1970; Sulfhydryl groups in meat proteins. Proc. 16th Eur. Meet. Meat Res. Workers, Varna, Bulgaria, 1970, p. 576–579.

Proc. int. Symp. Nitrite Meat Prod., Zeist, 1973. Pudoc, Wageningen.

Determination of volatile amines and amine oxides in food products

A. Ruiter

Institute for Fishery Products TNO, IJmuiden, the Netherlands

Abstract

A method for the analysis of trace amounts of volatile amines in foods is described. The amines are extracted and steam distilled; part of the concentrated distillate is fractionated by gas chromatography on Carbowax 400 + polyethylene imine. Amine oxides, after reduction with $TiCl_3$, can be determined in the same way. Under the conditions described, trimethylamine (TMA) and dimethylamine (DMA) could be easily separated. Traces of DMA were distilled quantitatively if ethylamine was added before distillation.

Some results of analysis in various foodstuffs are given.

Introduction

In many foodstuffs there are compounds that may be nitrosated if nitrite is present. In this study the possible presence of one group of substances that can be nitrosated: the volatile dialkyl and trialkylamines, in various foodstuffs was studied quantitatively. Amine oxides, which are easily reduced to the corresponding amines, were taken into consideration as well.

The rate of nitrosation for tertiary amines is much less than for secondary amines (Wolff, 1972). The nitrosation rate for amine oxides is not yet known, but, under certain conditions, these components tend to break down, e.g., to a secondary amine and formaldehyde (Sundsvold et al., 1969, 1971).

Until now, few concrete data have been available on the amount of these amines in food products, though it has been known for a long time that they are present in fish and fish products. Some years ago, Ito et al. analysed secondary amines in various foodstuffs. Their data were reviewed by Möhler (1972). The amines were converted into nitrosamines and analysed with the aid of a modified Griess reagent. Part of the nitrosamines were qualitatively identified by t.l.c.

Recently a method for determining both secondary and tertiary amines by gas chromatography was developed at our Institute. In fact, all volatile bases, with the exception of ammonia, can be analysed in this way. Amine oxides are determined by $TiCl_3$ reduction to the corresponding amines before gas chromatography.

20 g of fish + 80 ml of 5% trichloroacetic acid (TCA) solution

↓ homogenization

100 ml of crude extract

↓ filtration

clear extract

↓ aliquot (50 ml) → pH = 10
steam distillation

distillate (caught in excess HCl)

↓ concentration by evaporation (pH ⩽ 3)

concentrated distillate (vol. 5 – 20 ml; depends on amount of amine expected)

↓ pH → 10–11

solution ready for g.c. (2 μl injected)

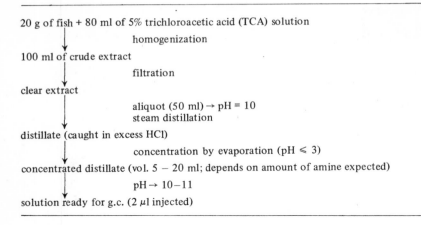

Fig. 1. Pre-treatment of fish samples for gaschromatographic determination of trimethylamine (TMA) and dimethylamine (DMA)

Description of the method

The method is a modification of the gas chromatographic determination of trimethylamine (TMA) and dimethylamine (DMA) in fish, and is described in detail elsewhere (Ritskes, 1973). The pretreatment of the sample is shown in Fig. 1. For the analysis a Hewlett-Packard Model 5700 gas chromatograph was used, the injector block temperature being 150°C and the detector temperature 200°C. Separation took place on a 1.83 m long glass column of 3.18 mm i.d. packed with 15% Carbowax 400 + 5% polythylene imine (Supelco, Inc.) on Chromosorb W-NAW (100-120 mesh), and provided with a ∿ 5 cm long pre-column of 3.18 mm i.d. (20% Dowfax 9N9 and 15% KOH on Chromosorb W-NAW 100–120 mesh). The column was maintained at 40°C and the pre-column at 150°C. The helium carrier gas flow rate was 20 ml/min. Retention times and peak areas were measured with a Hewlett-Packard 3370A electronic integrator.

A chromatogram is shown in Fig. 2. There is a satisfactory separation between the TMA and DMA peaks. Without the use of the pre-column the peaks are considerably broader and some overlap may occur.

If the procedure is not standardized thoroughly the injection of water on the column may cause some problems (e.g., splitting-up of TMA and particularly DMA peaks, and some loss of reproducibility). For the same reason it is necessary to keep the pre-column temperature at 150°C. The pH of the samples should be between 10 and 11.

It is worth mentioning that the use of an alkali flame ionisation detector (AFID) was found to be impracticable for several reasons (Ruiter & Ritskes, 1970):

– For optimum sensitivity (and for selectivity as well) the AFID needs a high detector temperature (400°C). For an acceptable ratio of signal : noise a relatively high (60 ml/min) flow rate is required. Both requirements are difficult to combine

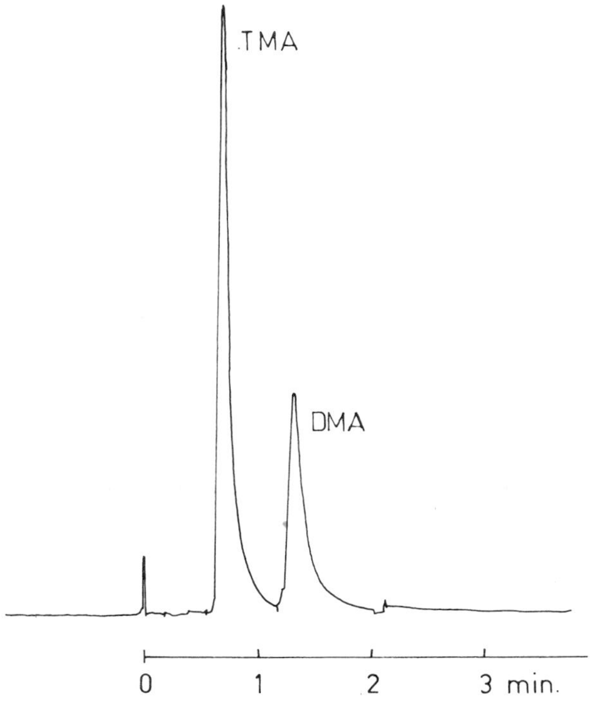

Fig. 2. Separation of TMA (45 ng) and DMA (30 ng). 15% Carbowax 400 + 5% polyethylene imine on Chromosorb W-NAW; FID; injection size 2 μl. Column temperature 40°C. He flow: 20 ml/min.

with the conditions necessary for optimum separation of the methylamines (column temperature 40–50°C; flow rate not exceeding 15–20 ml He/min).

– For amine separation, polyethylene imine was found to be an excellent column coating component, preventing tailing of the amines almost completely. However, it gave a large background signal with the AFID, thus making a quantitative determination of trace amounts impossible.

– We observed that injection of aqueous solutions strongly affect the base-line stability if an AFID is used.

TMA and DMA

With the method described, both TMA and DMA can be determined in concentrations down to about a concentration of 10 mg/kg in the product, which is sufficient for the analysis of these components in fish.

In this project, however, amine concentrations in the range of a few mg/kg had

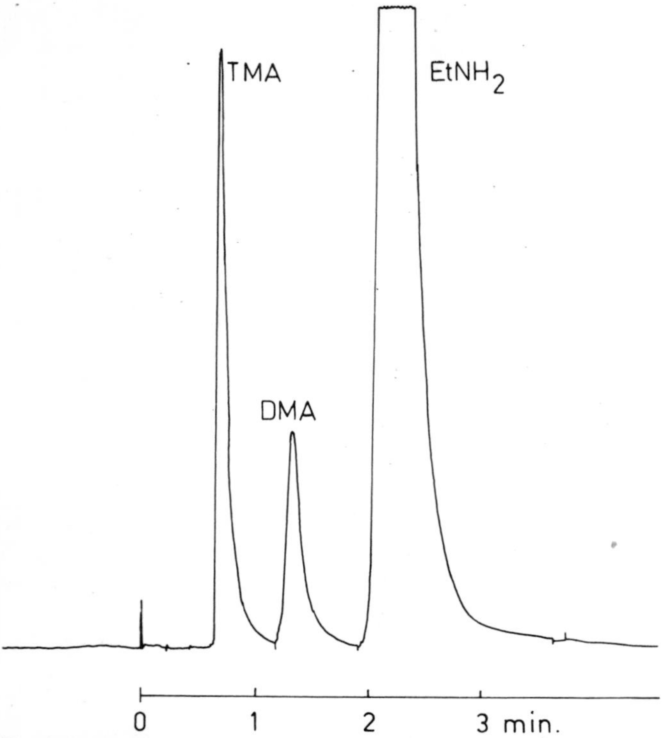

Fig. 3. Separation of TMA (45 ng), DMA (30 ng) and ethylamine. Conditions as in Fig. 2.

also to be taken into account and therefore the sensitivity of the method had to be increased. This increase could be easily achieved by a further concentration of the distillate (e.g., to 1 ml). Because of this concentration and the increased sensitivity of the detector system it proved possible to determine these amines even at a level under 1 mg/kg. However, steam distillation of microgram amounts of DMA in particular was found to be incomplete. Therefore a distillation aid had to be added to improve the yield of distilled DMA. It was found that the addition of another volatile nitrogen base before distillation satisfactorily solved the problem. Ammonia could be used for this purpose, but larger amounts of this component tend to influence the chromatogram. Ethylamine was found to be more suitable because of its good separation from the other amines, as is shown in Fig. 3.

Addition of 1.0 mg of ethylamine had the following effect on the yield of DMA:

Amount of DMA	Yield without $EtNH_2$	Yield with $EtNH_2$
10 μg	∿ 65%	∿ 115%
50 μg	∿ 75%	∿ 95%
100 μg	∿ 85%	∿ 95%

Obviously the method cannot distinguish between primary, secondary, and tertiary amine peaks. If unknown peaks are found beyond the TMA/DMA region, a further identification is necessary.

Amine oxides

The most common amine oxide, trimethylamine oxide (TMAO), was found to be easily reduced by titanous chloride. When small amounts of TMAO were left with 0.3% $TiCl_3$ (1 ml of 15% $TiCl_3$ in 50 ml of TCA extract) at 50°C for hours, more than 95% was reduced to TMA. Thus, amine oxides were reduced in a part of the TCA extract, before the pH was raised to 10 and the extract steam distilled. The results are compared with those obtained without the reduction step, and, from the differences observed, the amount of amine oxides can be calculated.

An alternative method, i.e., removal of volatile amines before reduction of the amine oxides, was not studied further.

Results and discussion

Some results of analysis are given in Table 1. It is shown that most of the products examined are extremely low in both TMA and DMA.

Other amines are not found in appreciable amounts except in orange juice which contains a volatile base with the same retention time as ethylamine. Calculated as $EtNH_2$ the concentration is about 5 mg/kg. I did not investigate this component further.

TMAO is found in all marine fish species. During loss of freshness it is partially reduced to TMA by bacterial action.

An enzymatic breakdown of TMAO to DMA and formaldehyde occurs in several gadoid species. This breakdown is not completely inhibited during frozen storage and, for this reason, frozen cod or coal fish may contain some DMA.

The DMA formation in some fish species was discussed, amongst other topics, in two recent reviews (Yamada, 1968; Ruiter, 1971).

Table 1. Results of analysis of TMA, DMA and TMAO (mg/kg) in various food products. The total amount of volatile bases (mainly NH_3) is also given, and expressed in mg nitrogen per 100 g.

	TMA	DMA	TMAO	Total volatile bases
Tomato juice	0.4	0.2	0.0	11
Orange juice	0.4	0.2	0.0	3
Old Gouda cheese	1.0	0.6	0.0	50
Banana	0.0	0.2	0.0	2
Minced meat (pork+beef 1:1)	2.5	0.4	0.0	5
Canned mackerel	550	145	265	50
Canned herring	345	100	520	36
'Maatjes' herring	50	20	1510	19
Fresh cod	8.4	16	3240	13.6
Cooked cod	195	83	2860	21
Fried whiting	840	67	570	39

The presence of DMA in some fishery products can be explained by partial TMAO breakdown during heat processing (Sundsvold et al., 1969, 1971).

Now the question arises whether because of the concentrations of DMA and TMA found in processed fish there is a risk that considerable amounts of dimethylnitrosamine will be produced in fishery products when nitrite is present.

To answer this question, it is important to look at the composition of the volatile bases as a whole. To these volatile bases TMA as well as DMA only contribute a small part. Most is ammonia, and this component has a strong affinity for nitrite. Ammonia and nitrite readily produce nitrogen and water under conditions that might permit nitrosation, thus possibly preventing, reducing, or retarding nitrosamine formation. Therefore the effect of ammonia in nitrosation should be studied both in model systems and in food before it is possible to give any opinion about the presence of DMA in heatprocessed fish products in relation to the nitrosamine problem.

References

Möhler, K., O. L. Mayrhofer & E. Hallermeyer, 1972. Das Nitrosaminproblem aus der Sicht des Lebensmittelchemikers. Z. Lebensmittel unters. u. -Forsch. 150: 1–11.

Ritskes, T. M., in preparation.

Ruiter, A. & T. M. Ritskes, 1970. The use of a thermionic nitrogen detector in the gas chromatographic analysis of volatile nitrogen compounds. Rep. 1st meet. res. workers W. Eur. Fish Technol. Inst., IJmuiden; p. 15–16.

Ruiter, A., 1971. Trimethylamine and the quality of fish. Voedingsmiddelentechnologie 2(43) 1–10.

Sundsvold, O. C., B. Uppstad, G. W. Ferguson, D. Feeley & T. McLachlan, 1969. The trimethylamine oxide content of Norwegian shrimps and its degradation to methylamines and formaldehyde. Tidsskr. HermetInd. 55: 94–97; 131–136.

Sundsvold, O. C., B. Uppstad, G. W. Ferguson, D. Feeley & T. McLachlan, 1971. The degradation of trimethylamine oxide to methylamines and formaldehyde in canned shrimps. J. Ass. pub. Anal. 9: 86–95.

Wolff, I. A., 1972. Nitrosamines. Paper presented before the Food Research Institute, University of Wisconsin, Madison, Wisconsin.

Yamada, K., 1968. Post-mortem breakdown of trimethylamine oxide in fishes and marine invertebrates. Bull. Jap. Soc. scient. Fish. 34: 541–551.

Discussion

Recovery of methylamines from fish

The recovery of TMA and DMA from meat products was not studied, but recovery from fish – even from fat fish – is complete, if the amines are allowed to protonize completely in the acid medium. In doing so, there is no risk that they are partially retained in the lipid phase.

Breakdown of TMAO

The breakdown of TMAO proceeds in the following way both chemically and enzymatically:

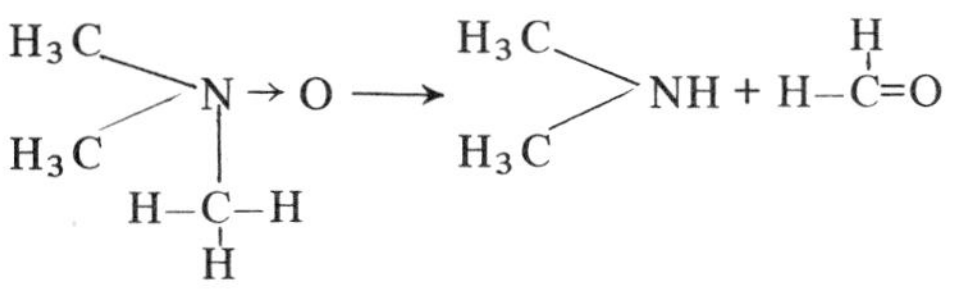

TMAO can also be reduced chemically to TMA under the same conditions.

TMA as an indicator for freshness

According to the author, TMA is not useful as a standard for the grading of fish freshness, although it has been proposed for this purpose several times. Too many factors influence the TMA value.

Amines of high molecular weight

Amines of higher molecular weight may be extracted more completely by perchloric acids, but their presence was not revealed in the trichloroacetic acid extract; they must be of minor importance.

Monoethylamine

Monomethylamine was detected in minute amounts in fish and fish products. Because of the addition of monoethylamine to the distillates of other foodstuffs including meat, it was impossible to detect monoethylamine. According to Mottram both DMA and TMA increase from 200 μg/kg in freshly slaughtered meat to about 600 μg/kg after curing, whereas monomethylamine reaches values of 2 mg/kg, which values decrease during curing.

Spores of nitrosamines

It was remarked that in fresh sablefish the presence of nitrosamines in amounts of some μg/kg has been confirmed twice by mass spectrometry.

Proc. int. Symp. Nitrite Meat Prod., Zeist, 1973. Pudoc, Wageningen.

Determination of volatile nitrosamines: a review

G. Eisenbrand

Institut für Toxicologie und Chemotherapie, Deutsches Krebsforschungszentrum, Postfach 449, D6900, Heidelberg, W-Germany

Abstract

Volatile N-nitrosamines are of considerable interest to environmental toxicologists and the potential danger represented by their presence in foods today is fully recognized. Their presence at low concentrations has been demonstrated in a limited range of foods, especially of animal origin, such as meat products, fish and cheese (Fazio et al., 1971; Wasserman et al., 1972; Sen, 1972).

Despite the wide range of techniques available to the analytical chemist, the problem of separation and analysis of nitrosamines has proved to be far more difficult than originally anticipated. The greatest difficulties in the development of reliable analytical techniques have been encountered in the enrichment and isolation of nitrosamines from samples in sufficient purity to allow an undisturbed determination. This situation is reflected in the wide range of different methods that have been developed for purification and final analysis. It makes it desirable to review and assess the whole field.

Methods of isolation

Most published methods rely on a few basic techniques which are usually modified slightly.

Solvent-extraction techniques

Because of its superior properties, dichloromethane has become the standard solvent for nitrosamine work. It has been used either for direct extraction of food samples in a Soxhlet or blender (Marquardt & Hedler, 1966; Thewlis, 1967; Eisenbrand, 1971; Sen, 1971; Sen et al., 1972) or for liquid – liquid extraction of steam distillates (Crosby et al., 1971; Alliston et al., 1972; Sen et al., 1972; Telling, 1972).

Methods depending on distillation

Many investigators have taken advantage of the volatility of nitrosamines when attempting their isolation from biological materials. The first to use this technique were Heath & Jarvis (1955) when they separated dimethylnitrosamine (DMNA) from animal tissues by steam distillation of the alkaline homogenate.

Eisenbrand et al. (1970) examined the distillation rates of 16 different nitrosamines. No significant differences in the respective yields were found on distillation at reduced or atmospheric pressure from alkaline, acidic or neutral media. Casselden et al. (1969) applied fractional distillation to increase the concentration of aqueous solutions of nitrosamines after the addition of methanol and salt. The nitrosamine-rich fraction was collected between the alcohol and aqueous phase, concentration factors of up to 30 could be obtained for nitrosamines of low molecular weight. Fractional distillation has also been used by Crosby et al. (1971) when examining East African spirits for the presence of nitrosamines. Vacuum distillation has been applied with different results by many workers for nitrosamine enrichment (Lydersen & Nagy, 1967; Devik, 1967; Heyns & Koch, 1971; Williams et al., 1971; Scanlan & Libbey, 1971). One of the most advanced modifications is probably the vacuum distillation technique described by Telling et al. (1971) and Telling (1972). They distilled from an only weakly alkaline medium (4% potassium carbonate) at moderate temperatures (65°C). Recoveries of even highly volatile nitrosamines have been found to be very satisfactory.

An enrichment technique that is frequently used consists of a combination of an alkaline-methanolic digestion of the sample with subsequent dichloromethane extraction of the digest, followed by distillation of the dichloromethane extract (Howard et al., 1970; Fazio et al., 1972). This technique has also been shown to be very efficient for nitrosamine isolation from samples.

Methods of further purification

Chromatographic techniques such as preparative thin-layer chromatography and column chromatography on various sorbents have been frequently used. Marquardt & Hedler (1966) applied multiple thin-layer chromatography on silica-gel plates to isolate nitrosamines from extracts of flour. Other workers have stated that losses of volatile nitrosamines under these conditions are extremely high (Schuller, 1969; Sen et al., 1969; Eisenbrand et al., 1970). Under certain conditions, t.l.c. can be successfully applied as a purification step, resulting in a preseparation of nitrosamines in groups of different chromatographic mobility. Separations which are carried out at 4°C under light protection on a silica-gel layer of 0.6 mm thickness result in recoveries of better than 90%, even for highly volatile nitrosamines (Eisenbrand et al., 1970). Howard and co-workers (1970), when they examined smoked fish for the presence of dimethylnitrosamine, used an acid-treated celite column for purification of concentrates. This technique has been abandoned in this experiment in favour of a more versatile silica-gel column purification (Fazio et al. 1972). Wasserman et al. (1972) used acid-treated florisil instead of celite. Sen, in some of his investigations (1972), tried polyamide for the same purpose. Telling (1972) obtained good results by purifying concentrates on a column of neutral alumina. In the purified eluates the nitrosamines were oxidized to nitramines and these again subjected to column chromatography on neutral alumina. The resulting final solutions were very pure. Recoveries after every chromatographic step were from 77–98% for different nitrosamines.

Methods of detection and estimation

In view of the high significance for public health, identification and quantitative determination of nitrosamines in food must be reliable and unequivocal even at a concentration of say, a μg/kg. Various methods have been developed and applied but only a few fulfill the necessary requirements. They can be divided into two classes: 1. methods for direct determination of intact nitrosamines and 2. methods for derivative formation.

Direct determination of intact nitrosamines

Because of its simplicity, t.l.c. has often been used for nitrosamine detection. Preussmann and co-workers (1964) developed this method as a simple and convenient tool for rapid nitrosamine analysis. Detection was carried out by u.v.-irradiation and spraying with Griess reagent or diphenylamine-palladium (II) chloride. Kröller (1967) localized substances with ninhydrin. Sen et al. (1972) applied Griess and ninhydrin reagents for semiquantitative determination of volatile nitrosamines in alcoholic beverages.

Many methods have been used for quantitative nitrosamine determination, among them polarography, u.v. spectrometry, colorimetry and gas chromatography. U.v. spectrometry can only be used for pure nitrosamine solutions since u.v. spectra are very easily disturbed by interfering compounds, especially in the range of 230–240 nm. Polarography as a very sensitive method has been used by Heath and Jarvis (1955) for dimethylnitrosamine estimation in animal tissues. Some workers have tried to improve the specificity of the method by differential polarography of unirradiated and irradiated samples (Walters et al., 1970). Colorimetric methods relying on photolytic splitting of the N-NO bond were developed by Daiber & Preussman (1964) and by Sander (1967). Eisenbrand & Preussmann (1970) took advantage of the facile acid-catalysed denitrosation of N-nitrosocompounds for the development of a simple, rapid and sensitive colorimetric technique. It has proved to be superior to any other colorimetric method and may even be used to determine molecular weights of pure unknown nitrosamines.

The final determination of nitrosamines in purified sample extracts has been carried out mostly by gas-liquid chromatography. Stationary phases were of medium to strong polarity such as carbowax 1540, (Rhoades & Johnson, 1970), Reoplex 400 (Sen et al., 1969), polyethylene glycol 4000 (Serfotein & Hurter, 1966), Carbowax 1000 (Du Plessis et al., 1969), diethylene glycol adipate (Saxby, 1970), Carbowax-KOH (Howard et al., 1970), Carbowax 20-M-TPA (Fiddler et al., 1971), Ucon LB 550-KOH and Marlophen 87 LWH-KOH (Heyns & Röper, 1970).

Although g.l.c. is a very powerful technique for the separation of mixtures, it should be applied with great care for the identification of substances. For example, Devik (1967) claimed to have identified diethylnitrosamine in roasted potato starch spiked with glucose and amino acids. His findings caused great concern. Heyns & Koch (1971) repeated the experiment of Devik and demonstrated by capillary g.l.c. and mass spectrometry that pyrazines had been confused with nitrosamines.

The introduction of nitrogen specific detectors was a great step forward in nitrosamine trace analysis. Howard et al. (1970) used a modified thermionic detector for DMNA analysis and Fazio et al. (1971) modified the procedure of Howard to a multidetection method for volatile nitrosamines, giving reproducible results with recoveries between 70–100% at a level of 10 μg/kg. Thermionic detectors have also been used by Fiddler et al. (1971), Crosby et al. (1972) and Wasserman et al. (1972).

Rhoades & Johnson (1970) used the Coulson electrolytic conductivity detector in the pyrolytic mode to analyse cigarette smoke condensates for the presence of nitrosamines by oxidation with peroxytrifluoroacetic acid. A very impressive the nitrosamines, but in the pyrolytic mode distinct nitrosamine peaks could be obtained. Sen (1972) also used the Coulson detector in the pyrolytic mode. He mentioned that although other nitrogen compounds also respond to the CECD, the technique is simple and useful for rapid screening. Crosby et al. (1972) used this detector in the reductive mode for detection of nitrosamines from fried bacon, fish and meat products. Despite the much better selectivity obtained with these specific detectors, all workers have emphasized the need for confirmation of positive results by mass spectrometry. Results have been confirmed in a number of cases, usually where the concentration of nitrosamine in the sample was at least in the range of 10 μg/kg (Fazio et al., 1972; Sen, 1972; Crosby et al., 1972; Wasserman et al., 1972). Concentrations far below this level were too low to be confirmed by Mass spectrometry.

Bryce & Telling (1972) reported on the use of high resolution mass-spectrometry for semiquantitative nitrosamine analysis. They monitored the g.c.-effluent for molecular ions or other characteristic ions of nitrosamines at a resolution of 10 000 and obtained much better sensitivities than which an earlier published method, where they monitored the NO^+ ion (Telling et al., 1971).

Formation of nitrosamine derivatives

Sen (1970) and Althorpe et al. (1970) developed procedures for the sensitive electron capture determination of N-nitro amines produced from the corresponding nitrosamines by oxidation with peroxytrifluoroacetic acid. A very impressive improvement of this method has recently been published by Telling (1972) who also utilized peroxytrifluoroacetic acid as the oxidizing agent and applied two column chromatographic steps for purification. Since the N-N bond of nitrosamines remains intact, the method ensures good selectivity. Another method to prepare suitable derivatives has been found by Eisenbrand and Preussman (1970) by the acid catalysed denitrosation of nitrosamines to secondary amines. These were reacted with heptafluorobutyryl chloride (HFB) and the resulting HFB-derivatives were detected by electron-capture gas chromatography (Eisenbrand, 1972). Since the amines could be selectively isolated by an ion-exchange step, interfering compounds were quantitatively removed. One advantage of the method is its suitability for gas chromatography – mass spectrometry measurements since HFB-derivatives can be easily detected in the mass spectrometer by monitoring the C_3F_7 fragment, which is abundant and characteristic for this class of compounds.

Alliston et al. (1972) split nitrosamines electrochemically under basic conditions and used essentially the same technique for HFB-derivative formation as Eisenbrand. Pailer and Klus (1971) when analysing tobacco smoke condensates for nitrosamine content, split nitrosamines with cuprous chloride in hydrochloric acid and detected corresponding amines by electron-capture gas chromatography of their trifluoracetyl derivatives.

Reduction to hydrazines has also been used for nitrosamine analysis. Neurath et al. (1965) reduced with lithium aluminium hydride and prepared the corresponding 5-nitro-2-hydroxybenzal derivatives which could be separated by t.l.c., but the yields obtained from tobacco smoke condensates were extremely low. A modification has recently been proposed by Yang & Brown (1972) who condensed the hydrazines with 9-anthraldehyde to form highly fluorescent compounds, but this has only been tested with pure compounds. Hoffmann & Vais (1972) obtained better results in the reduction step by the use of diboran instead of $LiAlH_4$. They condensed with 3,5-dinitrobenzaldehyde and determined the resulting hydrazones by electron-capture gas chromatography.

To summarize the data reviewed here we can say that considerable progress in the analysis of volatile nitrosamines has been achieved during the last years. Some of the methods seem to be well suited as standard procedures; these are mainly those relying on nitrogen specific detection and on formation of suitable derivatives. In this context, the *N*-nitro amine formation seems very promising because it ensures a high degree of specificity. Mass spectrometry nevertheless remains the method of choice for final confirmation.

References

Alliston, T. G., G. B. Cox & R. S. Kirk, 1972. The determination of steam – volatile *N*-nitrosamines in foodsstuffs by formation of electron capturing derivatives from electrochemically derived amines. Analyst 97: 915–920.

Althorpe, J., D. A. Goddard, D. J. Sissons & G. M. Telling, 1970. The gaschromatographic determination of nitrosamines at the picogram level by conversion to their corresponding nitramines. J. Chromat. 53: 371–373.

Bryce, T. A. & G. M. Telling, 1972. Semiquantitative analysis of low levels of volatile nitrosamines by gaschromatography – mass spectrometry. J. agric. Fd Chem. 20: 910–911.

Casselden, E. M., E. M. Johnson, N. Ray & C. L. Walters, (1969). Paper presented at IARC meeting on analytical problems in the estimation of traces of nitrosamines in food and other environmental media, with special references to clean-up methods. London, Oct. 1969.

Crosby, N. T., J. K. Foreman, J. F. Palframan & R. Sawyer, 1972. Estimation of steam-volatile N-nitrosamines in foods at the 1 μg/kg level. Nature (Lond.) 238: 342–343.

Crosby, N. T., J. K. Foreman, J. F. Palframan & R. Sawyer, 1972. Determination of volatile nitrosamines in food products at the μg/kg level. Proc. of a working Conf. held at Deutsches Krebsforschungszentrum (DKFZ), Heidelberg, 13–15 Oct. 1971. IARC Scient. Publs. No. 3, 38–42.

Daiber, D. & R. Preussman, 1964. Quantitative colorimetrische Bestimmung organischer N-nitroso-Verbindungen durch photochemische Spaltung der Nitrosaminbindung. Z. analyt. Chem. 206: 344–352.

Devik, O. G., 1967. Formation of N-nitrosamines by the Maillard Reaction. Acta chem. scand. 21: 2302–2303.

Eisenbrand, G. (1972). Determination of volatile nitrosamines at low levels in food by acid catalyzed denitrosation and formation of derivatives from the resulting amines. Proc. of a working Conf. held at the DKFZ, Heidelberg, 13–15 Oct. 1971, IARC Scient. Publ. No. 3: 64–70.

Eisenbrand, G., A. v. Hodenberg & R. Preussmann, 1970. Trace analysis of *N*-nitroso compounds. II. Steam distillation at neutral, alkaline and acid pH. Z. anal. Chem. 251: 22–24 (1970).

Eisenbrand, G. & R. Preussmann, 1970. Eine neue Methode zur kolorimetrischen Bestimmung von Nitrosaminen nach Spaltung der N-nitrosogruppe mit Bromwasserstoff in Eisessig. Arzneimittel-Forsch. 20: 1513–1517.

Eisenbrand, G. K. Spaczynski & Preussmann, 1970. Spurenanalyse von N-Nitroso-Verbindungen III. Quantitative Dünnschichtchromatographie von Nitrosaminen. J. Chromat. 51: 503–509.

Fazio, T., J. W. Howard & R. White, 1972. Multidetection method for analysis of volatile nitrosamines in food. Proc. of a working confer. held at DKFZ, Heidelberg, 13–15 Oct. 1971. IARC Scient. Publ. No. 3, 16–24.

Fiddler, W., R. G. Doerr, J. R. Ertel & A. E. Wasserman, 1971. Determination of N-nitrosodimethylamine in ham by gas-liquid chromatography with an alkali flame ionization detector. J. Ass. off. analyt. Chem. 54: 1160–1163.

Heath, D. F. & J. A. E. Jarvis, 1955. Polarographic determination of dimethylnitrosamine in animal tissues. Analyst 80: 613–615.

Heyns, K. & H. Koch, 1971. Zur Frage der Entstehung von Nitrosaminenen bei der Reaktion von Monosacchariden mit Aminosäuren (Maillard-Reaktion). Tetrahedron Lett. 10: 741–744.

Heyns, K. & H. Röper, 1970. Ein spezifisches analytisches Nachweisverfahren für Nitrosamine durch Kombination von Gaschromatographie mit der Massenspektrometrie. Tetrahedron Lett. 10: 737–740.

Hoffmann, D. & J. Vais, 1971. Analysis of volatile N-nitrosamines in unaged mainstream smoke of cigarettes. Presented at 25th Tobacco chem. Res. Conf. Lousville, Kentucky, Oct. 6–8 (1971).

Howard, J. W., T. Fazio & J. O. Watts, 1970. Extraction and gaschromatographic determination of N-nitroso dimethylamine in smoked fish: application to smoked nitrite-treated chub. J. Ass. off. Analyt. Chem. 53: 269–274.

Kröller, E., 1967. Untersuchungen zum Nachweis von Nitrosaminen im Tabakrauch und Lebensmitteln, Dt. Lebensmitt. Rdsch. 63: 303–305.

Lydersen, D. L. & K. Nagy, 1967. Polarographische Bestimmung von Dimethylnitrosamin in Fischprodukten. Z. anal. Chem. 230: 277–281.

Marquardt, P. & L. Hedler, 1966. Über das Vorkommen von Nitrosaminen in Weizenmehl. Arzneimittel Forsch. 16: 778–779.

Neurath, G., B. Pirmann, H. Wichern, 1964. Zur Frage der N-nitroso-Verbindungen im Tabakrauch. Beitr. Tabakforsch. 2: 311–319.

Neurath, G., B. Pirmann, W. Lüttich & H. Wichern, 1965. Zur Frage der N-Nitroso-Verbindungen im Tabakrauch. Beitr. Tabakforsch. 3: 251–262.

Pailer, M. & H. Klus, 1971. Die Bestimmung von N-nitrosaminen im Zigarettenrauchkondensat. Fachl. Mittlg. Austria Tabakwerke, 12. Heft: 203–211.

Preussmann, R., G. Neurath, G. Wulf-Lorentzen, D. Daiber & H. Hengy, 1964. Anfärbemethoden und Dünnschichtchromatographie von organischen N-nitroso-Verbindungen. Z. analyt. chem. 202: 187–192.

Pessis, I. S. Du, J. R. Nunn & W. A. Roach, 1969. Carcinogen in a Transkein Food additive. Nature (Lond.) 222: 1198.

Rhoades, J. W. & D. E. Johnson, 1970. Gas-chromatography and selective detection of N-nitrosamines. J. Chromat. Sci. 8: 616–617.

Rhoades, J. W. & D. E. Johnson, 1972. Method for the determination of N-nitrosamines in tobacco smoke condensate. J. natn. Cancer Inst. 48: 1841–1843.

Scanlan, R. A. & L. M. Libbey, 1971. N-nitrosamines not identified from heat induced D-glucose/alamine reactions. J. agric. Fd Chem. 19, 570–572.

Sen, N. P., 1970. Gas-liquid chromatographic determination of dimethylnitrosamine as dimethylnitramine at picogram levels. J. Chromat. 53: 301–304.

Sen, N. P., 1972. Multidetection method for determining volatile nitrosamines in foods. Proceedings of a working conference held at the DKFZ, Heidelberg, 13–15 Oct. 1971. IARC Scientif. Publ. No. 3: 25–30.

Sen, N. P. & C. Dalpe, 1972. A simple thin-layer chromatographic technique for the semiquantitative determination of volatile nitrosamines in alcoholic beverages. Analyst 97: 216–220.

Sen, N. P., D. C. Smith, L. Schwinghamer & J. Marleau, 1969. Diethylnitrosamine and other N-nitrosamines in foods. J. AOAC, 52: 47–52.

Sen, N. P., L. A. Schwinghamer, B. A. Donaldson & W. F. Miles, 1972. N-Nitrosodimethylamine in fish meal. J. Agr. Fd. Chem. 20: 1280–1281.
Schuller, P. L. Paper presented at IARC meeting on analytical problems in the estimation of traces of nitrosamines in food and other environmental media, with special references to clean up methods, London, Oct. 1969.
Serfontein, W. J. & P. Hurter 1969. Nitrosamines as environmental carcinogens II. Evidence for the presence of nitrosamines in tobacco smoke condensate. Cancer Res. 26: 575–579.
Saxby, M. J., 1970. The separation of volatile N-nitrosamines and pyrazines. Analyt. Lett. 3: 397–399.
Telling, G. M., 1972. A gas-liquid chromatographic procedure for the detection of volatile N-nitrosamines at the ten parts per billion level in food stuffs after conversion to their corresponding nitramines. J. Chromat. 73: 79–87.
Telling, G. M., T. A. Bryce, J. J. Althorpe, 1971. Use of vacuum distillation and gaschromatography mass spectrometry for determining low levels of volatile nitrosamines in meat products. J. agric. Fd Chem. 19: 937–940.
Thewelet, B. H., 1967. Testing of wheat flour for the presence of nitrosamines. Fd Cosmet. Toxicol. 5: 333–337.
Walters, C. L., E. M. Johnson & N. Ray, 1970. Separation and detection of volatile and non-volatile N-nitrosamines. Analyst 95: 485–489.
Wasserman, A. E., W. Fiddler, E. C. Doerr, S. F. Osman & C. J. Dooley, 1972. Dimethylnitrosamine in Frankfurters. Fd Cosmet. Toxicol. 10: 681–684.
Williams, A. A., C. F. Timberlake, O. G. Tucknott & R. L. S. Patterson 1971. Observations on the detection of dimethyl- and diethylnitrosamines in foods with particular reference to alcoholic beverages. J. Sci. Fd Agr. 22: 431–434.
Yang, K.W. & E. V. Brown, 1972. A sensitive analysis for nitrosamines. Analyt. Lett. 5: 293–304.

Discussion

Methods of nitrosamine analysis

The usufulness of the Coulson electrolytic conductivity detector in the pyrolytic method was demonstrated in a limited number of examples. Very clean gas chromatograms can be obtained from crude extracts of steam distillates without any further purification. In addition some different methods of derivative formation were demonstrated: the Dansyl method and the HFB method.

The need for confirmation of positive results by mass spectrometry was stressed.

Decomposition of nitrosamines

Decomposition of nitrosamines can occur in visible light. This is attributed to the near u.v. radiation, which is part of the sunlight and artificial light sources. The stability of standard solutions of nitrosamines depends greatly on the solvent. Expecially water can promote decomposition. A satisfactory procedure is to use dichloromethane rather than hexane and acetone as a solvent and to prepare the solutions afresh every week.

Handling nitrosamine waste

There are several ways to destroy nitrosamine waste products and to decontaminate glassware, e.g. HBr in glacial acetic acid, aqueous alkaline solutions in contact with aluminium foil, CuCl together with HCl and sunlight.

Groenen stressed the necessity of adequate labeling, safe shipment and handling of N-nitrosamines. In the USA is has been recommended to handle N-nitrosamines in a way comparable to that for radioactive materials.

Nitrosylation of prolylglycine

Prolylglycine in vitro is nitrosylated in the proline ring and presumably also on the N of the peptide bond. This could be confirmed by low resolution mass spectrometry, but not at high resolution, according to Walters.

Proc. int. Symp. Nitrite Meat Prod., Zeist, 1973. Pudoc, Wageningen.

The determination of total non-volatile nitrosamines in microgram amounts

C. L. Walters, D. G. Fueggle and T. G. Lunt

Biochemistry Section, BFMIRA, Leatherhead, Surrey, UK

Abstract

Nitrosamines and nitrosamides are denitrosated with thionyl chloride to form a product (presumably nitrosyl chloride) which can be volatilized in a stream of nitrogen and trapped in alkali as inorganic nitrite. At the 10 μg level, the yield of nitrite obtained usually approaches that obtained by denitrosation with hydrobromic acid in glacial acetic acid, which is virtually quantitative in most cases. Compounds tested in much higher amounts for their possible participation in this procedure have included a nitramine, nitramide, *C*-nitric and *C*-nitroso compounds, alkyl nitrites and nitrates, an oxime, azoxy compound, *S*-nitrosothiol, amine oxide and pyrazine; none gave significant release of nitrosyl chloride with thionyl chloride.

This procedure can be applied to the estimation in foods of total non-volatile nitrosamines. Water must be removed, preferably by freeze drying, and during the process volatile nitrosamines will be lost. Any non-volatile nitrosamines within the dried food matrix (for instance 500 g or more) suspended in dichloromethane will be denitrosated with thionyl chloride. Provided the suspension is not too viscous, the passage of nitrogen will volatilize nitrosyl chloride formed, though not always to a theoretical yield. In the absence of water, the percentage interference from inorganic nitrite is very small but detectable. The interference from nitrite increases with the addition of water up to a maximum of about 1.2% when about 1.0% water is present in the reaction medium.

Thus, this method provides a sensitive and selective procedure for the estimation of total non-volatile nitrosamines without the necessity for their extraction from the dried food or other matrix.

Introduction

Most secondary amines, tertiary amines and quaternary ammonium compounds recognized to be present in a biological matrix and to be potential precursors to nitrosamines after reaction with nitrite would give rise to non-volatile compounds. Unlike volatile nitrosamines, however, the non-volatile analogues possess no single physical or chemical property which permits their separation as a group, an essential requirement when estimating low levels of such compounds in the presence of a great excess of other aqueous and lipid soluble material.

An early attempt to separate non-volatile nitrosamines in aqueous extracts of foods, etc. took advantage of the fact that all such compounds examined were absorbed on to activated carbon (Walters et al., 1970). However, the efficiency of desorption of the adsorbed nitrosamines was by no means complete for all compounds even in model systems and the presence of lipid could reduce markedly the ability of the carbon to adsorb.

The introduction by Eisenbrand & Preussman (1970) of the denitrosation procedure of nitrosamines and nitrosamides using hydrobromic acid in glacial acetic acid solution and the subsequent demonstration (Johnson & Walters, 1971) of its specificity to these compounds commended this reaction in the estimation of total non-volatile nitrosamines, since little difficulty should be experienced in removing interfering water from such compounds. Its application, however, would necessitate either preliminary extraction of the nitrosamines from the food or other matrix, a difficult procedure at low levels likely to co-extract interfering substances such as inorganic nitrite. Alternatively, the reaction of nitrosamines in situ in the dried food would require a volume of acetic acid sufficient to cover the matrix effectively and this would restrict markedly the sensitivity of detection of any *N*-nitroso compounds present.

Accordingly, attempts were made to convert non-volatile nitrosamines and nitrosamides into (a) volatile derivative(s) which could be separated readily from a food matrix and interfering compounds such as inorganic nitrite and which could preferably be concentrated in some way to enhance the sensitivity of detection. Initially, the reaction investigated was that with hydrobromic acid followed by distillation of the glacial acetic acid solvent in the hope of carrying over any nitrosyl bromide formed for trapping in alkali but this proved impracticable by virtue of the excessive volume of acetic acid to be distilled and trapped. As a result, attention was directed towards denitrosation to the more volatile nitrosyl chloride.

Materials and methods

Estimation of total non-volatile nitrosamines and nitrosamides

After the suspension and/or solution of the nitrosamine or nitrosamide or of a dried matrix such as a food in dry CH_2Cl_2 (e.g. 200 ml) in a threenecked round bottom flask (usually 500 ml – 1 litre) equipped with a dropping funnel, inlet tube for nitrogen and an exit tube, dry nitrogen is bubbled through the system to displace damp air. If the suspension in use is at all viscous, a stirrer is desirable to ensure as complete a reaction as possible. The addition of 5 ml of a 25% solution of thionyl chloride in dry dichloromethane is completed rapidly from the dropping funnel and the mixture is left to stand for 15 minutes at room temperature. At the end of the reaction period, a steady stream of nitrogen is bubbled through the mixture for 15 minutes, the gas leaving the flask through the third neck and passing through a trap containing 5.0 ml 100% w/v aqueous sodium hydroxide. In order to determine the amount of nitrite produced in the alkali trap, 1.0 ml sulphanilamide solution (5.0% in concentrated HCl diluted with three volumes of water) is added, followed by 1.0 ml concentrated HCl to lower the pH for diazotization. Colour development occurs following the addition of 1.0 ml *N*-(1-naphthyl) ethylene-diamine dihydrochloride (0.1% in water) and is allowed to proceed for 15 minutes at room temperature. Then nitrite is determined by the optical density at 540 nm. Blank determinations without the presence of a nitrosamine should give no colour whatsoever.

Source of compounds used

N-nitrosoproline and -hydroxyproline and *N*-nitrososarcosine were prepared by a modification of the method of Stewart (1969). The preparation of *S*-nitrosothioglycolic acid was based upon that of Mirna & Hofmann (1969) for *S*-nitrosocysteine and -glutathione.

Results and discussion

Table 1 presents the percentage production of inorganic nitrite from a number of nitrosamines, nitrosamides and other related compounds after denitrosation with

Table 1. Yield of inorganic nitrite after denitrosation of compound with thionyl chloride in CH_2Cl_2.

Compounds	Amount	Yield(%)
Nitrosamine:		
N-nitrosoproline	10 μg	84
N-nitrosohydroxyproline	10 μg	90
N-nitrososarcosine	10 μg	54
N-nitroso-*N*′-phenylpiperazine	10 μg	81
N-nitrosodiphenylamine	10 μg	85
1,4-dinitrosopiperazine	10 μg	22
Nitrosamide:		
N-nitroso-*N*′-methyl urea	10 μg	40
Nitramine:		
N-nitrodimethylamine	100 μg	0.7
Nitramide:		
N-nitro urea	100 μg	0.0
C-nitro compounds:		
1-nitropropane	100 μg	1.2
C-nitroso compounds:		
1-nitroso-2-naphthol	100 μg	1.0
4-nitrosophenazone	100 μg	0.0
Oxime:		
cyclohexanone oxime	100 μg	0.0
Alkyl nitrites:		
n-bentyl nitrite	5.0 mg	0.5
isopropyl nitrite	5.0 mg	0.7
Alkyl nitrates:		
isopentyl nitrite	10 mg	0.0
ethyl nitrate	10 mg	0.0
Azoxy compound:		
cycasin	100 μg	0.0
Nitrosothiol:		
S-nitrosothioglycolic acid	1.0 mg	0.5
Inorganic nitrite:		
$NaNO_2$	10 mg	0.008
Amine oxide:		
3 hydroxy pyridine-1-oxide	100 μg	0.5
Pyrazine:		
methylpyrazine	100 μg	0.8

thionyl chloride. The percentage release of nitrite in the case of nitrosamines and nitrosamides is based upon that obtained by denitrosation in solution in glacial acetic acid with hydrobromic acid, which normally approaches the theoretical. That of compounds other than nitrosamines and nitrosamides is based upon the weight and nitrogen content.

At the 10 μg level, the recoveries of inorganic nitrite from nitrosamines and nitrosamides were not theoretical in all instances, but were greatly in excess of the yields obtained from the range of related compounds examined when tested at much higher levels. As can be seen from Table 1, virtually no response in terms of nitrite formation occurred with a nitramine, nitramide, *C*-nitro and *C*-nitroso compounds, oxime, alkyl nitrites and nitrates, azoxy compound, nitrosothiol, amine oxide or pyrazine. After drying thoroughly both dichloromethane and inorganic nitrite, the response from 1.0 mg $NaNO_2$ was equivalent to 0.009% falling slightly to 0.008% at 10 mg $NaNO_2$ and thence to 0.001% after the inclusion of 100 mg. The interference in the procedure provoked by an addition of 10 mg nitrite increased with the addition of water up to a value of 1.2% at a water concentration of 1.0% in the 500 ml volume of dichloromethane used. Above this water level, no further increase in the interference from 10 mg $NaNO_2$ was observed, but separation into two phases became apparent. No interference was observed in the Eisenbrand & Preussmann (1970) denitrosation procedure by 5% water but the reaction was completely prevented by 10%, probably by virtue of the easier protonation of water in preference to a nitrosamine.

No interfering chromogens have been found in the alkali trap after the displacement in a stream of nitrogen of nitrosyl chloride from the suspensions in dichloromethane of any of the dried food matrices so far examined. Unlike the denitrosation procedure with hydrobromic acid in glacial acetic acid solution, no interference has been detected in the spectrophotometric determination of liberated nitrite by compounds extracted from the food which sometimes cause asymetry of the peak at 540 nm through the formation of other colours absorbing at shorter wavelengths.

In studying the release of nitrosyl chloride from inorganic nitrite, use has been made of dried casein as a substrate. At a concentration of 0.3 g casein per ml dichloromethane, the suspension was found to be too viscous to allow the volatilization of nitrosyl chloride formed in a stream of nitrogen. However, a suspension of 0.1 g per ml CH_2Cl_2 was sufficiently mobile to permit ready volatilization of NOCl formed and, under these conditions, the estimation of 100 μg N-nitrosoproline was possible without interference from 500 mg sodium nitrite in 50 g casein.

One other method (Fan & Tannenbaum, 1971) has been proposed for the estimation of total non-volatile nitrosamines in a food extract and it has been applied to the autoanalyzer. This technique involves irradiation with ultraviolet light of an aqueous extract of the food etc., with release of inorganic nitrite. Many compounds other than nitrosamines release nitrite on photolysis but specificity has been claimed using monochromatic light of 360 nm wavelength in similar manner to Sander (1967), although some preliminary clean-up may be necessary to remove interference. Other compounds can also reduce or eliminate the formation of inorganic nitrite during photolysis, probably by acting as receptors of the nitrite released. No interference was, however, apparent from salts, organic and inorganic acids, sugars, nucleoside bases and amino acids.

The desirable features claimed for our estimation of total non-volatile nitrosamines and nitrosamides are as follows:

– Sensitivity: provided the denitrosation system is sufficiently fluid to permit the volatilization in nitrogen of nitrosyl chloride formed, the amount of food or other sample in suspension in dichloromethane is almost immaterial. A concentration factor is thereby introduced equivalent to the ratio of the weight of the original sample to the volume of liquid in which nitrite released is determined.

– Specificity: in the absence of water, little or no interference arises from inorganic nitrite and many other types of compounds from which it is potentially possible to obtain inorganic nitrite. Such interference would give rise to false positives rather than prevention of the observation of nitrosamines present.

– No extraction of non-volatile nitrosamines is necessary with this procedure, thus avoiding a stage of variable efficiency.

– The volatility of nitrosyl chloride formed is such that is can be volatilized readily in a stream of nitrogen, a procedure so mild that the possibility is remote of carry over of food components which could interfere with the determination of nitrite.

Probably the only drawbacks of the technique are the necessity for the drying of the food, with the accompanying possibility of the destruction of nitrosamines and nitrosamides, and the fact that the nitrosamines and nitrosamides detected are not characterized individually. Nevertheless, it should provide a ready screening procedure from which samples with nitrosamine levels of concern can be pinpointed and studied further by appropriate procedures.

References

Eisenbrand, G. & R. Preussmann, 1970. A new method for the colorimetric determination of nitrosamines after clearage of the N-nitroso group with hydrogen bromide in glacial acetic acid. ArzneimittelForsch. 20: 1513.

Fan, T. Y. & S. R. Tannenbaum, 1971. Automatic colorimetric determination of N-nitroso compounds. J. Agric. Fd Chem. 19: 1267–1269.

Johnson, E. M. & C. L. Walters, 1971. The specificity of the release of nitrite from N-nitrosamines by hydrobromic acid. Analyt. Letters 4: 383–6.

Mirna, A. & K. Hofmann, 1969. The behaviour of nitrite in meat products. Fleischwirtschaft 49: 1361–1364.

Sander, J., 1967. Detection of nitrosamines. Z. physiol. Chem. 348: 852–4.

Stewart, F. H. C., 1969. Peptide synthesis with the N-nitroso derivatives of sarcosine and proline. Aust. J. Chem. 22: 2451–61.

Walters, C. L., E. M. Johnson & N. Ray, 1970. Separation and detection of volatile and non-volatile N-nitrosamines. Analyst 95: 485–489.

Discussion

Labelling N-nitrosamines

In order to facilitate the observation of the release of non-volatile nitrosamines from meat products, studies have commenced using N-nitrosoproline and N-nitrososarcosine 'labelled' with ^{14}C. At an initial level of 20–50 $\mu g/kg$, good recoveries have been obtained from a number of cured meat products; the products were extracted by agitation in a phosphate buffer with or without the addition of protease; radioactivity entered the aqueous solution.

According to the experience of Dhont (Civo) who had sued the proposed method for total non-volatile nitrosamines, recoveries were sometimes very low and the blank values very high. The author could not confirm this and emphasized that the method was still being studied. The possible interference of chloride and sulphite was discussed. Chloride appeared to have no influence, but the effect of SO_2 has not yet been studied.

Non-volatile nitrosamines

The determination of individual non-volatile nitrosamines is not yet possible, because no reliable isolation and separation are available.

Conclusions and recommendations of the analytical session, 11th September 1973

1. Sufficiently precise methods are lacking for the determination of nitrate and of free and bound nitrite. A number of collaborative studies has and is being carried out on nitrate and free nitrite analyses (e.g. in ISO, EEC, IUPAC, AOAC). It is recommended that the efforts of these various groups be co-ordinated and extended to include the determination of bound nitrite.

2. Further and more specific information is needed on the determination and level in foods of precursors of nitrosamines and their relation to nitrosamine formation.

3. It is now possible to determine volatile nitrosamines in meat products at levels of a few μg per kg (ppb). Positive results have to be compared using adequate methods, such as m.r.

4. More research is recommended on:

a. methods for the determination of non-volatile *N*-nitrosocompounds in foods,
b. the range of compounds of interest, and
c. the nature of their combination in the food matrix.

Microbiological session

Reporters: J. H. Houben, S. J. Mulder, W. J. Olsman, P. S. van Roon

Proc. int. Symp. Nitrite Meat Prod., Zeist, 1973. Pudoc, Wageningen.

The microbiological effects of nitrite

M. Ingram

ARC Meat Research Institute, Langford, Bristol, BS18 7DY, UK

Abstract

The inhibitory effects of nitrite are discussed in relation to salt, pH, nitrate, number of bacteria, and storage temperature, in unheated and heated systems. The need for predictive generalisation is emphasized.

Introduction

Use and function of nitrite

Nitrite is used in meat curing, where it has at least the following four functions:
a. It is converted into nitrosomyoglobin which gives the characteristic pink colour to cured meat. It has been known for thirty years that a nitrite content as little as 5 mg/kg meat suffices to give a satisfactory colour for a limited time. It is believed that rather higher concentrations, perhaps up to 20 mg/kg, are necessary to provide commercially adequate colour stability, but detailed experimental confirmation of this is lacking.
b. Nitrite is essential for the characteristic flavour of cured meat; that which distinguishes ham and bacon for example from salt pork. The nature of the changes involved is still not known. To produce the characteristic flavour, some 50 mg of nitrite/kg are thought necessary; satisfactory quantitative evidence is lacking.
c. Curing meat provides an important degree of protection against botulism. It has been believed that more than 100 mg nitrite/kg is necessary to secure this protection, under commercial conditions. But the question proves to be very complex, because the necessary concentration of nitrite depends on several associated factors. Nitrite may provide similar protection against other food poisoning bacteria, e.g. *Clostridium welchii* or staphylococci, but the importance of this has never been assessed.
d. In raw fresh meat, there is protection against these kinds of food poisoning, since hazard does not usually arise until after the meat has gone putrid and the consumer rejects it. Cured meats, however, do not go putrid though they may carry equally large numbers of bacteria as putrid meat, because the curing salts inhibit putrefaction. It is not clear what are the relative contributions of nitrite and other curing

factors (especially salt) to the differences in spoilage; experience suggests that substantial concentrations of nitrite are involved, but again satisfactory experimental data are scarce. It seems important to understand these relations, because they make cured meat less 'perishable' than fresh.

Points (c) and (d) lead to the following general conclusions important to our present theme. In a situation where permissible concentrations of nitrite may be progressively reduced, the problems likely to arise first should be those involving the largest concentrations of nitrite, i.e. the microbiological problems. Though *Clostridium botulinum*, being the most dangerous, is the most important species, it is by no means the only one in question. We have to consider unheated systems, besides heated.

Chemical and biochemical properties

Nitrite has properties which seem important microbiologically:

– It reacts via nitrous avid with amino and other groups in proteins etc. Hence, undissociated form depending directly on the pH:
$\log_{10}\{(HNO_2)/(\text{-}NO_2)\} = pK - pH$, where pK = 3.4. The concentration of HNO_2 thus increases roughly ten-fold for every unit fall of pH.

– It reacts via nitrous acid with amino and other groups in proteins etc. Hence, nitrite disappears when added to meat or to culture media, usually roughly exponentially, at rates increased by acidity or high temperature; in meat, Nordin (1969) observed the relation:
$\log_{10}$ (half life in hours) = 0.65 – 0.025 × temp. (°C) + 35 × pH.
Because of this, it is necessary to add filter-sterilised nitrite in the cold, to obtain precisely defined concentrations in an experimental system. Because the nitrite gradually disappears, there might be opportunity for dormant organisms to grow out later (cf. Ashworth & Spencer, 1972); though it does not necessarily follow that the antimicrobial activity falls correspondingly, for nitrite might be converted into other substances with inhibitory properties.

– As a result of the balance between these two properties, there may be an optimum pH for the anti-microbial activity of nitrite, about 5.5.

– Nitrous acid decomposes to yield HNO_3 and NO, and the latter can be oxidised to HNO_3 by atmospheric oxygen in aqueous systems:

$$3HNO_2 \rightarrow HNO_3 + 2NO + H_2O$$
$$2NO + O_2 \rightarrow 2NO_2$$
$$2NO_2 + H_2O \rightarrow HNO_3 + HNO_2$$
$$\overline{}$$
$$3HNO_2 + O_2 \rightarrow 2HNO_3 + HNO_2$$

Besides helping to explain how nitrate appears in systems to which only nitrite was added, this may be why nitrite is more inhibitory to microorganisms when oxygen is absent, or in presence of reducing agents.

– Ascorbic acid decomposes nitrite in acid solution. If present during acidification of nitrite-containing extracts for colourimetric estimation, therefore, the estimated quantities of nitrite are too small. This may explain reports that a given microbiological effect is produced with less nitrite in the presence of ascorbate.

– Nitrite represents one step in a biochemical oxidation-reduction chain, potentially extending from nitrate to ammonia, and potentially reversible. Of the intermediate compounds, hydroxylamine and nitric oxide possess anti-microbial properties, the latter only weakly (Shank, Silliker & Harper, 1962).

$$\begin{array}{ccccccccc} HNO_3 & — & HNO_2 & — & H_2N_2O_2 & — & NH_2OH & — & NH_3 \\ & & NO_2 & & N_2O & & & & NH_4NO_2 \\ & & NO & & & & & & N_2 \end{array}$$

Nitrous oxide and nitrogen are common end-products of bacterial denitrification.

– The biochemical mechanism whereby nitrite inhibits bacteria is not known in detail. It appears to be a general metabolic inhibitor (Dainty, 1971–2).

Unheated systems

Interest in the anti-bacterial effects of nitrite developed in the 1920's, from its preservative effect in raw cured meat. But attempts failed to demonstrate a corresponding inhibitory action of nitrite; concentrations of the order ten times greater being needed, even in the presence of 3.5% sodium chloride (Lewis & Moran, 1928; Evans & Tanner, 1934). From subsequent knowledge, this failure is most reasonably explained by assuming that the workers of those days used massive inocula and media of pH above 6 incubated near 37 °C, the importance of those factors not being understood at that time.

It had been observed by Grindley (1929) that nitrite is more inhibitory under acid conditions, and he suggested that this might be due to the presence of nitrous acid. But his observation passed virtually unnoticed, and was re-discovered by Tarr (1941a; 1941b; 1942). Tarr showed that the preservative action of nitrite in fish was greatly increased by acidification; and correspondingly, that the inhibitory effect against several species of bacteria depended on the pH of the medium and increased markedly at levels below pH 6.0. Tarr did not explain the phenomenon, and it was left to Jensen (1945) to repeat Grindley's suggestion that the inhibition might be due to undissociated molecular nitrous acid. The expected ten-fold diminution in the inhibitory concentration of nitrite for unit drop of pH was clearly demonstrated by Castellani & Niven (1955) with *Staphylococcus aureus;* and later confirmed by Eddy & Ingram (1956) using a *Bacillus* species, and by Perigo, Whiting & Bashford (1967) using vegetative *Clostridium sporogenes* PA 3679.

The implication of Tarr's work, and of subsequent investigations, has been that the inhibitory action of nitrite is rather general. Indeed, Tarr listed several genera as being inhibited by 0.02% nitrite at pH 6.0: namely *Achromobacter, Aerobacter, Escherichia, Flavobacterium, Micrococcus,* and *Pseudomonas.* Later investigations have revealed that some bacteria are more resistant: for example salmonellas (Castellani & Niven, 1955); lactobacilli (Castellani & Niven, 1955; Spencer, 1971); and *Clostridium perfringens* among clostridia (Perigo & Roberts, 1968). There is however a lack of detailed information of this kind.

The effects of acidity are confused at low pH levels by the reaction of nitrite with constituents of the medium, so that its effect begins to diminish again at pH's approaching 5.5 (Henry, Joubert & Goret, 1954). This was confirmed by Shank, Silliker & Harper (1962) who observed that the anti-microbial effect increases with falling pH to a maximum near pH 5.0. Henry et al. (1954) also noted the importance of redox relations, and others have observed that nitrite is more inhibitory under anaerobic conditions (e.g. Castellani & Niven, 1955; Eddy & Ingram, 1956).

It has long been clear that pH influences the inhibitory effects of salt (sodium chloride) as well as of nitrite, smaller concentrations of salt being effective at lower pH (Ingram, 1948; Ingram, 1958). Because the sodium chloride can be partly substituted by other salts e.g. sodium nitrate, and because similar relations with pH apply to the inhibitory effects of sugars in pickling (e.g. Vas, 1957), it appears likely that the effect of sodium chloride relates mainly to water activity, and is only in minor degree specific. Such relations have recently been well discussed by Riemann, Lee & Genigeorgis (1972). A relation between the effects of sodium chloride and of sodium nitrite was implicit in some of the early work (e.g. Evans & Tanner, 1934; Riemann, 1963); but a clear demonstration was only given (so far a I know) some five years ago, when Pivnick, Barnett, Nordin & Rubin (1969) observed that their inhibitory effects were complementary. Though this demonstration was made with heated packs of canned pork luncheon meat containing *Clostridium botulinum* spores, recent research has amply confirmed similar relations in unheated systems. There are clear indications of a salt/nitrite interaction in a recent publication by Wood & Evans (1972) on their preservative effect in curing. Triple combined relations exist between the inhibitory effects of pH, nitrite, and sodium chloride, when inhibiting vegetative cells of several strains of *Clostridium botulinum,* or of *Clostridium perfringens* (Roberts & Ingram, 1973). Similar work by Bean & Roberts (about to be published) demonstrates similar inter-relations in inhibiting *Staphylococcus aureus;* and it appears that the phenomenon may be general.

The work of Roberts' group has further revealed a phenomenon like that observed with spores in heated systems, namely that the minimum inhibitory concentrations are smaller if they are challenged by a smaller number of vegetative cells. The differences have not been great until the number of vegetative cells fell below about 100 per gram; below this level, the effect of cell concentration has been marked. Little is known about the basis of this phenomenon, and more work is highly desirable.

A further important point, emerging in Bean & Roberts' experiments, is the very great influence of incubation temperature. Others have previously demonstrated that minimum inhibitory salt concentrations are less at lower growth temperatures (Segner, Schmidt & Boltz, 1966; Ohye & Christian, 1967; Alford & Palumbo, 1969; Matches & Liston, 1972); and that there is an analogous inter-relation between limiting pH and temperature (Baird-Parker & Freame, 1967). Similarly with the triple combined effects observed by Roberts & Ingram (1973) and Bean & Roberts (unpublished): the minimum inhibitory combinations of salt, nitrite and pH are similar at 25 and at 37 °C, but are much reduced at more normal temperatures of 20 °C and below. Even in unheated systems to which all the foregoing applies, quite high degrees of control are conceivable using practically realistic temperature conditions and combinations of salt, nitrite and pH. What we now need is some established basis for interpreting quantitatively the combined effects of these several factors, to measure the effect of changing one of them upon the critical levels of the others. The anti-microbial effect of nitrite clearly can no longer be considered in isolation.

Heated systems

Because vegetative cells are destroyed by heating, interest in heated systems is specially related to inhibitory effects upon bacterial spores, especially those of *Clostridium botulinum* and of the more heat-resistant 'indicator' strains of *Clostridium sporogenes* like PA 3679.

This phase of the subject effectively began with the work of Stumbo, Gross & Vinton in 1945, and continued through a series of similar investigations with inoculated packs, conveniently reviewed by Spencer (1966), which together gave the following indications. The major inhibition is due to salt, with some supplementary effect of nitrite; nitrate and sugar were not important. The supplementary effect of heating, to the order F_0[1] = 0.1 to 1.0, is of critical importance. Systems which are inhibitory with 1–10 spores per gram fail when challenged with numbers 100–1 000 times greater. (See Table 1).

This phase culminated in an experiment briefly described by Riemann (1963),

Table 1. Processes shown experimentally to give stable and/or safe systems (after Spencer, 1966).

Reference	Heat process	Sodium chloride concentration (% on water phase)	Initial nitrite concentration (mg/kg total)	Spore level/g	Comments
28	$F_0 = 1$	about 4.5	150	5	spoilage inhibited for up to one year
15	$F_0 = 0.05$ to 1.0	about 4.5–5.0	not stated	natural	viable spores detected after heat processes but no spoilage in 9–18 months
25	150°F/70 min	about 3–3.5	200	5000[1]	no toxin after 30 days at 30°C
31	176°F/20 min	3.6	150 (83)	natural, 1	pH 5.8 ± 0.2. Failed when challenged with 50 spores/g
27	$F_0 = 0.08$	3.5	78 (10–20)	natural, 1	failed when containing only 38 mg nitrite/kg
	$F_0 = 0.10$	3.5	78 (10–20)	added, 2.6	
12	$F_0 = 0.4$	4.02	not stated	3	

1. *Cl. botulinum*
The nitrite figures in brackets refer to post-processing levels
References:
28 – Stumbo, Gross & Vinton (1945)
15 – Vinton, Martin & Gross (1947)
25 – Steinke & Foster (1951)
31 – Bulman & Ayres (1952)
27 – Silliker, Greenberg & Schack (1958)
12 – Riemann (1963)

1. F_0 is a term normally used in process calculation work.
$F_0 = 0.5$ means a heat process in effect equal to a heating of 0.5 min at 121.1 °C (250 °F).

statistically planned on a factorial basis, the factors tested being salt, nitrate, nitrite, pH, F_0 value and spore level inoculated. After incubating six months at 30 °C, the number of spores developing was calculated from the numbers of swelled cans. Regarded singly, all the factors except nitrate had a statistically significant effect on the number of spores which developed. The following interactions also were statistically significant: $F_0 \times NaCl$; $F_0 \times NaNO_2$; $NaCl \times NaNO_3 \times NaNO_2$; $NaCl \times pH$; and $NaCl \times NaNO_3$. Several points are noteworthy. Though $NaNO_3$ was insignificant alone, it has a significant effect in interactions. pH alone had no effect, either in this, or in a parallel series of experiments reported at the same time. No interaction of pH with nitrite was revealed.

This kind of investigation, in which one observes spoilage or toxin formation in inoculated packs, has the advantage that it simulates realistic conditions. It has, on the other hand, the disadvantage that it provides little information about the nature of the inhibitory effects. In particular, it does not distinguish between four alternative possible explanations for the inhibitory effect of nitrite, first coherently enunciated by Johnston, Pivnick & Samson (1969):

The enhanced destruction of spores by heat

This possibility was suggested by Jensen & Hess in 1941. It is now largely eliminated, as a result of experiments in which spores have been heated in the absence or presence of curing salts and then transferred to separate growth media with or without salts. So far as sodium chloride and nitrate are concerned, the effect of heating in the presence of realistic concentrations is small, the important element being the presence of the salt in the medium on which the heated spores attempt to develop (Roberts & Ingram, 1966; Duncan & Foster, 1968). As regards nitrite, results are ambiguous; Duncan & Foster claimed that the presence of nitrite during heating increased the heat sensitivity of spores of PA 3679, especially at pH 6.0 (though this is not clear from the data presented); whereas Ingram & Roberts (1971), working with *Clostridium botulinum* strain 33 A at pH 6.0, failed to observe any corresponding phenomenon.

An increased rate of germination of spores during the heat process followed by death of the germinated spores from the heat process

This possibility was soon negatived by observations showing that some spores are in fact not killed as a result of the heating (e.g. Silliker, Greenberg & Schack, 1958; Riemann, 1963). Indeed, Silliker, Greenberg & Schack calculated that 20% of the inoculated spores survived in cured meat packs which nevertheless remained stable; and they then concluded that the failure of those spores to develop was probably the result of an inhibitory effect of the curing salts dependent on heat damage.

Prevention of germination of spores that survived the heat process

This third possibility, stated by Johnston and co-workers, was plainly demonstrated in the already cited work of Roberts & Ingram (1966) and of Duncan & Foster (1968). Large inhibitory effects were observed when spores, heated in whatever system, were plated out on media containing salt or nitrite. With nitrite, this was especially so at acid pH values, and Roberts & Ingram remarked that the relations suggested the involvement of undissociated HNO_2, by analogy with similar phenomena with unheated bacteria.

Production of more inhibitory substances from nitrite

Up to this point, everybody had ignored the fourth possibility. Then Perigo, Whiting & Bashford (1967) observed that the inhibitory concentrations of nitrite in the experiments of Roberts & Ingram (1966) were of the order 100–500 mg/kg, some ten-fold greater than the concentrations effective in canning practice; and they noted that Roberts & Ingram added filtersterilized cold nitrite, instead of heating nitrite with the supporting medium as is done in canning. They therefore speculated that the nitrite might, when heated in the presence of the supporting medium, be converted into some substances which is a much more powerful inhibitor. To test this hypothesis, Perigo et al. developed an artificial culture medium, portions of which they could adjust to various pH values; to one set they added nitrite without heating, and to another they added nitrite and then heated, then challenged both with inocula of 8×10^6 vegetative cells of PA 3679. Much smaller concentrations of nitrite were indeed effective in the heated system. There were two other notable observations: first, that with medium at pH 6.0 heated for 20 minutes, it was necessary to heat in the temperature range 95–125 °C, but that in the range 100–110 °C the enhancement became ten-fold or more; and second, this effect is much less pH-dependent than that of nitrite. The conclusion was that a small proportion of the added nitrite turns into a different substance at least ten times more inhibitory than nitrite; and this presumed substance has since come to be called the Perigo Factor or Inhibitor (PF).

Not long after, Johnston, Pivnick & Samson (1969) addressed themselves to the same possibility, but in a different manner. They blended meat with the culture medium and heated with the various concentrations of nitrite up to 200 mg/kg, then challenged with *Clostridium botulinum.* The inhibition was found to be no greater than that expected from the residual nitrite in the substrate; in addition, the inhibitory substance in meat was dialysable, while PF in culture medium was not; these facts suggested that the inhibitory substance in the meat system was nitrite itself. Further, using a culture medium in which PF was formed by heating with nitrite for 20 min at 110°C, they found that additions of 20% or more of meat prevented the development of PF and that as little as 1% of meat interfered with its development. Finally, they observed that addition of meat to culture medium in which PF was already present had the effect of neutralising its activity. These observations, on the absence of PF in the meat system, and the ability of meat to neutralise pre-formed PF, have been confirmed independently by Dr T. A. Roberts. The conclusion seems to be that the Perigo Inhibitor is not produced during the heat treatment of meat containing nitrite, and cannot explain the role of nitrite in the stability of canned cured meats.

Since that time, there has been an apparent dichotomy between work in culture media and work in meat systems, and the two are best considered separately, because it is not clear what may be the relation of one with the other.

Table 2. Inhibitory concentrations (mg/kg) at pH 6 of sodium nitrite and heated sodium nitrite (Roberts, 1971).

	Spores/ml	Unheated nitrite			Heated nitrite		
		A	E	S	A	E	S
Unheated	10^6	240	160	200	20	15	10
	10^3	240	240	200	15	7.5	10
Heated	10^6	200	160	200	40	5	15
	10^3	200	240	200	10	5	15
Irradiated	10^6	240+	240+	240+	40	15	20
	10^3	240	200	200	40	7.5	20

A = *Cl. botulinum* Type A
E = ,, ,, Type E
S = *Cl. sporogenes*

Culture media Immediately after the demonstration of PF with vegetative cells it was demonstrated by Perigo & Roberts (1968) that a similar effect was observable with a variety of clostridia in the vegetative state. In subsequent work, which has been reported only briefly, Roberts (1971) went on to demonstrate a PF effect with spores of several types of *Clostridium* added to the Perigo medium heated with nitrite. As with vegetative cells, there was an about ten-fold enhancement of the effect of nitrite as a result of heating (Table 2). It is noteworthy that the number of spores (10^6 or 10^3 per ml), and heating or irradiation of the spores beforehand, made little difference to the inhibitory concentrations. In the meantime, there has been a good deal of work on the Perigo effect in artificial media, but it is difficult to discuss because little of it has been published, perhaps because of inconsistencies in the work of individuals and differences from the results of others. Almost the only point of general agreement is that a reducing agent is necessary in the medium: the thioglycollate used by Perigo et al. (1967) can be replaced by ascorbic acid or cystein. A protein hydrolysate is necessary, preferably of casein, but it is not decided which fractions are important. There have been suggestions that iron is necessary, to form Roussin salts, but my colleaques have had no consistent benefit from the addition of iron, and it seems that two inhibitors may be involved. There is however little profit in discussing these complexities, because of doubt whether they are applicable to the situation in the presence of meat.

Meat systems Since the work of Johnston, Pivnick & Samson (1969) demonstrated that the Perigo Inhibitor could be neutralised by meat, there have been two reports that an extra inhibitory effect does occur when nitrite is heated in the presence of meat. A positive response was observed by Pivnick's group, in work reported by Chang & Pivnick to the 1973 Meeting of the Canadian Institute of Food Technologists and to be published in their Journal. Chang & Pivnick canned commercially formulated luncheon meat with varying concentrations of nitrite and, after autoclaving, stored the cans until all residual nitrite had disappeared. Then, the cans were challenged with an inoculum of washed spores of *Clostridium botulium* previously heated to $F_0 = 0.4$ in a meat homogenate containing 4% NaCl and

150 mg/kg $NaNO_2$. In the now nitritefree meat, an inhibition was observed which was greater the greater the concentration of nitrite initially. This was interpreted as indicating the presence of an inhibitor derived from the original nitrite, as a result of reaction during autoclaving or storage. The inhibition was greater with smaller numbers of spores; but, with only 10 spores per can, and relatively high concentrations of nitrite initially (up to 200 mg/kg), one out of 16 cans showed growth. The other report was from Ashworth & Spencer (1972), who added nitrite directly to minced pork and heated, observing a greater inhibitory effect when nitrite was added before rather than after heating. The effect was observed in the range 150–300 mg nitrite/kg initially. But the enhancement factor was unusually small, less than 1.5; and practically independent of pH, unlike the situation in culture medium where Perigo et al. (1967) showed that the enhancement factor is much greater at higher pH values. Using more resistant strains of clostridia isolated from spoiled cured meats, Roberts (personal communication) has confirmed that there is an effect in pork, but only by using much higher initial nitrite concentrations up to 800 mg/kg.

The outcome of a substantial amount of work on the Perigo effect seems disappointing. In culture media, it is possible to get complete inhibition in heated nitrite with residual concentrations of the order 50–10 mg/kg which are frequently found in safe situations in practice. Yet it seems that the Perigo Inhibitor developed in culture media may not be the same as that in meat. In culture media, the enhancement factor is much greater, and sensitive to pH; its effect is little influenced by the number of spores or by heating or irradiating them beforehand, factors which are known to have a large influence in realistic meat systems; and finally, this inhibitor is neutralised by the presence of only quite small proportions of meat. For these reasons, it seems premature to speak of the Perigo Inhibitor or Perigo Factor in meat, and such terminology should be avoided. As to the inhibitor actually produced by heating nitrite in meat, to attain effective concentrations seems to require initial nitrite concentrations much greater than those observed to be effective in actual practice; and, indeed, Ashworth & Spencer (1972) themselves conclude 'from the work reported here no evidence has been obtained which would indicate that a Perigo effect is likely to be involved, under practical conditions, with the safety and stability of canned cured meats.'

Recent inoculated pack experiments

There has been administrative pressure in many countries recently to decide what concentrations of nitrite are necessary to provide protection against botulism. This has led in U.S.A. to a resumption of inoculated pack experiments, some of which have been described by Greenberg (1972) and by Christiansen, Johnston, Kautter, Howard & Aunan (1973). Ground pork ham was prepared with eight levels of nitrite (0–500 mg/kg as formulated), and inoculated with a mixture of spores from 5 Type A and 5 Type B strains of *Clostridium botulium*, at two levels namely 90 and 5 000 per gram. The canned materials were cooked in water to an internal temperature of 155°F (68.5°C); and the cans were then stored at 80°F (26.7°C) and observed over 6 months. The essential data are in Tables 3 and 4. Greenberg

Table 3. Effect of nitrite on botulinum toxin formation in canned ham (light inoculum). (Greenberg, 1972).

Nitrite added, mg/kg	Days until toxin first detected	Total number of toxic cans
0	7	45/80
50	7	32
100	28	16
150	68	2
200	>180	0
300	>180	0
400	>180	0
500	>180	0

Table 4. Effect of nitrite on botulinum toxin formation in canned ham (heavy inoculum). (Greenberg, 1972).

Nitrite added, mg/kg	Days until toxin first detected	Total number of toxic cans
0	7	63/80
50	12	18
100	13	33
150	42	13
200	70	5
300	168	1
400	21	2
500	>180	0

states that the observations correlated better with the initial concentrations of nitrite than with (calculated) residual concentrations. The problem is, how to interpret such observations?

First, it cannot be too strongly emphasized that they have no general validity. The limiting concentrations of nitrite refer only to the particular circumstances of the experiment: for example the salt concentration (2.5% or 3.7% brine); the particular numbers of spores (90 or 5 000 per gram), which were relatively high; and the particular heating regime, relatively low and not very precisely specified. Dextrose and ascorbate were present, but no polyphosphate; their likely effect is uncertain. The effect of pH (6.24) is uncertain. It does seem certain that, had the salt content, spore number of heating treatment been different, the limiting concentrations of nitrite would have been different. But we do not, unfortunately, know how the inter-relations observed in model systems, for example between salt, nitrite and pH, will remain valid in commercially heated packs; there are already indications (e.g. from Riemann, 1963) that the pH relations may not apply, which might perhaps be expected if a large part of the inhibition were due to substances other than nitrite. What we need at the present time, in my opinion, is not more inoculated pack experiments but a rationale for interpreting them.

References

Alfrod, J. A. & S. A. Palumbo, 1969. Interaction of salt, pH, and temperature on the growth and survival of salmonellae in ground pork. Appl. Microbiol. 17(4): 528–532.

Ashworth, J. & R. Spencer, 1972. The Perigo effect in pork. J. Fd Technol. 7(2): 111–124.

Baird-Parker, A. C. & B. Freame, 1967. Combined effect of water activity, pH and temperature on the growth of *Cl. botulinum* from spore and vegetative cell inocula. J. appl. Bact. 30(3): 420–429.

Bean, P. G., 1972. Growth and recovery of damaged *Staphylococcus aureus* NCTC 10652. M.Sc. thesis Bristol University.

Castellani, A. G. & C. F. Niven Jnr., 1955. Factors affecting the bacteriostatic action of sodium nitrite. Appl. Microbiol. 3: 154.

Chang, P.-C. & H. Pivnick, 1973. Effect of sodium nitrite on *Cl. botulinum* in commercially produced canned luncheon meat. Can. Inst. Fd Technol. J. (in press).

Christiansen, L. N., R. W. Johnston, D. A. Kautter, J. W. Howard & W. J. Aunan, 1973. Effect of nitrite and nitrate on toxin production by *Cl. botulinum* and on nitrosamine formation in perishable canned comminuted cured meat. Appl. Microbiol. 25(3): 357–362.

Dainty, R. H., 1971–2. Mechanisms of inhibition of growth of bacteria by nitrite. Meat Res. Inst. Ann. Rep. 1971–2: 80.

Duncan, C. L. & E. M. Foster, 1968. Role of curing agents in the preservation of shelf stable canned meat products. Appl. Microbiol. 16: 401.

Eddy, B. P. & M. Ingram, 1956. A salt-tolerant, denitrifying *Bacillus* strain which 'blows' canned bacon. J. appl. Bact. 19: 62.

Evans, F. L. & F. W. Tanner, 1934. The effect of meat curing solutions on anaerobic bacteria. IV. The effect of mixed curing solutions. Zentbl. Bakt. ParasitKde (Abt. 11), 91: 135.

Greenberg, R. A., 1972. Nitrite in the control of *Cl. botulinum*. Proc. American Meat Inst. Found. Meat Ind. Res. Conf., March 1972: p. 25.

Grindley, H. S., 1929. The influence of potassium nitrate on the action of bacteria and enzymes. Studies in Nutrition, Univ. of Illinois, Urbana, Ill., 2: 359.

Henry, M., L. Joubert & P. Goret, 1954. Méchanisme biochimique de l'action du nitrite dans la conservation des viandes. Conditions physicochimiques favorables à son action bactériostatique. C. R. Séanc. Soc. Biol. 148: 819.

Ingram, M., 1958. Fatigue musculaire, pH, et prolifération bactérienne dans la viande. Annls Inst. Pasteur 75: 139–147.

Ingram, M., 1958. L'importance du pH pour la microbiologie de la viande. Rev. Ferment. Ind. aliment. Bruxelles 13: 139–146.

Jensen, L. B., 1945. Microbiology of Meats. 2nd edn. Garrard Press, Champaign, Ill.

Jensen, L. B. & W. R. Hess, 1941. A study of the effects of sodium nitrate on bacteria in meat. Fd Mf. 16: 157.

Johnston, M. A., H. Pivnick & J. M. Samson, 1969. Inhibition of *Clostridium botulinum* by sodium nitrite in a bacteriological medium and in meat. Can. Inst. Fd Technol. J. 2: 52.

Lewis, W. L. & J. A. Moran, 1928. The present status of our knowledge of ham souring. Amer. Inst. Meat Packers, Bull. No. 4, Chicago, Ill.

Matches, J. R. & J. Liston, 1972. Effect of pH on low temperature growth of *Salmonella*. J. Milk Fd Technol. 35(1): 49.

Nordin, H. R., 1969. The depletion of added sodium nitrite in ham. Can. Inst. Fd Technol. J. 2(2): 79–85.

Ohye, D. F. & J. H. B. Christian, 1967. Combined effects of temperature, pH and water activity on growth and toxin production by *Cl. botulinum* types A, B and E. 'Botulism 1966', Proc. 5th Int. Symp. Fd Microbiol., Moscow, July 1966. ed. M. Ingram & T. A. Roberts. Chapman & Hall.

Perigo, J. A., E. Whiting & T. E. Bashford, 1967. Observations on the inhibition of vegetative cells of *Clostridium sporogenes* by nitrite which has been autoclaved in a laboratory medium, discussed in the context of sub-lethally processed cured meats. J. Fd Technol. 2: 377.

Perigo, J. A. & T. A. Roberts, 1968. Inhibition of clostridia by nitrite. J. Fd Technol. 3: 91.

Pivnick, H., H. W. Barnett, H. R. Nordin & L. J. Rubin, 1969. Factors affecting the safety of canned, cured, shelf-stable luncheon meat inoculated with *Clostridium botulinum*. Can. Inst. Fd Technol. J. 2: 141–148.

Riemann, H., 1963. Safe heat processing of canned cured meats with regard to bacterial spores. Fd Technol., Champaign 17: 39.

Riemann, H., W. H. Lee & C. Genigeorgis, 1972. Control of *Cl. botulinum* and *Staph. aureus* in semi-preserved meat products. J. Milk Fd Technol. 35(9): 514.

Roberts, T. A., 1971. The inhibition of bacteria in canned pasteurized hams by sodium nitrite. Proc. 17th Eur. Meeting of Meat Res. Workers, Bristol, Sept. 1971.

Roberts, T. A. & M. Ingram, 1966. The effect of sodium chloride, potassium nitrate and sodium nitrite on recovery of heated bacterial spores. J. Fd Technol. 1: 147–163.

Roberts, T. A. & M. Ingram, 1973. Inhibition of growth of *Cl. botulinum* at different pH values by sodium chloride and sodium nitrite. J. Fd Technol. (in press).

Segner, W. P., C. F. Schmidt & J. K. Boltz, 1966. Effect of sodium chloride and pH on the outgrowth of spores of type E *Cl. botulinum* at optimal and suboptimal temperatures. Appl. Microbiol. 14(1): 49–54.

Silliker, J. H., R. A. Greenberg & W. R. Schack, 1958. Effect of individual curing ingredients on the shelf stability of canned comminuted meats. Fd Technol., Champaign 12: 551.

Shank, J. L., J. H. Silliker & R. H. Harper, 1962. The effect of nitric oxide on bacteria. Appl. Microbiol. 10: 185.

Spencer, R., 1966. Processing factors affecting stability and safety of non-sterile canned cured meats. Fd Mf. 41(3): 39.

Spencer, R., 1971. Nitrite in curing – microbiological implications. Proc. 17th Eur. Meet. Meat Res. Workers, Bristol, Sept. 1971.

Stumbo, C. R., C. E. Gross & C. A. Vinton, 1945. Bacteriological studies relating to thermal processing of canned meats. III. Influence of meat curing agents upon growth of a putrefactive anaerobic bacterium in heat processed meat. Fd Res. 10: 293.

Tarr, H. L. A., 1941a. The action of nitrites on bacteria, J. Fish. Res. Bd Can. 5: 265.

Tarr, H. L. A., 1941b. Bacteriostatic action of nitrites. Nature, Lond. 147: 417.

Tarr, H. L. A., 1942. The action of nitrites on bacteria; further experiments. J. Fish. Res. Bd Can., 6: 74.

Vas, K., 1957. How to improve the keeping qualities of fruit and vegetable concentrations. Fd Mf. 32: 71–76.

Wood, J. M. & G. G. Evans, 1972. Curing limits for mild cured bacon. Proc. Inst. Fd Sci. Technol. Symp. on Meat Curing, Meat Res. Inst., March 1972: p. 6–20.

Discussion

Added or residual nitrite

The references to nitrite in the paper usually deal with the added nitrite. Some authors suggest a correlation of the inhibiton of clostridia with the amount of added nitrite other with the residual nitrite; there is no agreement on this point. Perhaps there is a relation with some reaction products of nitrite, i.e. with the nitrite which disappears.

The necessity for nitrite

meat

The control of *Cl. botulinum* would be no problem without nitrite, if the storage temperature were always lower than 10°C or the pH lower than 4.5 or the NaCl content were higher than about 12%, but these are not realistic conditions in practice. The type of spoilage of raw and cured meats depends largely on the initial microflora, the curing conditions, the heating temperature and the storage conditions of a product.

Realistic numbers of spores in raw meat are about 10 per kg meat, and of these about 1 on 10^4 is a *Cl. botulinum* spore, according to American investigations. At retail levels this may be somewhat higher.

In pasteurized comminuted products these numbers are much higher, counts of about 100 *Bacillus* and 10 *Clostridium* spores per gram are reported in Europe. Other authors have even found higher numbers occasionally.

Recent outbreaks of botulism in man due to unheated meat products without nitrite have been reported. No substitute for nitrite which is less hazardous has so far been found. If we knew the specific effect of nitrite on micro-organisms it would be easier to suggest a replacement.

Specific inhibitory action of nitrite

In the USA, experiments have demonstrated the ability of nitrite, but not nitrate, to retard or prevent botulinal toxin development in a series of cured meat products as manufactured currently.

Action of nitrosamines

Dr Walters examined the anti-clostridial effect of about a dozen nitrosamines; none of these had any effect at relevant concentrations. Others specifically looked for DMNA and DENA in culture media with which the Perigo effect had been demonstrated, but these nitrosamines were not found.

Action of aerobic sporeformers

Whether aerobic sporeformers can inhibit clostridia is not known. Bacilli can metabolize nitrite under certain conditions.

Proc. int. Symp. Nitrite Meat Prod., Zeist, 1973. Pudoc, Wageningen.

The inhibition of Clostridium botulinum by nitrite and sodium chloride

A. C. Baird-Parker and M. A. H. Baillie

Unilever Research Laboratory, Colworth/Welwyn, Colworth House, Sharnbrook, Bedford, UK

Abstract

In a bacteriological growth medium (MRCM) most *Clostridium botulinum* type A and proteolytic type B and F strains will grow at pH 6.0 in the presence of at least 150–200 mg/kg sodium nitrite or 6% sodium chloride at 25°C. Under the same growth conditions (medium, incubation temperature and inoculum level), *Cl. botulinum* type E and non-proteolytic strains of types B and F are mainly inhibited by 4.5% sodium chloride or 100–150 mg/kg sodium nitrite. Less nitrite and salt are required to inhibit growth at lower storage temperatures. Sodium chloride and sodium nitrite are apparently synergistic in inhibiting the growth of *Cl. botulinum* in bacteriological media and 200 mg/kg nitrite in the presence of 3% sodium chloride inhibits the growth of almost all strains at 20 and 30°C as also does 100 mg/kg sodium nitrite in the presence of 4.5% sodium chloride at pH 6.0; these results were obtained in a bacteriological medium. However, in pork macerate broth (PMB) higher concentrations of nitrite and salt are required to inhibit the growth of the same number of cells as that used in the bacteriological growth medium. If smaller numbers of cells (10 or 10^3) are used to inoculate PMB containing nitrite most strains are inhibited by the presence of 200 mg/kg nitrite at pH 6.0 and a temperature of 20°C. L-Ascorbic acid (0.1 or 1.0%) in MRCM (pH 6.0) reduces the minimum inhibitory concentration (MIC) of sodium nitrite for all strains to less than 100 mg/kg. However, in PMB containing 0.1% L-ascorbic acid this enhancement of nitrite activity is not observed and the MIC of nitrite to cells of *Cl. botulinum* was the same in the presence as in the absence of ascorbic acid.

Preliminary experiments done in a laboratory cured bacon stored at 25°C show that the addition of 0.1% L-ascorbic acid to bacon containing 500 mg/kg sodium nitrate, 4.5% sodium chloride and 100 or 200 mg/kg nitrite (pH 5.6–5.8) does not affect the antibotulinum activity of nitrite.

Introduction

The microbial stability of a cured meat or fish product cannot be ascribed to any single attribute but depends on the interacting effects of a number of parameters. These include the type and quantity of the contaminating microflora, the a_W, pH, E_h and composition of the product, the degree of heat injury induced to micro-organisms surviving any thermal process used during manufacture, the packaging and storage conditions and the amounts and types of curing salts they contain (Riemann et al., 1972, Christiansen et al., 1973, Leistner et al., 1973, Pace & Krumbiegel, 1973).

There is a considerable amount of published data on the separate effects of most of the above parameters on the growth of *Clostridium botulinum* in bacteriological media and some foods (Baird-Parker, 1969) and quite a lot of data on the effects of two or more combinations of these (Segner et al., 1966; Baird-Parker & Freame, 1967; Spencer, 1967; Ohye & Christian, 1967; Emodi & Lechowich, 1969; Pivnick et al., 1969 and Roberts & Ingram, 1973).

The present study was done to obtain further data on the individual and combined effects of curing salts, in particular sodium chloride and nitrite, on *Cl. botulinum* at concentrations which are relevant to production and sale of raw cured meat and fish products. Sodium chloride was included in these studies as it is believed to interact strongly with nitrite in preventing the growth of *Cl. botulinum* in cured foods.

The studies have been based on the use of a large collection of *Cl. botulinum* strains isolated from foods and natural environments in many parts of the world. The use of a large collection of strains was regarded as particularly important in these studies as it was considered that a major deficiency in almost all previous studies was that only a few strains had been studied. This is important, as it is well recognised that individual strains and toxigenic types of *Cl. botulinum* vary enormously in their tolerance to curing salts and other factors important to the safety and stability of cured meat and fish products.

Materials and methods

Sources of organism

Strains representing all toxigenic types of *Cl. botulinum* were used including the recently described type G (Gimenez & Ciccarelli, 1970). These were mainly obtained from workers and collections in North America, Europe and the United Kingdom but included a number of our own isolates. Most of the strains studied were toxigenic types A, B, E and F as only these have been incriminated unequivocally in human botulism.

Inoculum and inoculum levels

Vegetative cell inocula were used in all studies except for those involving bacon. In previous studies we have not observed any differences between the ability of spores and vegetative cells to initiate growth under growth limiting conditions (Baird-Parker & Freame, 1967; A. Baillie, unpublished). The vegetative cell inocula used were from overnight cultures (30°C) grown in modified Reinforced Clostridial Medium (see basal media) and subcultured from refrigerated stock cultures maintained in Robertson's Cooked Meat Medium. An inoculum level of approximately 10^6 cells/ml was used in all tests except those on the effect of inoculum level on growth. For the latter Thoma slide counts were done, and the cultures appropriately diluted in freshly steamed and cooled 0.1% peptone water to give inoculum levels of about 10^6, 10^3 and 10 cells/bottle of test medium.

Basal media

All media were dispensed in 20 or 25 ml amounts in 1 oz (30 g) screw-capped bottles and freshly steamed before making additions to the heat sterilized media. All pH adjustments were done with HCl or NaOH. The following two basal media were used.

a. Modified Reinforced Clostridial Medium (MRCM). The medium was that described by Baird-Parker & Freame (1967) except that ascorbic acid was omitted. It supports good growth of all toxigenic types of *Cl. botulinum* except for the fastidious type C strains which grow only moderately well in this medium.

b. Pork Macerate Broth (PMB). This medium was used to simulate a meat situation. Lean pork was diced and as much fat as possible removed. It was than minced, weighed into 1 kg aliquots and 1 litre of distilled water added to each aliquot. The macerates were then steamed for 30 minutes, with occasional stirring, and when cool the liquor was strained off through muslin bags; as much free liquor as possible was removed from the minced pork by squeezing the bags tightly. The meat was dispensed in 10 g amounts in wide necked 30 g screw-capped bottles. The liquor was clarified by adding about 50 g Celite 545 to each litre and filtering through Whatman No. 1 filter paper. It was then adjusted to pH 6.0 and 10 ml amounts were added to each bottle of minced pork. Appropriate additions and further pH adjustments were made after sterilizing the PMB at 121°C for 15 minutes. This medium supports excellent growth of all toxigenic types of *Cl. Botulinum*.

Additions

Sodium nitrite (Analar grade) was used for all experiments involving nitrite. This was made up as a stock solution in distilled water containing 50 times the highest final concentration of sodium nitrite required. It was sterilized by membrane filtration and diluted so that when 0.4 or 0.5 ml was added to the appropriate basal medium the required final concentration of sodium nitrite was obtained; all pH adjustments and other additions were made to the freshly steamed media prior to the addition of nitrite. Nitrite concentrations of 50, 100, 150 and 200 mg/kg were used. All nitrite additions were checked by chemical analysis in the final media. Nitrite was very stable in MRCM and analysed initial levels were within 5% of the calculated levels and after 28 days storage at the highest temperature used (30°C) the nitrite levels were still within 10% of the initial analysed levels. As would be expected nitrite was rather unstable in PMB. Analysed levels were all 10–20% below the calculated amounts added and during storage at 20°C for 28 days this dropped to 40–50% of the initial level.

Sodium chloride was added prior to sterilization of the basal media; the pH was adjusted prior to sterilization such that it was correct after sterilization. Concentrations of sodium chloride used were 1.5, 3.0, 4.5 and 6.0% w/v. All additions were checked analytically in the final media and were within 2.0% of the calculated amounts added.

L-Ascorbic Acid (Analar grade) was made up as stock solutions which were 50 times the required final concentrations in PMB or MRCM. The stock solutions were adjusted to pH 6.0 and with NaOH and filter sterilized; 0.4 or 0.5 ml of each stock solution was added to the freshly steamed and cooled medium to give the final required concentration. The presence of L-ascorbic acid in MRCM and PMB resulted in a loss of the order of 75–85% of the initial nitrite concentration over the storage period of 28 days at 20°C.

Inoculation and storage

Five replicates of each combination of test medium and test organism were set up. In most experiments, 20°C was used as the storage temperature. To test the effect of temperature, experiments were also done at 15, 25 and 30 °C. Storage was done in jacketed incubators (accuracy ± 1°C) and inoculated media were stored for up to 28 days except for the 15°C storage experiments when the storage time was extended to 56 days.

Assessment of results

Throughout the storage periods bottles were examined at weekly intervals for visible signs of growth, i.e. turbidity, digestion of meat or gas production. At the end of the storage period contents of all bottles showing growth at the most severe test conditions were checked for purity by aerobic and anaerobic plating on MRCM agar; the contents of representative bottles were also tested for botulin by mouse toxicity tests (Baird-Parker, 1969). As it was often difficult to decide visually whether growth had occurred in PMB the contents of bottles were streaked out aerobically and anaerobically to assess growth. Growth is recorded in the tables, if one or more bottles of the 5 replicates tested showed growth.

Laboratory cured bacon

Bacon was slice cured in bags (3 rashers/bag; wt. c. 100 g) using a solution containing sodium chloride, sodium nitrate, and where appropriate sodium nitrite and L-ascorbic acid (adjusted to pH 6.0). The concentrations of curing salts and ascorbic acid used were calculated to give final concentration in the bacon of 4.5% sodium chloride on water phase, 500 mg/kg nitrate, 1000 mg/kg L-ascorbic acid, and 100 or 200 mg/kg nitrite. The bacon was then vacuum-packed and placed at 4°C for 2 days. Packs were opened and surface and deep inoculated (total inoculum 0.01 ml) with either a mixture of vegetative cells and spored of 7 proteolytic strains of *Cl. botulinum* (2, type A; 2, Type B; 2, Type F; 1, Type G) or 6 non-proteolytic strains (2, Type B; 2, Type E; 2, Type F). The inocula contained equal numbers of spores and vegetative cells. The inoculum levels were: proteolytic type A, B, F and G strains c. 260 cells/g, non-proteolytic type B, E and F strains c. 8 cells/g. The strains used were chosen to represent those strains which in PMB were most resistant to sodium chloride or sodium nitrite. The inoculated packs were revacuum-packed and then stored at 25°C. After 14 and 28 days storage 5 replicates of each combination of inoculum type and cure were examined for cells and toxins of *Cl. botulinum*.

Results and discussion

Inhibition by sodium nitrite

Results obtained in MRCM (pH 6.0) stored for 28 days at 20 and 25°C and 56 days at 15°C shown in Table 1. They show that although individual strains of *Cl. botulinum* vary markedly in their resistance to sodium nitrite they can be divided into two main groups on the basis of nitrite tolerance. The most resistant strains, generally growing in upto 150–200 mg/kg sodium nitrite in these tests at 25°C, occur amongst the heat resistant and physiologically closely related proteolytic and mesophilic types, i.e. *Cl. botulinum* type A and proteolytic types B and F. The most sensitive strains which are mainly inhibited by 100–150 mg/kg sodium nitrite are found amongst the heat sensitive, non-proteolytic and psychrotrophic types, i.e. *Cl. botulinum* type E and non-proteolytic types B and F. The mesophilic, non-proteolytic type C and D strains belonged to the more sensitive group. There was also a marked difference in the tolerance to nitrite of the proteolytic, mesophilic strains growing at 20°C and 25°C but virtually no difference in the nitrite tolerance of the non-proteolytic, psychrotrophic strains growing at these two temperatures. This difference may be a reflection of the fact that at 20 and 25°C the mesophilic strains are further from their optimum growth temperatures than the psychrotrophic strains and are therefore more subject to an influence of a small

Table 1. Effect of sodium nitrite on the growth of *Clostridium botulinum* in MRCM (pH 6.0) at 15, 20 and 25°C.

Toxigenic type	Storage temperature (°C)	No. of strains	Sodium nitrite (mg/kg)[3]			
			50	100	150	200
A	15[1]	6	4	4	3	2
B (proteolytic)	15	6	6	6	2	3
B (non-proteolytic)	15	6	4	2	0	0
E	15	3	3	3	0	0
F (proteolytic)	15	3	2	1	1	0
F (non-proteolytic)	15	3	0	0	0	0
A	20[2]	51	46	40	27	14
B (proteolytic)	20	12	11	8	2	1
B (non-proteolytic)	20	9	6	2	1	0
E	20	64	24	14	5	0
F proteolytic	20	3	2	2	1	0
F (non-proteolytic)	20	3	1	0	0	0
A	25[2]	50	50	50	46	30
B (proteolytic)	25	11	11	11	10	5
B (non-proteolytic)	25	9	7	5	3	2
E	25	64	51	19	5	2
F (proteolytic)	25	3	3	3	3	1
F (non-proteolytic)	25	3	3	0	1	0

1. Stored for 56 days.
2. Stored for 28 days.
3. Calculated amount of sodium nitrite added.

Table 2. Effect of sodium nitrite on the growth of *Clostridium botulinum* at 20°C in MCRM (pH 5.5, 6.0, 6.5 and 7.0).

Toxigenic type	pH	No. of strains	Sodium nitrite (mg/kg)[1]		
			50	100	200
A	5.5	6	0	0	0
	6.0	6	6	6	2
	6.5	6	6	6	6
	7.0	6	6	6	6
B (proteolytic)	5.5	6	0	0	0
	6.0	6	5	3	1
	6.5	6	6	5	5
	7.0	6	6	6	6
B (non-proteolytic)	5.5	6	0	0	0
	6.0	6	4	1	0
	6.5	6	5	5	2
	7.0	6	6	6	6
E	5.5	6	0	0	0
	6.0	6	1	1	0
	6.5	6	2	2	0
	7.0	6	5	5	5
F (proteolytic)	5.5	3	0	0	0
	6.0	3	2	2	1
	6.5	3	3	3	1
	7.0	3	3	3	3
F (non-proteolytic)	5.5	3	0	0	0
	6.0	3	1	0	0
	6.5	3	3	3	0
	7.0	3	3	3	3
G	5.5	1	0	0	0
	6.0	1	1	0	0
	6.5	1	1	0	0
	7.0	1	1	1	1

1. Calculated amount added.

temperature change. At 15°C, the proteolytic strains still remain most resistant to nitrite.

The results in Table 2 demonstrate the effect of pH on the inhibition of *Cl. botulinum* by nitrite. Thus at pH 5.5, none of the strains grew in the presence of 50 mg/kg nitrite whereas at pH 7.0 all strains grew in the presence of 200 mg/kg nitrite.

The results shown in Table 3 of the effect of nitrite on the growth of *Cl. botulinum* in PMB indicate, similar trends and differences to that noted for strains grown in MRCM although all strains grew to higher nitrite concentrations. This could either have been the result of the different growth substrate, redox potential, lower analysed nitrite initially and about a 50% drop in nitrite concentration during storage for 28 days or most likely to the production of the inhibitory substance

Table 3. Effect of sodium nitrite on the growth of *Clostridium botulinum* at 20°C in PMB (pH 6.0) without sodium chloride.

Toxigenic type	No. of strains	Sodium nitrite (mg/kg)[1]			
		50	100	150	200[2]
A	6	6	6	6	6
B (proteolytic)	6	6	6	5	6
B (non-proteolytic)	6	6	5	5	1
E	6	6	6	3	4
F (proteolytic)	3	3	1	2	0
F (non-proteolytic)	3	3	2	2	1
G	1	1	1	1	1
Total strains	31	31	27	24	19

1. Calculated amount added.
2. Analytical results of 200 mg/kg sample after storage at 20°C: 0 days 185 mg/kg $NaNO_2$, 7 days 164 mg/kg $NaNO_2$, 14 days 109 mg/kg $NaNO_2$, 28 days 92 mg/kg $NaNO_2$.

Table 4. Effect of inoculum size on the inhibition of growth of *Clostridium botulinum* by sodium nitrite in PMB (pH 6.0) at 20°C without sodium chloride.

Toxigenic type	No. of strains	Inoculum	Sodium nitrite (mg/kg)[1]		
			50[2]	100[2]	200[2]
A	2	10^1	2	0	0
	2	10^3	2	1	1
	2	10^6	2	2	2
B (proteolytic)	2	10^1	2	0	0
	2	10^3	2	2	0
	2	10^6	2	2	2
B (non-proteolytic)	2	10^1	0	0	0
	2	10^3	2	0	0
	2	10^6	2	2	1
E	2	10^1	0	0	0
	2	10^3	2	0	0
	2	10^6	2	2	2
F (proteolytic)	2	10^1	1	0	0
	2	10^3	1	0	0
	2	10^6	2	2	1
F (non-proteolytic)	2	10^1	0	0	0
	2	10^3	2	0	0
	2	10^6	2	2	1
G	1	10^1	1	1	0
	1	10^3	1	0	0
	1	10^6	1	1	1

1. Calculated amount added.
2. Analytical results (mg/kg):

	50 mg/kg	100 mg/kg	200 mg/kg
0 day	38	93	168
28 days	0	45	91

(reported by Van Roon, this symposium) in MRCM containing nitrite which enhanced the antibotulinum activity of nitrite in MRCM.

Table 4 demonstrates the effect of inoculum size on the ability of nitrite to inhibit the growth of *Cl. botulinum.* The effects are most marked. Inoculum size is an important consideration in determining the antibotulinum activity of nitrite, but results of high inoculum levels should not be ignored as such inocula contain the small number of resistant cells which occur in any population of cells and might also be expected to occur in nature.

Inhibition by sodium chloride

Results obtained in MRCM and PMB are shown in Tables 5 and 6 respectively. They show precisely the same differences as were noted for sodium nitrite. Thus *Cl. botulinum* type A and proteolytic type B and F strains generally grow in the presence of at least 4.5–6.0% sodium chloride at 20 or 25°C while most of the non-proteolytic and psychrotrophic types are inhibited by 4.5% sodium chloride in MRCM but some grow in PMB in the presence of 6% sodium chloride. At 20 and 25°C the effect of temperature on the level of sodium chloride inhibiting growth was only marked for the mesophilic proteolytic strains and was not apparent for the psychrotrophic strains; similar differences of the effect of temperature on the a_W tolerance of type A, B and E strains can be observed in the results reported by

Table 5. Effect of sodium chloride on the growth of *Clostridium botulinum* in MRCM (pH 6.0) at 15, 20 and 25°C.

Toxigenic type	Storage temperature (°C)	No. of strains	Sodium chloride (% w/v)			
			1.5	3.0	4.5	6.0
A	15[1]	6	1	1	0	0
B (proteolytic)	15	6	3	2	0	0
B (non-proteolytic)	15	6	4	1	0	0
E	15	6	6	2	0	0
F (proteolytic)	15	3	3	0	0	0
F (non-proteolytic)	15	3	3	0	0	0
A	20**	49	49	45	38	6
B (proteolytic)	20	10	10	10	7	0
B (non-proteolytic)	20	9	9	8	0	0
E	20	61	60	54	3	0
F (proteolytic)	20	4	4	2	1	0
F (non-proteolytic)	20	4	4	2	0	0
A	25[2]	50	50	50	50	38
B (proteolytic	25	11	11	11	10	7
B (non-proteolytic)	25	9	9	9	0	0
E	25	61	61	61	2	0
F (proteolytic)	25	4	4	4	2	0
F (non-proteolytic)	25	4	4	4	1	0

1. Stored for 56 days.
2. Stored for 28 days.

Table 6. Effect of sodium chloride on the growth of *Clostridium botulinum* at 20°C in PMB (pH 6.0).

Toxigenic type	No. of strains	Sodium chloride (% w/v)			
		1.5	3.0	4.5	6.0
A	6	6	6	5	5
B (proteolytic)	6	6	6	5	3
B (non-proteolytic)	6	6	6	2	0
E	6	6	6	4	3
F (proteolytic)	3	3	3	2	2
F (non-proteolytic)	3	3	3	2	0
G	1	1	1	0	0
Total strain	31	31	31	20	13

Table 7. The effect of sodium chloride on the growth of *Clostridium botulinum* at 20°C in MRCM (pH 5.0, 5.5, 6.0 and 7.0).

Toxigenic type	pH	Sodium chloride (% w/v)				
		0	1.5	3.0	4.5	6.0
A	5.0	–[1]	–	–	–	–
	5.5	+[2]	+	–	–	–
	6.0	+	+	+	+	–
	7.0	+	+	+	+	+
B (proteolytic)	5.0	–	–	–	–	–
	5.5	+	+	+	–	–
	6.0	+	+	+	+	–
	7.0	+	+	+	+	+
E	5.0	–	–	–	–	–
	5.5	+	+	–	–	–
	6.0	+	+	+	–	–
	7.0	+	+	+	–	–

1. No growth.
2. Growth.

Ohye & Christian (1967). At 15°C, 4.5% sodium chloride was inhibitory to all strains and 3% sodium chloride inhibitory to most.

Table 7 which is based on data presented in the paper by Baird-Parker & Freame (1967) shows that like nitrite, the inhibitory effect of sodium chloride against *Cl. botulinum* is highly dependent on pH.

Inhibition by sodium nitrite in the presence of sodium chloride

A synergistic effect of combinations of sodium chloride and sodium nitrite against *Cl. botulinum* is apparent from the results presented in Table 8. At both 20 and 30°C only one strain out of the 61 strains tested was able to grow in a

Table 8. Effect of sodium chloride in the presence of sodium nitrite on the growth of *Clostridium botulinum* at 20°C and 30°C in MRCM (pH 6.0).

Toxigenic type	Storage temperature (°C)	No. of strains	3% w/v sodium chloride + sodium nitrite (mg/kg)[1]				4.5% w/v sodium chloride + sodium nitrite (mg/kg)[1]			
			0	50	100	200	0	50	100	200
A	20	40	36	7	4	1	33	5	0	0
	30	10	10	7	4	1	10	4	1	0
B (proteolytic)	20	8	8	1	0	0	7	1	0	0
	30	4	4	3	0	0	4	1	0	0
B (non-proteolytic)	20	1	1	0	0	0	0	0	0	0
E	20	11	10	1	0	0	3	0	0	0
	30	6	6	0	1	0	1	0	0	0
F (non-proteolytic)	20	1	0	0	0	0	0	0	0	0
	30	1	1	0	0	0	1	0	0	0

1. Calculated amounts added.

combination of 3% sodium chloride and 200 mg/kg sodium nitrite in MRCM at pH 6.0. In the presence of 4.5% sodium chloride only one strain grew at 30°C in the presence of 100 mg/kg sodium nitrite. Very similar results are reported by Roberts & Ingram (1973) who in a study of single strains of *Cl. botulinum* types A, B, E and F found that at pH 6.0 a combination of 4% sodium chloride and 200 mg/kg sodium nitrite inhibited the growth of all strains at 35°C. However, these tests are on bacteriological media and preliminary data (unpublished) shows that higher concentrations of salt and nitrite will be required to inhibit the growth of high inocula (10^6/ml) of *Cl. botulinum* at high ambient temperatures in PMB.

Inhibition by sodium nitrite and sodium chloride in the presence of ascorbic acid

We have noted in previous unpublished studies that at pH 7.0 and an incubation temperature of 37°C 100–200 mg/kg sodium nitrite is inhibitory to all strains growing in MRCM containing 1% L-ascorbic acid; in MRCM and other media not containing ascorbic acid the inhibitory level is 1000–1500 mg/kg (unpublished and Perigo & Roberts, 1968). Results obtained at pH 6.0 in MRCM with and without 1% ascorbic acid are shown in Table 9. There was no effect of L-ascorbic acid on

Table 9. Effect of sodium nitrite and sodium chloride on the growth of *Clostridium botulinum* at pH 6.0 in MRCM and in MRCM + 1.0% L-ascorbic acid at 20°C.

Toxigenic type	No. of strains	Growth medium[1]	Sodium nitrite (mg/kg)					Sodium chloride (% w/v)			
			0	50	100	150	200	1.5	3.0	4.5	6.0
A	4	1	4	0	0	0	0	4	4	1	0
	4	2	4	4	4	1	1	4	4	2	0
B (Proteolytic)	4	1	4	0	0	0	0	4	4	3	0
	4	2	4	4	4	1	0	4	4	2	0
B (Non-proteolytic)	4	1	4	0	0	0	0	4	4	0	0
	4	2	4	2	1	0	0	4	4	0	0
E	4	1	4	0	0	0	0	4	4	1	0
	4	2	4	3	1	0	0	4	4	0	0
F (Proteolytic)	1	1	1	0	0	0	0	1	1	0	0
	1	2	1	1	1	1	1	1	1	0	0
F (Non-proteolytic)	2	1	2	0	0	0	0	2	2	0	0
	2	2	2	1	0	0	0	2	2	0	0

1. 1 = MRCM + 1.0% L-ascorbic acid, 2 = MRCM.

Table 10. Effect of sodium nitrite on the growth of *Clostridium botulinum* in the presence of ascorbic acid after 20 days at 20°C in MRCM (pH 6.0) without sodium chloride.

Toxigenic type	No. of strains	0 mg/kg ascorbic acid				1 000 mg/kg ascorbic acid			
		0	50	100	200	0	50	100	200
A	2	2	2	1	0	2	1	0	0
B (proteolytic)	2	2	2	1	0	2	1	0	0
B (non-proteolytic)	2	2	1	0	0	2	0	0	0
E	2	2	1	0	0	2	1	0	0
F (proteolytic)	1	1	1	0	0	1	0	0	0
F (non-proteolytic)	2	2	0	0	0	2	0	0	0
G	1	1	1	1	0	1	0	0	0
Total	12	12	8	3	0	12	3	0	0

sodium chloride tolerance but a very marked effect on tolerance to sodium nitrite, i.e. no strain out of the 38 strains tested grew in the presence of 50 mg/kg nitrite which was the lowest concentration tested. There can be little doubt that the principle reason for this pronounced inhibition of *Cl. botulinum* in MRCM containing ascorbic acid is due to the formation of a complex of the type described by Van Roon in this symposium. Similar results are obtained in MRCM containing 0.1% L-ascorbic acid (Table 10) but not in PMB containing the same level of L-ascorbic acid (Table 11). In the latter medium, the stimulatory effect of L-ascorbic acid on the antibotulinum activity of nitrite is not observed. This would be expected if the inhibitory factor formed in MRCM containing L-ascorbic acid and nitrite was a Van Roon type inhibitor as such inhibitors are inactivated by meat particles, a main component of PMB. It is important to note that whereas there is no stimulatory effect of ascorbic acid on the antibotulinum activity of nitrite in PMB there is no reduction in activity in the presence of ascorbic acid.

Table 11. Effect of ascorbic acid on the inhibition of *Clostridium botulinum* by sodium nitrite after 21 days at 20°C in PMB (pH 6.0 – 6.2).

Toxigenic type	No. of strains	No ascorbic acid, sodium nitrite (mg/kg)[1]					Ascorbic acid 1 000 mg/kg, sodium nitrite (mg/kg)[1]			
		0	300	400	500	600	0	300	400	500
A	3	3	3	3	3	3	3	3	3	3
B (Proteolytic)	2	2	2	2	2	2	2	2	2	2
B (Non-proteolytic)	2	2	2	2	1	0	2	2	2	0
E	2	2	2	1	1	1	2	2	2	1
F (proteolytic)	2	2	2	2	2	1	2	2	2	1
F (Non-proteolytic)	2	2	2	2	0	0	2	2	2	0
G	1	1	1	1	0	0	1	1	1	0
Total strains	14	14	14	13	9	7	14	14	14	7

1. Calculated amount added.

Inoculation trial with laboratory cured bacon

Results obtained of a storage trial at 25°C in which *Cl. botulinum* was inoculated into packs of bacon containing different levels of nitrite in the presence and absence of L-ascorbic acid are shown in Table 12. Bacon prepared with a level of c. 4.5% sodium chloride on water phase and an initial level of 500 mg/kg nitrate supported good growth and toxin formation by *Cl. botulinum,* i.e. 9 out of 10 packs were toxic after storage for 14 and also 28 days. This bacon was judged by the bacteriologists assessing the results as spoiled (putrid) by 14 days which was the first time the bacon was examined after inoculation. In contrast to this bacon containing the same levels of salt and nitrate but in addition 100 or 200 mg/kg nitrite did not readily support growth or toxin formation by *Cl. botulinum* either in the presence or absence of L-ascorbic acid. The bacon was also much more shelf-stable than the nitrite-free bacon and only slight souring of a few packs was observed after 28 days storage. Only one out of 40 packs examined after storage for 14 days and the same after 28 days, were toxic and contained cells of *Cl. botulinum*. The two toxic packs contained only 3.0% salt on water phase when tested and were therefore low in curing salts. *Cl. botulinum* type B cells were present at the 10^5-10^6/g level in the toxic packs but no cells of any toxigenic type were detected in the non-toxic packs; limit of detection was a most probable number count of 0.36 cells/g. Thus we conclude from this preliminary experiment that 0.1% ascorbic acid does not affect the antibotulinum activity of curing salts in bacon.

Table 12. Effect of L-ascorbic acid on *Clostridium botulinum* in vacuum packed bacon stored at 25°C.

Cure[1] additive	Number of days stored at 25°C	
	14	28
0 mg/kg ascorbic acid 0 mg/kg $NaNO_2$	9[2]	9
0 mg/kg ascorbic acid 100 mg/kg $NaNO_2$	0	1[3]
1 000 mg/kg ascorbic acid 100 mg/kg $NaNO_2$	0	0
0 mg/kg ascorbic acid 200 mg/kg $NaNO_2$	0	0
1 000 mg/kg ascorbic acid 200 mg/kg $NaNO_2$	1[3]	0

1. Bacon contained an average of 4.4% salt on water phase (range 3.0 – 5.8%) and an initial level of 500 mg/kg sodium nitrate (pH of bacon 5.6 – 5.8).
2. Number of toxic packs out of 10 tested (5 inoculated with proteolytic and 5 with non-proteolytic strains of *Cl. botulinum*).
3. Contained type B toxin (non-proteolytic inoculum).

References

Baird-Parker, A. C., 1969. In: The Bacterial Spore, G. W. Gould & A. Hurst (eds), p. 517–548. Academic Press, New York and London.

Baird-Parker, A. C. & B. Freame, 1967. Combined effect of water activity, pH and temperature on the growth of *Clostridium botulinum* from spore and vegetative cell inocula. J. appl. Bact. 30: 420–429.

Christiansen, L. N., R. W. Johnston, D. A. Kautter, J. W. Howard & W. J. Aunan, 1973. Effect of nitrite and nitrate on toxin production by *Clostridium botulinum* and on nitrosamine formation in perishable canned comminuted cured meat. Appl. Microbiol. 25: 357–362.

Emodi, A. C. & R. V. Lechowich, 1969. Low temperature growth of type E *Clostridium botulinum* spores. 1. Effect of sodium chloride, sodium nitrite and pH. J. Fd Sci. 34: 78–81.

Emodi, A. C. & R. V. Lechowich, 1969. Low temperature growth of type E *Clostridium botulinum* spores. 2. Effects of solutes and incubation temperature. J. Fd Sci. 34: 82–87.

Gimenez, D. F. & A. S. Ciccarelli, 1970. Another type of *Clostridium botulinum*. Zentbl. Bakt. I. Abt. Orig. 215: 221–226.

Leistner, L., 1973. Welche Konsequenzen hätte ein Verbot oder eine Reduzierung des Zusatzes von Nitrat and Nitritpökelsalz zu Fleischerzeugnissen? –Aus mikrobiologischer Sicht. Fleischwirtschaft p. 371–375.

Ohye, D. F. & J. H. B. Christian, 1967. In: Botulism 1966. Proc. 5th Int. Symp. Fd Microbiol, July 1966, M. Ingram & T. A. Roberts (eds). p. 217–223. Chapman & Hall, London.

Pace, P. J. & E. R. Krumbiegel, 1973. *Clostridium botulinum* and smoked fish production: 1963–1972. J. Milk Fd Technol. 36: 42–49.

Perigo, J. A. & T. A. Roberts, 1968. Inhibition of clostridia by nitrite. J. Fd Technol. 3: 91–94.

Pivnick, H., H. W. Barnett, H. R. Nordin & L. J. Rubin, 1969. Factors affecting the safety of canned cured shelf-stable luncheon meat inoculated with *Clostridium botulinum*. Can. Inst. Fd Technol. J. 2: 141–147.

Roberts, T. A. & M. Ingram. 1973. Inhibition of *Clostridium botulinum* at different pH values by sodium chloride and sodium nitrite (submitted to J. Fd Technol).

Segner, W. P., C. P. Schmidt & J. K. Baltz, 1966. Effect of sodium chloride and pH on the outgrowth of spores of type E *Clostridium botulinum* at optimal and suboptimal temperatures. Appl. Microbiol. 4: 49–54.

Spencer, R., 1967. In: Botulism 1966. Proc. 5th Int. Symp. Fd Microbiol., July 1966. M. Ingram & T. A. Roberts (eds) p. 123–135. Chapman and Hall, London.

Discussion

Experimental details

The end pH in the experiments (initial pH 6.0) was 6.0–6.1. The end nitrite concentrations in MRCM + 1% ascorbic acid were 15–25% of the initial concentrations. At the 0.1% ascorbic acid level, the nitrite concentrations after 20 days at 20°C and initial levels of 50, 100 and 200 mg/kg nitrite were 4.4, 13 and 44, respectively and the corresponding ascorbic acid levels were 44, 37, and 65 mg/kg.

The experimental effect of ascorbic acid

The author does not think that one can explain the differences observed in MRCM with and without ascorbic acid solely in terms of differences in oxidation and reduction potential. It is likely that an inhibiting substance of the type II that van Roon described in his paper is formed.

The inoculum level

The author agrees that it is necessary to repeat experiments at lower inoculum levels. However, it is important not to disregard results obtained at high inoculum levels as populations of cells may contain only a few number of resistant organisms.

Dying out of spores

In the experiment reported on bacon no *Cl. botulinum* cells (limit of detection < 0.36 organisms/g) were detected in all non-toxic packs examined after storage of the inoculated bacon for 14 and 28 days. All toxic packs contained $10^5 - 10^6$/g cells of *Cl. botulinum*. The author believes that the *Cl. botulinum* cells (spores and vegetative cells) had died out in the non-toxic packs during the first few days of storage when the nitrite levels were high. We do not know what could have happened if further spores of *Cl. botulinum*, had been inoculated into the bacon after it had been stored and nitrite levels were low. Similar die out of spores has been observed in experiments with canned meat products containing curing salts.

Proc. int. Symp. Nitrite Meat Prod., Zeist, 1973. Pudoc, Wageningen.

Inhibition of bacterial growth in model systems in relation to the stability and safety of cured meats

T. A. Roberts

ARC Meat Research Institute, Langford, Bristol, BS18 7DY, UK

Abstract

The interaction of inhibitory factors present in cured meats are known to be responsible for the bacteriological safety and stability of those products. Methods for the quantitative evaluation of these interactions are discussed with particular reference to pH, sodium chloride, nitrite and nitrate. Storage temperature is also considered.

Introduction

It is not the purpose of this paper to review the literature on the inhibition of growth of *Clostridium botulinum* in relation to cured meats: such reviews have already been provided by Ingram and Baird-Parker (this Symposium). I propose instead to describe the ways in which we have attempted to evaluate the extent of interactions between factors known to be important in inhibiting *Cl. botulinum* and *Staphylococcus aureus*. When an additional factor is added to a system its effect may be independent of those already acting, in which case it will be additive; or it may display synergistic or antagonistic interactions with one or more of the other factors. All such cases are included in the description 'interaction' in this paper.

Riemann (1963) demonstrated clearly that interactions of factors were significant in maintaining the bacterial stability of cured meat products, but made no attempt to determine their extent. Even a cursory consideration reveals the complexity of the subject. Within the commercially acceptable ranges of concentration of sodium chloride and sodium nitrite, higher values are known to inhibit bacterial growth more effectively than low, and nitrite is more effective in acid than in alkaline conditions (Roberts & Ingram, 1966). At given pH and concentrations of salt and nitrite, poorer growth results at sub-optimal storage temperatures, or if very low numbers of cells are used in the inoculum. Heat damaged spores are more sensitive to inhibition by curing salts than are unheated spores (Roberts & Ingram, 1966; Roberts, Gilbert & Ingram, 1966).

Already these five factors, if investigated at four levels each, lead to 1024 conditions without considering replication.

Medium

In addition the medium in which the tests are to be made is crucially important. Pork macerate closely resembles the practical situation and is excellent in respect of growth, but problems arise in handling large amounts and in obtaining uniform batches. Partial spoilage of raw pork macerate during, or even before, the test can be a real problem, and cooked macerate is in some respects easier to handle. If cooked macerate is used, the absence of the natural bacterial flora of pork might influence the effects of curing salts since no bacterial competition will be present. Different commercial products formulations should be considered, since surprisingly little is known of the effects on bacterial growth in the presence of curing salts of polyphosphates (which are difficult to study precisely because of variability in chemical composition), ascorbic acid, ascorbate, erythorbate or sugars, all of which may be used commercially and add to the interactions to be tested. Hence, for ease of experimentation, we initially used a bacteriological medium shown to give results very similar to pork macerate and are currently testing our conclusions in pork macerate itself.

Bacteria

A further problem arises in the choice of bacteria for the test. Clearly the toxigenic types of *Cl. botulinum* causing human botulism, types A, B, E and F should be tested, although type E may be omitted since it is associated with the marine environment and is already known to be much more sensitive than the other types to curing salts (Roberts & Ingram, 1973). However, within each toxigenic type the choice of bacterial strain could be important, since there is no guarantee that the strain most resistant under one combination of screening conditions will also be the most resistant under other conditions. Virtually no information is available on the variability of resistance of *Cl. botulinum* to salt and nitrite among strains and this all leads to considerable effort on screening. If spores are to be used, the manner in which they have been prepared and stored may be important, and, ideally it should be established that 'laboratory' spores are identical in relevant properties to those which occur in nature as contaminants of meat products. Again remarkably little relevant information is available.

It would also be prudent to make similar studies on other bacteria causing food poisoning e.g. *Cl. welchii, Staphylococcus* and *Salmonella,* although such studies could initially be a simple test of whether conditions inhibitory to *Cl. botulinum* are capable of inhibiting them. A range of strains of 'putrefactive anaerobes' could also be tested to determine whether they behave as 'indicators' of possible growth of *Cl. botulinum,* as they do in heat processed foods.

The problem is, therefore, enormous, and unlikely to be covered fully by any one laboratory, and it is time that a serious attempt was made to investigate inter-laboratory variation in experiments of this nature, with a view to sharing the work load. At present workers in different laboratories are using different media, methods and strains, and consequently it is difficult to correlate their results.

We have concentrated our effort on relatively simple systems which seemed more likely to be reproducible and suited to the evaluation of the complex of interactions.

Evaluation of the inhibitory effects of mixtures of curing salts: method

Counts of viable bacteria in agar media

The simplest approach is to count the viable bacteria under test in agar media with a range of pH values and containing different concentrations and combinations of NaCl and $NaNO_2$ (or other relevant additives) and to compare them with counts in the same media containing no additives. We have used this method to show that a 12-D concept equivalent to that used in heat processing could be developed for pasteurized cured meats (Ingram & Roberts, 1971). These data also showed the absence of any effect of a mixture of NaCl and $NaNO_2$ on the heat resistance of spores of *Cl. botulinum* and demonstrated that the absence of growth was due to the inhibition of viable spores by the curing salts in the growth medium. This method is convenient when the incubation temperature is near optimal and growth consequently rapid, but at lower incubation temperatures, the slower growth necessitates prolonged incubation which is inconvenient when using anaerobe jars. The opening of anaerobe jars to examine the contents during prolonged incubation may cause changes in the redox conditions detrimental to spores surviving but remaining dormant, which could result in the test conditions being adjudged more inhibitory than in fact they are.

Growth – no growth tests

We have investigated in some detail the interactions of pH, NaCl and $NaNO_2$ in a medium based on a casein hydrolysate ('Trypticase', BBL) using *Cl. botulinum* types A, B, E and F. These tests were made in airtight screw-capped bottles, in which prolonged incubation without breaking anaerobiosis was possible.

Results and discussion

pH, NaCl, $NaNO_2$, storage temperature

In an initial experiment using spores of *Cl. botulinum* type B growth (at 35°C) was sporadic and not evidently related to the concentration of salt or nitrite. The reason for this was not discovered, and, although subsequent spore crops behaved more reproducibly and in a less unexpected manner, we eventually decided to work with inocula of vegetative cells arguing that, since toxin is elaborated during vegetative growth, inhibition of such growth would be an adequate safeguard against toxin production.

Results using vegetative cells of *Cl. botulinum* types A, B, E and F are in press (Roberts & Ingram, 1973). The four diagrams of that paper have been drawn together as one, representing the greatest growth of *Cl. botulinum* type A, B, E or F (Fig. 1). The same data may be expressed in tabular form (Table 1). In less extensive experiments, addition of 0.3% of a polyphosphate in common commer-

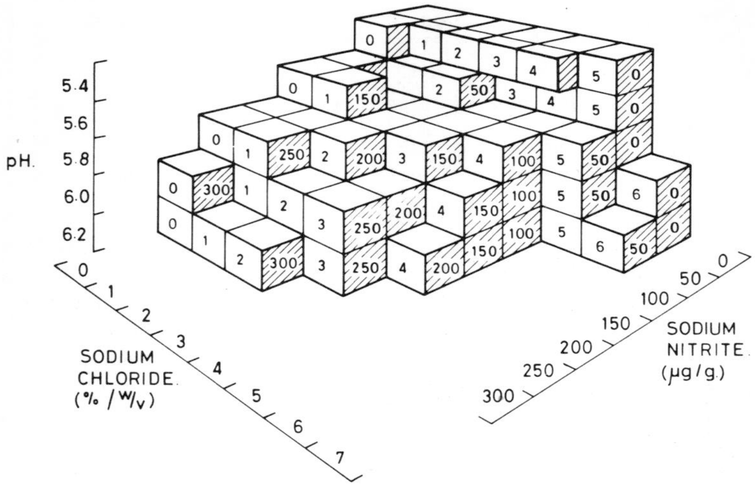

Fig. 1. The effect of pH, NaCl and $NaNO_2$ on growth of *Cl. botulinum* at 35°C. Composite diagram of types A, B, E and F. Blocks represent growth at the concentrations and pH indicated.

cial usage did not greatly affect the inhibitory levels of NaCl and $NaNO_2$. It is evident from Fig. 1 that the limit of the possibility of bacterial growth is expressed by a surface which bounds all effective concentrations of acid, NaCl and $NaNO_2$. Once this surface is defined, it is a simple matter to test the effect on it of other factors. After suitable mathematical transformation of the units of the axes, the surface describing the bulked data for types A, B, E and F approximates to $\frac{1}{8}$ of a sphere (Fig. 2) (although this might not be so when more data from current experiments at pH 7 are included). Although a diagram of this nature is of little

Table 1. Effect of sodium nitrite, sodium chloride and pH on vegetative inoculum of *Cl. botulinum* types A, B, E and F at 35°.

NaCl (% w/v)	a_W	Highest concentration of nitrite (mg/1) in which growth of any *Cl. botulinum* occurred at pH:				
		6.2	6.0	5.8	5.6	5.4
0	0.999	300+	300+	250	150	50
1	0.994	300	250	250	150	–
2	0.989	300	250	200	50	–
3	0.983	250	250	150	–	–
4	0.977	200	150	100	–	–
5	0.971	50	50	50	–	–
6	0.965	50	–	–	–	–

Note: The levels of nitrite are those initially added to the medium. – no growth with 50 mg/l sodium nitrite.

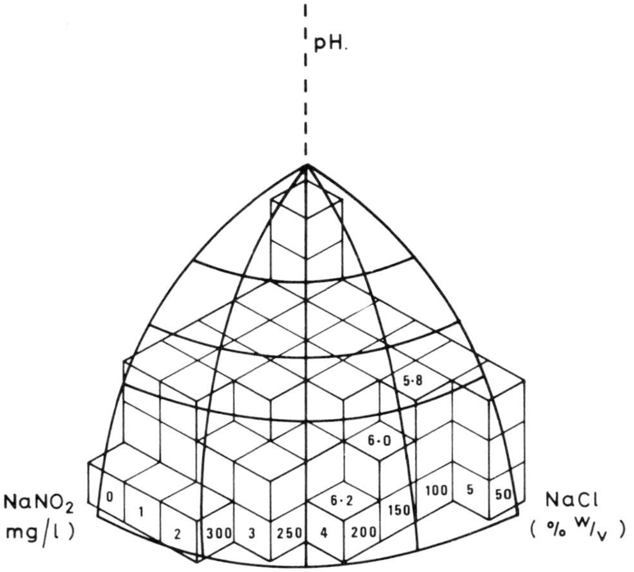

Fig. 2. Plot of growth zone of *Cl. botulinum* inside a $\frac{1}{8}$ sphere (using transformed co-ordinates).

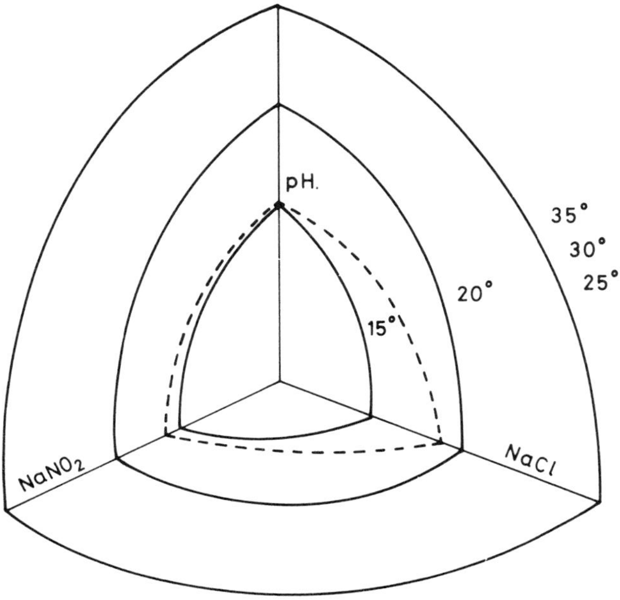

Fig. 3. Inhibition of *Cl. botulinum* by NaCl × $NaNO_2$ × pH. Effect of incubation temperature.

value to research or commerce, it may readily be converted into a nomogram which is simple to use.

We know that *Cl. botulinum* is inhibited more readily by curing salts when the incubation temperature is reduced, and under such conditions new definitions of the limiting surfaces will be needed. Preliminary tests indicate that temperatures from 25 to 35°C give essentially the same results; below 25°C temperature becomes increasingly important, and below 20°C inhibition of growth of *Cl. botulinum* may be achieved with commercially acceptable concentrations of NaCl and $NaNO_2$ provided that the pH is near 6.0. Such changes are represented diagrammatically in Fig. 3.

Similar detailed data on the effect of incubation temperature are available for the inhibition of *Staphylococcus aureus* by NaCl and $NaNO_2$ and resemble those described above (Bean, 1972; Bean & Roberts, in preparation) i.e. differences between 35 and 25°C are relatively small, but incubation at lower temperatures results in markedly greater inhibition.

All the above indicate that concepts of 'limiting water activity' are erroneous in not taking into account at least the pH of the growth medium.

$NaNO_3$

The effect of nitrate in the stabilizing of pasteurized cured meats has been little investigated, it is generally considered to be equivalent, weight for weight, to sodium chloride; however, Riemann (1963) suggested that it could be more than this.

Unpublished data kindly supplied by J. A. Perigo (Metal Box Co. Ltd., London) have been grouped and re-plotted in a manner resembling the data for *Cl. botu-*

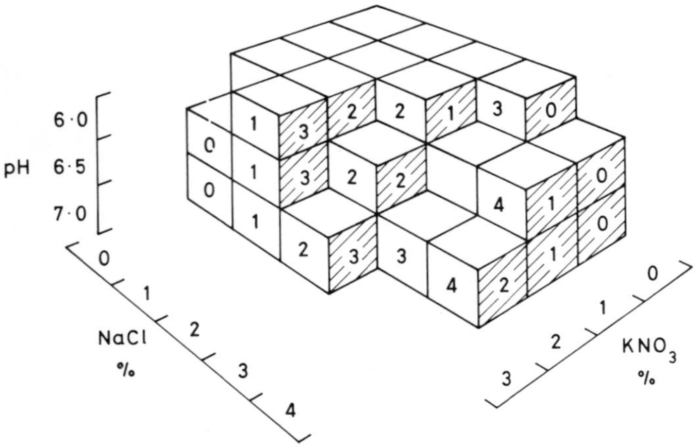

Fig. 4. The effect of pH, NaCl and KNO_3 on growth of *Cl. sporogenes* spores at 30°C ($NaNO_2$ absent).

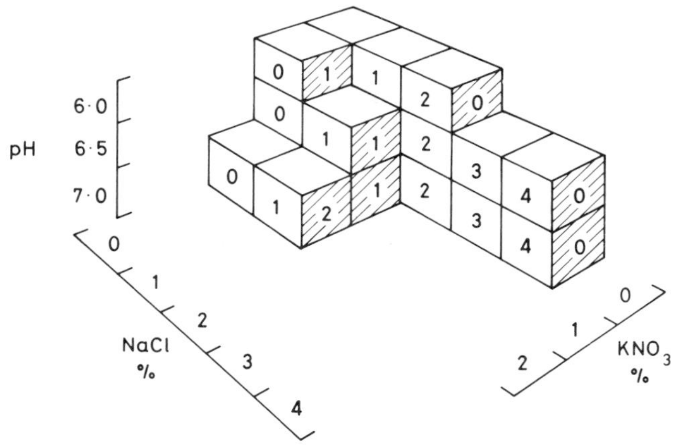

Fig. 5. The effect of pH, NaCl and KNO_3 on growth of *Cl. sporogenes* spores at 30°C ($NaNO_2$ 10 mg/l).

linum. The basal growth medium was that used in Perigo's studies of the heated nitrite inhibitor, and in these experiments, too, the nitrite was heated in the medium. Hence the concentrations limiting growth cannot be compared directly with those presented in Fig. 1 and Table 1 for *Cl. botulinum.* Three so far unconfirmed conclusions may be drawn:

1. the interaction of NaCl × KNO_3 reported by Riemann (1963) is demonstrated clearly both in the absence of nitrite (Fig. 4) and in the presence of 10 mg/1 of heated nitrite (Fig. 5).

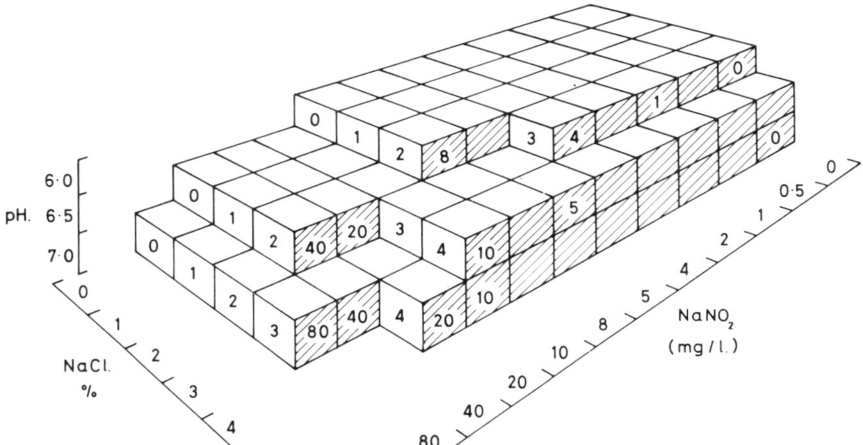

Fig. 6. The effect of pH, NaCl and $NaNO_2$ on growth of *Cl. sporogenes* at 30°C (KNO_3 absent).

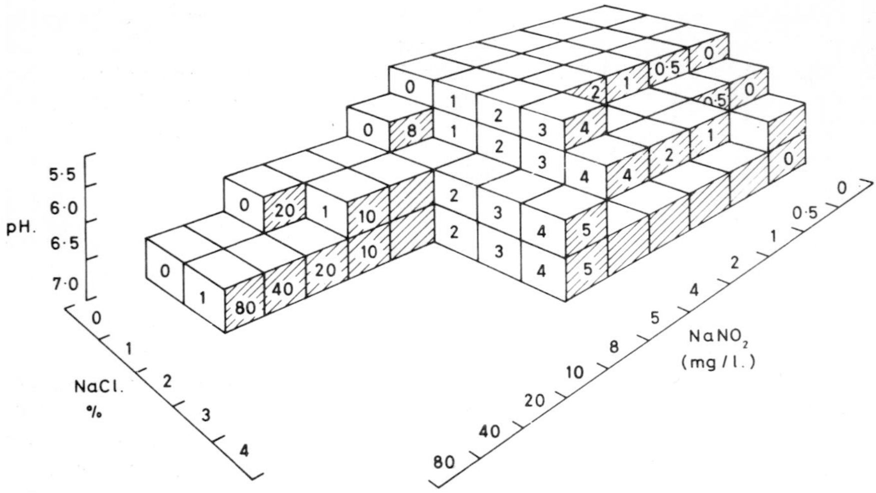

Fig. 7. The effect of pH, NaCl and $NaNO_2$ on growth of *Cl. sporogenes* at 30°C (KNO_3 1%).

2. if the data are arranged in 3-dimensions(pH × NaCl × $NaNO_2$)with and without 1% KNO_3, growth occurred, at constant pH, in higher concentrations of NaCl and $NaNO_2$ when nitrate was absent than when present, indicating an additional inhibitory effect of nitrate (Figs. 6 and 7).
3. further plotting of the data assuming 1% KNO_3 to have an effect equivalent to 1% NaCl (i.e. comparing 2% NaCl with 1% NaCl + 1% KNO_3), less growth occurred when nitrate was present (Fig. 8). The effect of 1% KNO_3 on water activity (a_W) is much smaller than 1% NaCl, hence a system containing 1% NaCl + 1% KNO_3 will

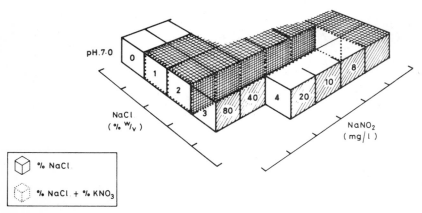

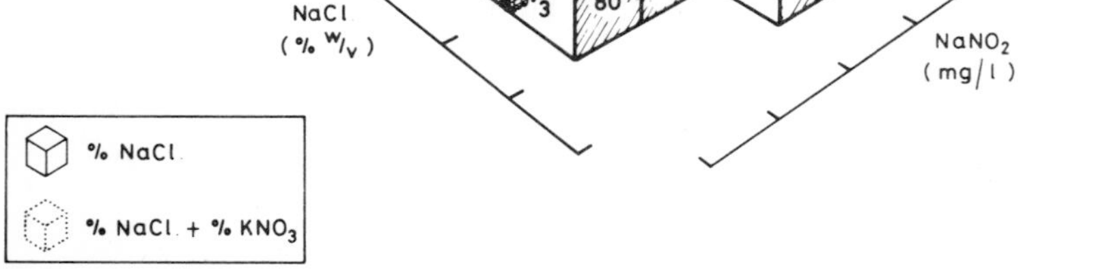

Fig. 8. The effect of KNO_3 on inhibition of *Cl. sporogenes* at 30°C by NaCl and $NaNO_2$.
Solid line = KNO_3 absent.
Broken line = sum of the concentration (%) NaCl + KNO_3.

have a higher a_W than a system containing 2% NaCl. Yet the former is the more inhibitory, indicating an effect of KNO_3 greater than can be accounted for by a_W alone. This occurred at pH 6.0, 6.5 and 7.0, although responses were not identical at these pH values. These data suggest that it might be unwise to insist upon the absence of nitrate from cured meats, at least until it has been established that products without nitrate are in no way less safe than those which contain it in low concentration.

Final remark

Viable counts by Most Probable Number (MPN) method

An alternative to viable counts in agar media is to make MPN counts in media containing different concentrations and combinations of curing salts. Although the method is inherently less accurate than plate counts, it has the advantage that incubation can be prolonged almost indefinitely, and examination for growth does not change the conditions of incubation. The results may be expressed conveniently to indicate the number of spores which are inhibited by a particular system. Results are accumulated as illustrated in Fig. 9, and when sufficient data become available, a surface may again be drawn covering the numbers of spores inhibited by different combinations of NaCl and $NaNO_2$. If experiments are made in media of different pH values, or at constant pH but incubated at different temperatures, a diagram similar to Fig. 10 will evolve, and it seems possible that they may be used to predict the safety margin of cured meat systems, in much the same way that the 12-D concept led to a safe canning process. A disadvantage of the method is that relatively large numbers of cells (or spores) are required to avoid unrealistic extrapolation to the level of 10^{12} spores. We are now concentrating on this type of

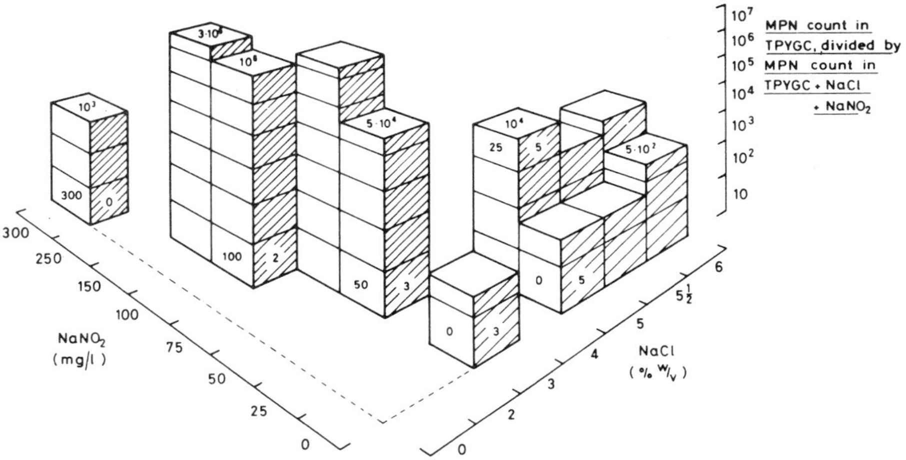

Fig. 9. Inhibition of *Cl. botulinum* in TPYGC (pH 6.2).

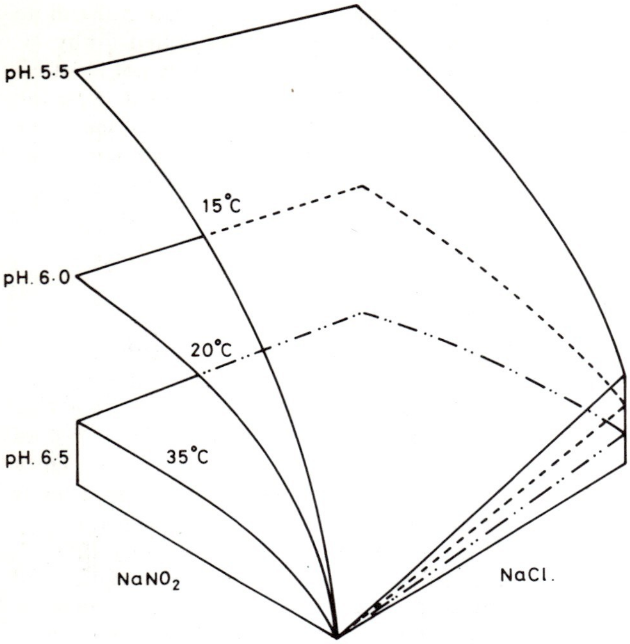

Fig. 10. Effect of pH or incubation temperature on inhibition of bacterial growth.

experiment, and particularly upon means of automating the rather tedious process of making very large numbers of MPN counts.

These investigations are beginning, for the first time, to lead to an understanding of the quantitative relationships between factors important in cured meats.

References

Bean, P. G., 1972. Growth and recovery of damaged *Staphylococcus aureus* NCTC 10652. M.Sc. thesis Bristol University.

Bean, P. G. & T. A. Roberts, 1973. The effect of incubation temperature and damage by heat and gamma-radiation on growth of *Staph. aureus* NCTC 10652 in a medium containing curing salts (in preparation).

Gonzalez, C. & C. Gutiérrez, 1972. Intoxication botulique humaine par *Clostridium botulinum* B.

Ingram, M. & T. A. Roberts, 1971. Application of the D-concept to heat treatment involving curing salts. J. Fd Technol. 6: 21–28.

Riemann, H., 1963. Safe heat processing of canned cured meats with regard to bacterial spores. Fd Technol., Champaign 17: 39.

Roberts, T. A. & M. Ingram, 1966. The effect of sodium chloride, potassium nitrate and sodium nitrite on recovery of heated bacterial spores. J. Fd Technol. 1: 147–163.

Roberts, T. A. & M. Ingram, 1973. Inhibition of growth of *Cl. botulinum* at different pH values by sodium chloride and sodium nitrite. J. Fd Technol. (in press).

Roberts, T. A., R. J. Gilbert & M. Ingram, 1966. The effect of sodium chloride on heat resistance and recovery of heated spores of *Cl. sporogenes* (PA $3679/S_2$). J. appl. Bact. 29: 549–555.
Sebald, M., 1970. Sur le botulism en France de 1956 à 1970. Bull. Acad. Nat. Méd. 154: 703–707.

Discussion

Petri dishes in plastic pouches

The use of Petri dishes sealed in plastic pouches containing alkaline pyrogallol as an absorbent was suggested as an alternative incubation method. Prolonged incubation and periodic examination is then possible without disturbing anaerobiosis.

Duration of incubation

The diagram obtained at near-optimal growth temperatures (35°C) using vegetative cell inocula (Fig. 1) did not alter greatly after 2 weeks' incubation, although incubation was continued for 3 months.

Nitrate concentrations

The nitrate concentrations in Figs. 4–8 are high, and tests should be made with commercially realistic concentrations to determine whether it has any antimicrobial effect prior to considering its prohibition.

Temperature dependence of pK of HNO_2

Experiments at different incubation temperatures must take into account that the pK of HNO_2 (presuming this to be the active agent) rises considerably with temperature.

Proc. int. Symp. Nitrite Meat Prod., Zeist, 1973. Pudoc, Wageningen.

Minimum nitrite concentrations for inhibition of clostridia in cooked meat products

A. B. G. Grever

Central Institute for Nutrition and Food Research TNO, Utrechtseweg 48, Zeist, the Netherlands.
Dept.: Netherlands Centre for Meat Technology

Abstract

Emulsions of cooked sausage and of liver sausage prepared in a factory or a laboratory were tested for the amount of nitrite required to prevent growth of clostridia after pasteurization at 80°C and storage for 5 weeks at 24°C.

Endogenous and inoculated spores, and spores from contaminating spore-bearing materials were involved. It was concluded that 200 mg/kg nitrite must be added, provided that the maximum value of pH is 6.2.

Also the amount of nitrite needed for sterilized sausage emulsions was determined. Inoculated sausage emulsions heated at F_0[1] = 0.5 often showed inhibition without nitrite. This could be due to the inhibitory action of salt on spores damaged by heat.

Thus for meat products, heated in F_0 = 0.5, addition of 100 mg/kg sodium nitrite is sufficient to prevent growth of clostridia.

Simultaneously, growth possibilities for bacilli in pasteurized and sterilized sausage were studied. To prevent growth of bacilli more nitrite is required than to prevent growth of clostridia. The investigation also provided information on the degree of nitrite depletion during heating and storage.

Introduction

In order to restrict the use of nitrite in food as far as possible there is a need for information on the minimum nitrite concentration needed in meat products to prevent growth of clostridia. This was investigated for pasteurized factory-made sausages.

At least 200 mg/kg nitrite must be added when the pH is not above 6.2. This was verified with laboratory-made liver sausages contaminated in different ways. It was studied whether it was possible to decrease the required amount of nitrite by lowering the pH.

It could be expected that in appertized and sterilized meat products less nitrite is needed to prevent growth of clostridia. This was investigated for two meat products: liver paste and luncheon meat.

1. See p. 67.

Experimental

Factory-made emulsions

Six factories were requested to prepare emulsions of cooked sausage and of liver sausage in the usual way but without nitrite and salt. These were added at the Institute: the salt to a brine percentage of 3.5% and the nitrite to amounts of 0, 50, 100 and 200 mg/kg.

Small cans (size 76 × 35 mm) were filled with the emulsions, pasteurized (in the centre 10 min at 80°C) and stored at 24°C. After 1, 3 and 5 weeks the cans were tested for growth of spore forming bacteria. Residual nitrite content was estimated at the same time.

Laboratory-made liver sausage emulsions

Liver sausage emulsions were prepared in the usual way. The brine percentage was 3.5% and the amounts of added sodium nitrite were respectively 75, 125, 150 and 200 mg/kg. The pH was approximately 6.2. The emulsions were inoculated with various strains of clostridia and with spore bearing materials such as spices, soil, well-decayed cow-dung and dust. The emulsions were canned, pasteurized and stored just as for the factory-made products. After pasteurization the spore number was 30–300 per gram.

Non-inoculated emulsions were also included in the experiments. In spite of careful preparation under sanitary conditions the emulsions also contained some spores. The growth possibilities of these spores in the nitrite-containing emulsions were examined as well.

Laboratory-made cooked sausage with pH 5.8

A cooked sausage was prepared with an adjusted pH-value of 5.8. The brine percentage was again 3.5%. Inoculation, pasteurization and storage conditions were the same as for liver sausage emulsions.

Sterilized sausage emulsions

Luncheon meat and liver paste emulsions were prepared (brine percentage 3.5%) and 100 and 200 mg/kg nitrite added, respectively. They were inoculated with spore-bearing materials to 100 and 500 spores per gram. The emulsions were packed in small cans (76 × 35 mm) and heated at 95°C and 105°C with F_0 values of 0.05 and 0.5, respectively. The cans were incubated for 5 weeks at 30°C and examined for growth of spore-forming bacteria.

Nitrite depletion

To obtain information about the nitrite depletion in the various sausage products, nitrite concentrations were determined immediately after pasteurization and sterilization and during storage. The analysis was carried out simultaneously with the bacteriological examination. To determine the non-bacteriological depletion of nitrite only the results for cans without observed bacterial growth were taken into account.

Results

Pasteurized sausage

Factory-made sausage emulsions As shown in Table 1, five out of twelve lots of sausage emulsions, failed to show growth of clostridia after pasteurization, even when no nitrite was added. Apparently the emulsions either did not contain clostridia, or, if present, the conditions were unfavourable for growth. One of the other lots was inhibited by 50 mg/kg nitrite, others only by 100 or 200 mg/kg nitrite. In two lots of liver sausage emulsions no inhibition was found even when 200 mg/kg nitrite had been added. The pH value of one of these emulsions was as high as 6.9 as is shown in Table 1. The table also demonstrates that the counts of *Clostridium* spores determined immediately after pasteurization were low.

Laboratory-made sausage emulsions In all experiments with liver sausage emulsions (pH = 6.2) inoculated spores were inhibited by 200 mg/kg nitrite. Growth occurred if the nitrite addition was less than 200 mg/kg.

For cooked sausage emulsions, 50 mg/kg nitrite did not inhibit inoculated spores in spite of the low pH-value of 5.8. The spores, originating from the spore-bearing materials, were not even inhibited by 200 mg/kg nitrite.

Nitrite depletion The degree of nitrite loss during pasteurization varied consider-

Table 1. Inhibition of clostridia in pasteurized, factory-made sausage emulsions.

Product	Factory code	Amount of nitrite needed for inhibition of clostridia (mg/kg)	pH after pasteurization	Count of *Clostridium* spores per gram after pasteurization
Cooked sausage emulsion	I	50	6.5	< 10
	II	100	6.3	2×10
	III	0	5.9	2×10
	IV	0	5.9	< 10
	V	100	6.4	2×10
	VI	200	6.5	< 10
Liver sausage emulsion	I	200	6.9	1×10
	II	> 200	6.9	2×10
	III	0	6.5	1×10
	IV	0	6.8	< 10
	V	0	5.8	1×10
	VI	> 200	6.5	1×10

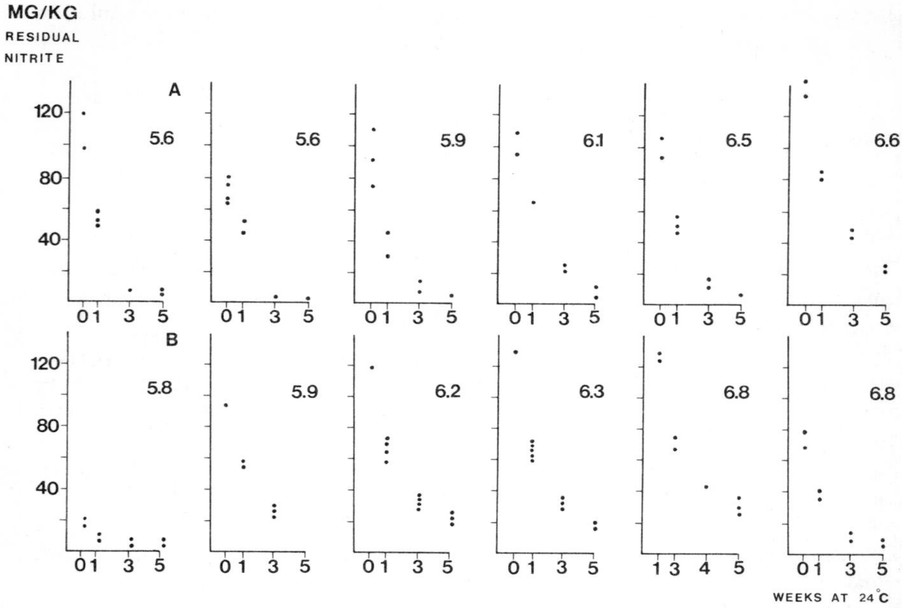

Fig. 1. Decrease of nitrite during storage at 24°C of pasteurized canned sausage emulsions made in the lab or from the factory, containing 200 mg/kg added nitrite. The figures refer to cans showing no growth of bacteria. A: cooked sausage; B: liver sausage.

ably. Nitrite losses were found to vary from 30 to 90%. The non-microbiological depletion of nitrite during storage is shown in Fig. 1.

Sterilized sausage emulsions

Inoculated *Clostridium* spores, after having survived a heat treatment in the meat product at 95°C ($F_0 = 0.05$), were not inhibited by 200 mg/kg nitrite during incubation at 30°C. However, the few spores in the non-contaminated meat products were inhibited by this amount of nitrite.

The contaminated meat products with 100 and 200 mg/kg of added nitrite which were processed at 105°C to an F_0 value of 0.5 did not show growth during incubation at 30 °C. Surprisingly, growth often did not occur in the absence of nitrite either. Closer examination revealed this inhibition to be related to the action of salt on heat-damaged spores; the same meat product without salt did show growth.

This salt-sensivity has been established by Roberts & Ingram (1966), Roberts et al. (1966), Duncan & Foster (1968), Pivnick & Thacker (1970), Ingram & Roberts (1971), Van der Laan (1971) and Holwerda (1973).

Bacillus spores, surviving the heat processing of the meat product at 95°C (F_0-value 0.05), were not inhibited by 200 mg/kg nitrite. Inhibition by nitrite only occurred sometimes in the meat product which was heated at an F_0 value of 0.5.

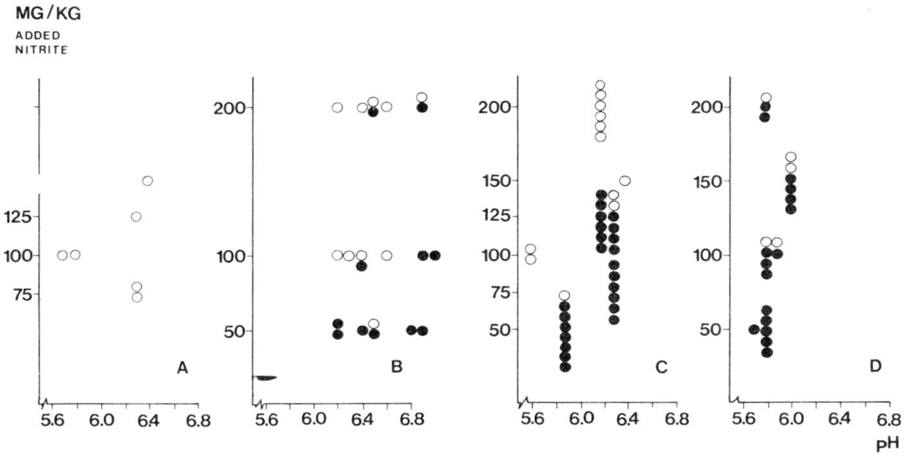

Fig. 2. Growth of clostridia after storage (5 weeks at 24°C) of pasteurized, canned sausage emulsions made in the lab or from the factory arranged according to added amounts of nitrite and pH values. A: endogenous spores from laboratory-made sausage emulsions; B: endogenous spores from factory-made sausage emulsions; C= inoculated spores; D: spores from contaminating spore-bearing materials. ●= growth; ○= no growth.

Discussion

The results of all 350 pasteurization/storage experiments are graphically represented in Fig. 2.

It should be emphasized that all these experiments refer to those cases in which growth of clostridia occurred without added nitrite.

To guarantee complete prevention of growth of clostridia 200 mg/kg sodium nitrite should be added. Even then – with a heavy contamination – growth of clostridia might occur. For still better security, a maximum pH-value of 6.2 should be maintained.

This conclusion is also based on the work of others. Steinke & Foster (1951) established inhibition of 5 000 *Cl. botulinum* spores in liver sausage with 200 mg/kg nitrite when heated ag 70 °C. Bulman & Ayres (1952) reported the prevention of spoilage of pork trimmings, heated at 80 °C, containing 150 mg/kg of added nitrite and a natural level of spores below 1 per gram. However, no inhibition was osberved at an inoculation level of 50 *Clostridium* spores per gram. Mol (1970) inhibited growth of inoculated Clostridium spores in a meat mix, heated at 78 °C, by adding 144 mg/kg sodium nitrite as a pH-value of 6.2. Pivnick & Barnett (1965) inoculated Bologna sausage with *Clostridium botulinum* spores and obtained inhibition with approximately 100 mg/kg nitrite (10^4 spores per gram before cooking at 71 °C).

Bacilli are less inhibited by nitrite than clostridia as can be seen from Fig. 3. Only in the non-inoculated emulsions were bacilli always inhibited by 200 mg/kg nitrite.

As far as sterilized meat products are concerned, the present investigation indicates that less nitrite may be sufficient.

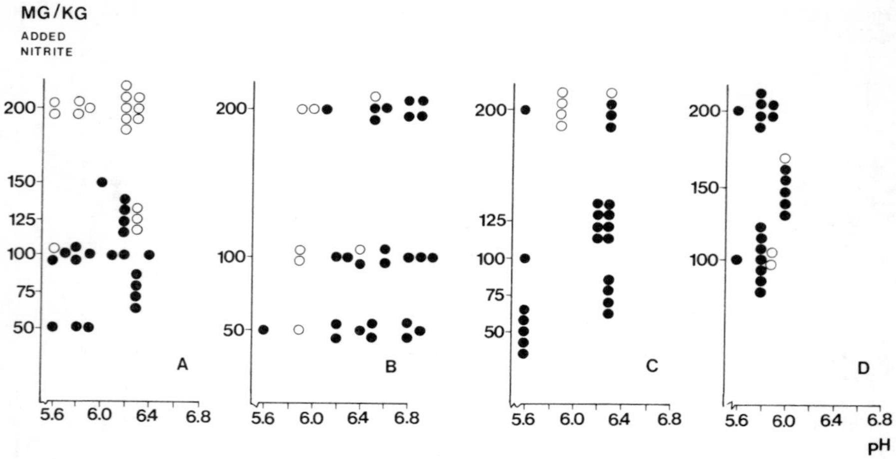

Fig. 3. Growth of bacilli after storage (5 weeks at 24°C) of pasteurized, canned sausage emulsions made in the lab or from the factory arranged according to added amounts of nitrite and pH values. A: endogenous spores from laboratory-made sausage emulsions; B: endogenous spores from factory-made sausage emulsions; C: inoculated spores; D: spores from contaminating spore-bearing materials. ●= growth; ○= no growth.

For a heat treatment resulting in a F_0 value of 0.5, clostridia were found to be inhibited by 100 mg/kg nitrite, provided that the brine percentage is 3.5 %.

The question arises whether this also applies to somewhat lower salt percentages and F_0 values.

References

Bulman, C. & J. C. Ayres, 1952. Preservative effect of various concentrations of curing salts in comminuted pork. Fd Technol. 6: 255–259.

Duncan, C. L. & E. M. Foster, 1968. Role of curing agents in the preservation of shelf stable canned meat products. Appl. Microbiol. 16: 401–405.

Holwerda, E., 1973. Invloed van natriumchloride op het verloop van de afsterving van Bacillus sporen. CIVO rapport Nr R4009.

Ingram, M. & T. A. Roberts, 1971. Application of the 'D-concept' to heat treatments involving curing salts. J. Fd Technol. 6: 21–28.

Laan-van Daalen, H. C. van der, 1971. Winning van gezuiverde sporen en bepaling van afstervingskrommen. CIVO rapport Nr R3668.

Mol, J. H. H. & C. A. Timmers, 1970. Assessment of the stability of pasteurized comminuted meat products. J. appl. Bact. 33: 233–247.

Pivnick, H. & C. Thacker, 1970. Effect of sodium chloride and pH on initiation of growth by heat-damaged spores of Clostridium botulinum. Can. Inst. Fd Technol. 3: 70–75.

Pivnick, H. & H. Barnett, 1965. Effect of salt and temperature on toxicogenesis by Clostridium botulinum in perishable cooked meats vacuum-packed in air-impermeable plastic pouches. Fd Technol. 19: 140–143.

Roberts, T. A., R. J. Gilbert & M. Ingram, 1966. The effect of sodium chloride on heat resistance and recovery of heated spores of Clostridium sporogenes (P A 3679/S_2). J. appl. Bact. 29: 549–555.

Roberts, T. A. & M. Ingram, 1966. The effect of sodium chloride, potassium nitrate and sodium nitrite on the recovery of heated bacterial spores. J. Fd Technol. 1: 147–163.
Steinke, P. K. W. & E. M. Foster, 1951. Botulinum toxin formation in liver sausage. Fd Res. 16: 477–484.

Discussion

Experimental details

The addition of salt and nitrite and the heat treatments were done in the pilot plant of the Institute, with the highest accuracy.

The data in Figure 2 refer to cans in which clostridia and in many cases also bacilli had developed.

Values of pH in practice

The pH range found in cooked, cured meat products is between 5.8 and 6.8. The growth of clostridia is better at the higher pH values. However the residual nitrite content after heating and during storage will remain higher.

Proc. int. Symp. Nitrite Meat Prod., Zeist, 1973. Pudoc, Wageningen.

Perigo effect in pork

H. Pivnick and P-C. Chang

Food Research Laboratories, Food Directorate, Health Protection Branch, Health and Welfare Canada, Ottawa

Abstract

An inhibitor against *Clostridium botulinum* was formed when canned pork luncheon meat was processed to F_0=0.4. The meat was manufactured with 0–300 mg/kg of $NaNO_2$ and all of it was held at 35°C after processing until the highest concentration of nitrite declined to less than 2 mg/kg. Meat in cans was then inoculated with spores of *Cl. botulinum* that had survived a heat treatment of F_0=0.4 in a solution of raw meat juice, 4.5% salt and 150 mg/kg of $NaNO_2$. The inhibitory effect could be demonstrated by an increase in time required for inoculated cans to swell and by the number of spores required to initiate growth and cause swelling. However, the inhibitory effect was relatively small; meat made with 3.8% salt in the water phase and 200 mg/kg of nitrite inhibited 3.3 $\log_{10}$ of spores (3.3 units of inhibition). Salt contributed 1.86 units and the inhibitor formed from nitrite contributed 1.43 units.

Introduction

The value of nitrite as an inhibitor of *Cl. botulinum* in canned meat has been recognized for several decades, but is was thought that nitrite per se caused the inhibition. The discovery that nitrite could form an antimicrobial inhibitor (PF) when heated with some components of bacteriological media (Perigo et al., 1967; Perigo & Roberts, 1968) suggested that this inhibitor might be important in the safety of canned cured meat.

However, PF added to meat was inactivated: mixtures of PF and meat were not inhibitory to *Cl. botulinum* (Johnston et al., 1969). This discovery does not preclude the possibility that nitrite heated with meat might form a Perigo-type (inhibiting) factor (PTF) that is not identical with PF. Johnston et al. (1969) and Ashworth & Spencer (1972) showed that meat heated with nitrite could form an inhibitor that was not formed in meat heated without nitrite. The inhibitor was not very potent and, to demonstrate it, they had to add sterile nitrite to the heated meat or have appreciable amounts of free nitrite present after heating. Only recently, an inhibitor has been demonstrated in meat in the absence (<2 mg/kg) of free nitrite (Chang & Pivnick, 1973). The purpose of this report is to describe the experiments and to demonstrate a method for quantitating the inhibitory effect.

Materials and methods

Pork luncheon meat containing 0–300 mg/kg of sodium nitrite and 2.3% salt (3.8–3.9% in the water phase) was stuffed in cans of 3.00 x 3.08 inches (76.2 x 78.2 mm) and processed at 110°C for 85 min to F_0=0.4. All cans in a single experiment were then held at 35°C until the highest concentration of nitrite had decreased to < 2 mg/kg. Spores of *Cl. botulinum* 13983B were used to inoculate the meat. The spores were heated at 100°C to F_0=0.4 in an non-particulate extract of raw, sterile (γ-irradiated) pork supplemented after irradiation with 150 mg/kg $NaNO_2$ and 4.5% NaCl. Cans were pierced aseptically, and 0.1 ml of suitably diluted inoculum was introduced into the centre of the meat with a hypodermic syringe and needle. Thereafter, the cans were resealed, incubated at 30°C, and examined weekly for swelling. Meat in all cans that swelled contained botulinum toxin. No toxin was found in non-swollen cans.

Estimates of the numbers of spores that grew after introduction into the canned meat were made by the method of Fisher (1922) and Fisher & Yates (1963).

Results and discussion

Two main experiments were carried out. For each, canned meat and spore preparations were freshly prepared. In Experiment 1, the sodium nitrite (nitrite) added to the raw product varied from 0–300 mg/kg, and the heat damaged spores inoculated per can varied from $2.2 \times 10^1 - 2.2 \times 10^5$. Table 1 shows that with 22 spores per can there was little difference in the proportion of cans with toxic meat regardless of whether 0 or 200 mg/kg of nitrite was used; unfortunately we did not inoculate 22 spores into meat made with 300 mg/kg. In meat processed without nitrite and inoculated with 22 spores per can (440 spores in 20 cans), the estimated number of spores that grew was 11.86 and the ratio of spores added/ spores that grew was 37.08 ($\log_{10}$ = 1.57); 97.3% of the spores failed to grow. This is about the amount of inhibition against spores heat damaged to F_0=0.4 that one would expect from 3.8% salt (Pivnick & Thacker, 1970); the brine % in the meat was 3.8%.

Table 1. Toxigenesis by Cl. botulinum in canned meat depleted of nitrite, Experiment 1[1]. Recorded are: number of cans toxic/number of cans inoculated.

$NaNO_2$ added (mg/kg)	Number of spores added per can				
	2.2×10	2.2×10^2	2.2×10^3	2.2×10^4	2.2×10^5
0	9/20	18/20	20/20	.	.
100	4/20	20/20	20/20	10/10	.
200	2/10	20/20	20/20	20/20	.
300	.	19/20	20/20	20/20	20/20

1. 11.8% protein; 59.7% water; 4.0% dextrose equivalent; 2.4% salt; 3.8% brine (salt in water phase). All cans held 15 weeks at 35°C until meat with 300 mg nitrite per kg contained less than 2 mg/kg.

Table 2. Weeks at 30°C for 50% of inoculated cans to swell (t_{50}). Experiment 1.

$NaNO_2$ [1] (mg/kg)	Number of spores added per can				
	10^1	10^2	10^3	10^4	10^5
0	.	1.5	1.2	.	.
100	.	2.0	1.3	1.2	.
200	.	2.5	1.7	1.3	.
300	.	3.4	3.0	1.7	1.6

1. See footnote for Table 1.

Although we had used too many spores for most of the cans in Experiment 1, we did observe that the time for 50% of the cans to swell (t_{50}) increased with the concentration of nitrite added to the meat (Table 2). Thus, there was considerable indication that the nitrite, although no longer present at time of inoculation, had been converted into an inhibitor. As expected, the t_{50} decreased as the inoculum was increased. Christiansen et al. (1973) have observed similar results with canned meat inoculated before pasteurization, but their results, although suggesting the presence of a PTF, are far from conclusive: free nitrite was available after pasteurization.

Experiment 2 was similar to Experiment 1, except that we used smaller numbers of heat-damaged spores. Table 3 shows the composition of the meat and the nitrite level at time of inoculation (after 14 weeks at 35°C). Table 4 shows the number of cans that became toxic at each level of nitrite and spores. The numbers of heat-damaged spores added per can varied from 8 to 128 and the total numbers of heat-damaged spores added to the 80 cans at each nitrite level was 3968. The estimates (Fisher & Yates, 1963) of spores that grew at each level of nitrite is based on the total number of cans that became toxic divided by the total number of cans that were inoculated. These estimates were in good agreement with most probably

Table 3. Analysis of canned luncheon meat. Experiment 2[1].

Sodium nitrite (mg/kg)					Brine (%)
added	in raw product	after F_0=0.4	after storage at 35°C		
			6 weeks	14 weeks	
0	0	0	0	0	3.7
50	32	9	1.6	0.6	3.8
100	80	17	3.4	0.8	3.8
150	136	31	4.7	1.1	3.9
200	192	54	7.5	2.1	3.9

1. 24% fat; 12.1% protein; 58.2% water; 4.0 dextrose equivalent.

Table 4. Toxigenesis by *Cl. botulinum* in canned meat depleted of nitrite. Recorded are: number of cans toxic per 16 cans inoculated[1]. Experiment 2.

$NaNO_2$ added (mg/kg)	Number of spores added per can					Total toxic/ inoculated	Estimate[2] of total number of spores growing
	8	16	32	64	128		
0	1	6	5	9	12	33/80	54.25
50	1	2	6	7	9	25/80	35.34
100	0	1	2	5	10	18/80	22.63
150	0	3	4	7	7	21/80	27.59
200	1	0	0	1	0	2/80	2.01

1. 10 cans inoculated with sterile water did not swell.
Cans incubated 14 weeks at 35°C until $NaNO_2$ was < 2 mg/kg; then inoculated and incubated at 30°C until swollen, or maximum of 11 weeks.
2. 16(8+16+32+64+128) = 3968 spores added per level of nitrite. Estimate of number growing by method of Fisher and Yates (1963).

number calculations (Stumbo et al., 1950) for the various combinations of spores and nitrite (data not presented).

The time required for 50% of the cans that eventually swelled to be detected as incipient swells was calculated for those inoculated with 128 spores per can. The t_{50}'s were as follows: 0 mg/kg nitrite, 1.3 weeks; 150 mg/kg, 2.9 weeks. It was not possible to calculate the t_{50} for 200 mg/kg of nitrite.

Quantitative assessment of the Perigo effect

In studying inhibition by PF or Perigo-type factors (PTF), various workers have used fixed inocula while varying the concentration of nitrite added to the raw meat before processing plus supplemental nitrite added after processing. Additionally, they have used high or low numbers of vegetative cells, spores, and mixtures of the spores and vegetative cells. Their experimental approach, however, has not offered much aid in assessing one of the most pressing problems concerned with nitrite in canned, shelf-stable meat, i.e., how much does the PTF contribute to inhibition of the outgrowth of spores of *Cl. botulinum* that have survived thermal processing.

In an attempt to provide a simple quantitative approach to inhibitory properties of cured meats with respect to *Cl. botulinum*, we (Pivnick 1970a; Pivnick, 1970b; Pivnick & Petrasovits, 1973) have proposed an equation:

$$xPr=y\,Ds+z\,In$$

where Pr=Protection against *Cl. botulinum,* Ds=destruction of spores by heat (or other means), and In=Inhibition of those spores that survive the destructive process. Pr, Ds and In are expressed on the basis of 1 $\log_{10}$ i.e., protection against, or destruction or inhibition of, 90% of the spores.

In the experiments described here, Ds (destruction) is not considered; we are concerned only with inhibition. Inhibition can be expressed quantitatively by the ratio: spores added/spores that grew. For example, meat processed without nitrite permitted growth of 54.25 spores out of the total of 3968 spores added to the 80

Table 5. Number of spores inhibited by salt and by Perigo-type factor. Experiment 2[1].

$NaNO_2$ added (mg/kg)	Spores that grew	Ratio spores added[2] / spores that grew	Total units of inhibition in $\log_{10}$	Ratio spores growing in salt/ spores growing in salt and nitrite	Units of inhibition due to nitrite added
0	54.25	73.14	1.86		
50	35.34	112.28	2.05	1.54	0.19
100	22.63	175.34	2.24	2.40	0.38
150	27.59	143.82	2.15	1.97	0.29
200	2.01	1974.13	3.30	26.99	1.43

1. See footnotes for Table 4.
2. Spores added: 3968.

inoculated cans (Table 5); the ratio is 73.14 ($\log_{10}$=1.86 or 1.86 units of In). This could also be stated as follows: meat containing salt, but made without nitrite, inhibited outgrowth of 98,63% of heat damaged spores (compared with 97.3% in Experiment 1). The increased inhibition due to PTF is relatively small for meat made with 50–150 mg/kg nitrite, i.e., 0.19–0.38 units of In. With 200 mg/kg of nitrite, however there was some additional increase, 1.43 units of In above that obtained with meat made without nitrite. Assuming that cooked meat is the optimum medium for *Cl. botulinum* and that the only inhibitors were salt and PTF, then the total number of units of In for salt and PTF (in meat made with 200 mg/kg of nitrite) is 3.30.

We do not claim that the approach described above answers all of the questions concerned with inhibition of surviving spores, e.g., the combined action of nitrite, PTF, salt, etc., the long delay in swelling of cans inoculated before processing (Pivnick et al., 1969) which could be due to loss of PTF with time, etc. However, we do believe that our work: (1) demonstrates a PTF in the absence of nitrite and (2) presents a useful method of measuring the extent of inhibition in terms that are of use to scientists concerned with the safety of canned meat.

References

Ashworth, R. & R. Spencer, 1972. The Perigo effect in pork. J. Fd Technol. 7: 111–124.

Chang, P-C. & H. Pivnick. Effect of sodium nitrite on *Clostridium botulinum* in commercially produced canned meat. 16th Ann. Meeting, Can. Inst. Fd Sci. Technol., Vancouver, May, 1973.

Christiansen, L. N., R. W. Johnston, D. A. Kautter & J. W. Howard, 1973. Effect of nitrite and nitrate on toxin production by *Clostridium botulinum* and on nitrosamine formation in perishable canned comminuted cured meat. Appl. Microbiol. 25: 357–62.

Fisher, R. A., 1962. On the mathematical foundation of theoretical statistics. Phil. Trans. Soc., A. 222: 309–68.

Fisher, R. A. & F. Yates, 1963. Statistical tables for biological, agricultural and medical research. 6th edn, Oliver and Boyd, Edinburgh.

Johnston, M. A., H. Pivnick & J. M. Samson, 1969. Inhibition of *Clostridium botulinum* by sodium nitrite in a bacteriological medium and in meat. Can. Inst. Fd Technol. J. 2: 52–55.

Perigo,J. A. & T. A. Roberts, 1968. Inhibition of clostridia by nitrite. J. Fd Technol. 3: 91–94.
Perigo, J. A., E. Whiting & T. Bashford, 1967. Observations on the inhibition of vegetative cells of *Clostridium sporogenes* by nitrite which has been autoclaved in a laboratory medium discussed in the context of sub-lethally processed cured meats. J. Fd Technol. 2: 377–397.
Pivnick, H. The inhibition of heat-damaged spores of *Clostridium botulinum* by sodium chloride and sodium nitrite. Presented at the Symposium on the Microbiology of Semi-preserved Foods. Prague, Oct., 1970.
Pivnick, H. & Petrasovits, 1973. A rationale for the safety of canned shelf-stable cured meat: protection = destruction + inhibition. Presented at the XIX Réunion européenne des chercheurs en viande. Paris, Sept. 1973.
Pivnick, H. & C. Thacker, 1970. Effect of sodium chloride and pH on initiation of growth by heat-damaged spores of *Clostridium botulinum*. Can. Inst. Fd Technol. J. 3: 70–75.
Pivnick, H., H. W. Barnett, H. R. Nordin & L. J. Rubin, 1969. Factors affecting the safety of canned, cured, shelf- stable luncheon meat inoculated with *Clostridium botulinum*. Can. Inst. Fd Technol. J. 2: 141–148.
Stumbo, C., J. R. Murphy & J. Cochrane, 1950. Nature of thermal death time curves for P. A. 3679 and *Clostridium botulinum*. Fd Technol. 4: 321–326.

Discussion

Inhibitory action of nitrite on spores

Earlier experiments showed that there was no inhibitor formed due to the presence of nitrite during heating of the spores in the raw meat extract, because spores heated in the same material devoid of nitrite behaved similarly.

In other experiments (Pivnick et al., 1969) spores were added to meat and then the meat was canned and heated; in some cans that never swelled or showed evidence of growth, viable spores were found 18 months later. Other workers have also observed these viable, but dormant spores in canned meat. However, in one experiment of the type described in the report spores could not be recovered from a few unswelled containers.

Nitrite the active compound?

Some American inoculated pack experiments with ham (Christiansen et al., 1973) showed that the antibotulinal effect during incubation at 27°C was fairly well correlated with the initial level of nitrite. This suggests the possibility of an inhibitor being formed during processing. However, because considerable free nitrite was also present after processing the inhibitory effect of nitrite per se could not be excluded.

The 'protection' equation

The equation $x\mathrm{Pr} = y\mathrm{Ds} + z\mathrm{In}$ can simply be written as $\mathrm{Pr} = \mathrm{Ds} + \mathrm{In}$. This equation can also be applied to irradiation preserved products.

Proc. int. Symp. Nitrite Meat Prod., Zeist, 1973. Pudoc, Wageningen.

Inhibitors in cooked meat products

P. S. van Roon

Institute of Food Hygiene, Department of Meat Technology, Faculty of Veterinary Sciences, University of Utrecht, Biltstraat 172, Utrecht, the Netherlands

Abstract

For several reasons it seems possible that iron-nitrosyl coordination complexes can be formed in canned, cured meat products during heating. Two complexes, the black Roussin salt, $[Fe_4S_3(NO_7)]^-$, and a nitrosyl-cysteyl-ferrate, were studied. Both inhibit the growth of clostridia spores. The Roussin salt is only produced when free Fe(II) ions and hydrogen sulphide are present. The second complex needs cysteine in place of hydrogen sulphide.

With a tentative assay method for the Roussin salt in sterilized meat products, this complex was not detectable in a luncheon meat. An assay method for the second complex is being studied.

The likelihood that the second complex is produced during heating of meat products is surveyed. It is also pointed out why the presence of such an inhibiting complex does not always imply that it inhibits growth.

Introduction

Perigo et al. (1967) and Perigo & Roberts (1968) described the inhibition of vegetative cells of several clostridia by very low amounts of nitrite, present in the growth medium during sterilization. The authors concluded that a potent growth inhibitor might have been formed with properties differing from nitrite.

Johnston et al. (1969); Johnston & Loynes (1971) and Roberts (1971) studied this Perigo effect in sterilized meat products and could not find such an effect. Addition of rather high amounts of ascorbic acid or cysteine before sterilization of the meat induced growth inhibition with small amounts of nitrite. The inhibiting effect which Perigo discovered in his medium, does not seem probable in meat under normal curing and sterilizing conditions. Johnston et al. (1969) were not able to find growth inhibition with Perigo's medium after addition of coagulated meat-protein. This protein effect was also found by Grever (1973).

These findings do not rule out the formation of an inhibiting agent in sterilized, cured meat, but tend to indicate that the activity of such a substance is perhaps inhibited by adsorption on to meat protein.

Grever (1973) studied the conditions for inhibiting growth of spores of *Clostridium sporogenes* after addition of a liquid containing low amounts of nitrite and cysteine in water, which was heated in closed cans for 20 min at 110°C. The inhibition by this fluid appeared to be irreproducible. The reason for this lack of reproducibility may be found in trace metal ions like Fe, Sn and Pb originating from the can.

For several reasons the effect of Fe ions seems to be most important:
- Fe ions are present in the Perigo medium,
- Fe, being a transition metal, is able to form coordination complexes with nitric oxide and cysteine (Mc Donald et al., 1965; Vanin, 1967; Woolum et al., 1968),
- addition of small amounts of Fe^{2+} to a heated beef-product increases the rate of nitrite depletion (Olsman & Krol, 1972).

Another possible inhibitor might be the black Roussin salt, a complex of Fe, NO and S^-, which is formed from nitrite, sulphide and ferrous sulphate at 100 °C (Pawel, 1882). It appears to be a potent inhibitor for several microbes including *Bacillus* species (Dobry & Boyer, 1945; Candeli, 1949). In sterilized meat products hydrogen sulphide is produced (Hofmann & Hamm, 1966), which might react with nitrite or nitric oxide and Fe^{2+}, resulting in black Roussin salt.

The experiments described in this paper deal with:
- the synthesis of cysteyl-nitrosyl complexes,
- their growth inhibiting effect on clostridia,
- the occurrence and detection of one of these complexes in a meat product.

Experimental

All reagents were of analytical grade. Nitrogen was purified by washing with alkaline pyrogallol, nitric oxide by passing through a solution of potassium carbonate.

Black Roussin salt

The synthesis was according to the method of Pawel (1882).

Experiments with Na_2S, $FeSO_4$ and $NaNO_2$; cysteine, $FeSO_4$ and $NaNO_2$

The reactions were carried out in Thunberg vessels under vacuum in distilled water, presaturated with nitrogen. The added components were allowed to react for 30 min at 90 °C, after which the vacuum was replaced by nitrogen. The pH of the reaction mixture was 7.0.

Experiments with cysteine, $FeSO_4$ and NO

The reaction took place at room temperature for 15 min by introduction of nitric oxide (P_{NO} approx. 120 mm Hg) into the evacuated Thunberg vessel. The pH was 7.0.

Procedure for isolation and assay of black Roussin salt

The meat product (approx. 10g) was extracted twice with 60 ml of methanol in a Waring blender. The extract was separated from the mixture by filtration through a Buchner funnel. 10 g of anhydrous sodium sulphate was added to the filtrate and it was placed for 1 h in a cold room (−20°C) to solidify the fat. The Na_2SO_4 and separated fat were removed by filtration. The dried defatted extract was concentrated in a rotating evaporator until a volume of V_1 was obtained. V_2 ml of this solution was applied to a column of aluminium oxide (activity state II–III) suspended in chloroform. The adsorbed Roussin salt was eluted with methanol and collected in a 50-ml fraction which was concentrated by evaporation in vacuo to a volume of V_3 ml. The Roussin salt concentration was estimated by spectrophotometry at 350 mm ($\epsilon_{10\ mm}$ in methanol 15.5×10^3. litre mol^{-1} cm^{-1}). This tentative method has a recovery of 87% with a pure solution of black Roussin salt. Losses mainly occurred during the chromatography.

For the meat product with a high fat content, a control without nitrite is required.

A_s and A_b being the absorption of the sample and the control respectively, the concentration of Roussin salt in the meat product is calculated as follows, x being the weight of the sample in grams:

$$C_{RS} = \frac{1000}{x} \cdot \frac{V_1 \cdot V_3}{V_2} \cdot \frac{A_s - A_b}{15.5} \ \mu mol/kg \ product$$

Preparation of the meat product (luncheon meat)

The emulsion was prepared in the usual way except that sometimes the salt containing nitrite was replaced by NaCl. The products were prepared in two main series with either 1.5 kg meat emulsion and 150 ml solution or 1.0 kg and 50 ml. All solutions contained ascorbic acid 10 mmol/litre. In the first series, one treatment also contained black Roussin salt 5 mmol/litre. In the second series, solutions contained also cysteine 10 mmol/litre or $FeSO_4$ 20 mmol/litre or both cysteine 20mmol/litre and $FeSO_4$ 20 mmol/litre. When a salt containing nitrite was used the initial nitrite content was approximately 1.6 mmol per kg (110 mg/kg). After mixing the emulsion was put into 57 x 75 cans and heated for 65 min at 112°C ($F_0 = 0.5$)

Isolation and assay of the cysteyl-nitrosyl-Fe coordination complex

The study of a method of analysis has recently been started.

Results

Inhibition of growth of Clostridium sporogenes by black Roussin salt

No growth of these spores in the medium used by Grever (1973) was found at concentrations of 50 μM (29 mg/l) and higher; no inhibition occurred with concentrations lower than 12.5 μM (15 mg/l).

Inhibition by the combined action of $NaNO_2$ (0.6 mmol/l), Na_2S (10 mmol/l) and $FeSO_4$ (1.0 mmol/l) after heating at 90°C

This combination gives growth inhibition after dilution (1:1) with the growth medium used by Grever (1973). Without heating no inhibition was detected. Therefore it is likely that the inhibition is due to formation of the black Roussin salt.

Inhibition by a solution containing $NaNO_2$ (0.6 mmol/l), cysteine (10 mmol/l) and $FeSO_4$ (1.0 mmol/l) after heating at 90°C

Only this combination of substances gives inhibition after a 1 : 1 dilution with growth medium. The solution becomes faintly yellow. Replacement of cysteine by cystine results in a colourless solution without inhibiting properties. Other experiments indicate that for complete growth inhibition the amounts of cysteine and Fe(II) can be reduced by a factor of 10. The yellow component which is very sensitive to oxidation seems to be a growth inhibitor.

Inhibition by a mixture of NO (P_{NO} approx. 120 mm Hg), cysteine (10 mmol/l) and $FeSO_4$ (1.0 mmol/l)

The reaction product with a bright yellow colour, appears to be an inhibiting agent. Neither nitric oxide alone, nor replacement of cysteine by cystine, giving a colourless solution, produced inhibition.

Detection of black Roussin salt in sterilized luncheon meats

Estimation of Roussin salt – 445 μmol added to 1 kg of the luncheon meat – by the assay method described above, resulted in a recovery of 215 μmol per kg product, i.e. 47%. In spite of the tentative character of this method it was used for analysis of normal luncheon meat and products with addition of $FeSO_4$ and/or cysteine as described above under preparation of the meat product. In all these cases no black Roussin salt could be detected.

Discussion

In the introduction it was suggested that one of the possible nitrosyl-Fe coordination complexes that might be formed during heating in cured meat is the black Roussin salt (I),

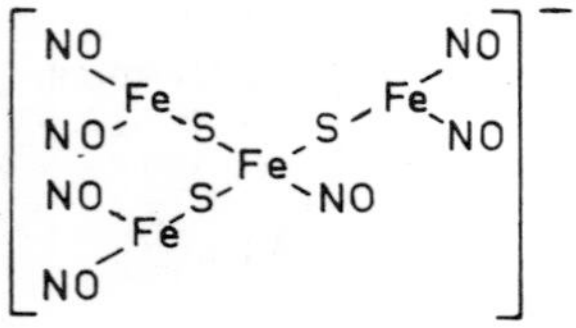

I

It appears to be an inhibitor for growth of Clostridium spores but it could not be detected in sterilized luncheon meat. Reasons for its absence could be:
- the rather low amount of H_2S developed during heating,
- the adsorption of the non-heme iron to myofibrillar protein (Hamm & Bünnig, 1972).

Addition of free Fe^{2+} and cysteine – being the main source for the pyridoxal catalysed production of H_2S (Gruenwedel & Patnaik, 1971) – or the combination of both additives gave no indication for production of complex I. Thus it does not seem very likely that the black Roussin salt contributes to growth inhibition.

Another possible growth inhibitor in cooked, cured meat products could be the yellow component formed by reaction of Fe^{2+}, cysteine and NO. According to McDonald et al. (1965) the probable chemical formula (II) is,

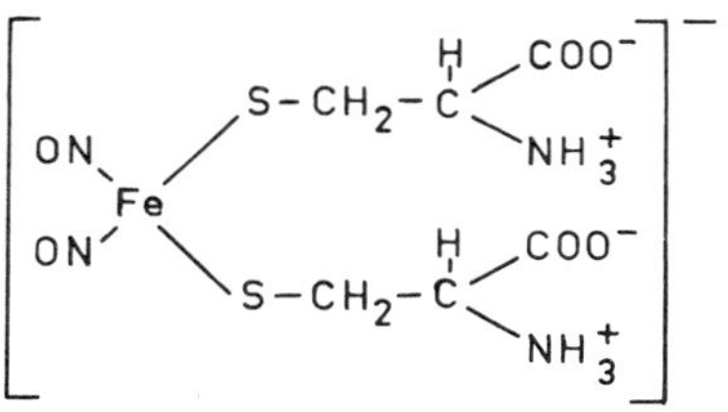

II

This inhibiting agent can be produced at room temperature, if nitric oxide, free Fe^{2+} and accessible SH-groups are present. Perhaps cooking temperatures are necessary to produce,
- higher amounts of NO,
- more available free Fe^{2+},
- more accessible free SH-groups, obtained by protein denaturation.

Vanin (1967) showed direct reaction of NO and Fe^{2+} with cysteyl groups of actomyosin. Because the inhibiting complex is tightly bound to the protein matrix, its inhibiting capacity may be unfavourably affected. More important for the formation of an inhibiting factor is the presence in meat of endogenous low-molecular weight substances, containing SH-groups like glutathione. Addition of Fe^{2+} and cysteine will act favourably on the production of complex II.

These considerations indicate why no Perigo-like effect could be detected in heated cured meat, assuming the Perigo factor is a complex like complex II.

References

Candeli, A., 1949. Meccanismo dell'azione antibatterica del ferro-epta-nitroso-trisulfuro di pottasia. Boll. Soc. Biol. sper. 25: 495–497.

Dobry, A. & F. Boyer, 1945. Sur le nitrososulfure de fer ou sel de Roussin. Action antiseptique. Annls Inst. Pasteur 71: 455–462.

Grever, A. B. G., 1973. Vorming van de Perigo factor in voedingsbodems. Voedingsm. Technol. 4(30/31)105.

Gruenwedel, D. W. & R. V. Patnaik, 1971. Release of hyrogen sulfide and methylmercaptan from sulfur containing amino acids. J. agric. Fd Chem. 19: 775–779.

Hamm, R. & K. Bünnig, 1972. Myoglobin, hemoglobin and iron in bovine and porcine muscle. Proc. 18th Meet. Meat Res. Workers, Guelph, Canada, p. 156–161.
Hofmann, K. & R. Hamm, 1969. Einflusz der Erhitzung auf Struktur und Zusammensetzung von Muskeleiweisz. Fleischwirtschaft 49: 1180–1182; 1184.
Johnston, M. A., H. Pivnick & J. M. Samson, 1969. Inhibition of Clostridium botulinum by sodium nitrite in a bacteriological medium and in meat. Can Inst. Fd Technol. J. 2: 52–55.
Johnston, M. A. & R. Loynes, 1971. Inhibition of Clostridium botulinum by sodium nitrite as affected by bacteriological media and meat suspensions. Can. Inst. Fd Technol. J. 4: 179–184
McDonald, C. C., W. D. Phillips & H. F. Mower, 1965. An electron spin resonance study of some complexes of iron, nitric oxide and anionic ligands. J. Am. Chem. Soc. 87: 3319–3326.
Olsman, W. J. & B. Krol, 1972. Depletion of nitrite in heated meat products during storage. Proc. 18th Meet. Meat Res. Workers, Guelph, Canada, p. 409–415.
Pawel, O., 1882. Ueber Nitrososulfide und Nitrosocyanide. Berichte 15: 2600–2615.
Perigo, J. A., E. Whiting & T. E. Bashford, 1967. Observations on the inhibition of vegetative cells of Clostridium sporogenes by nitrite which has been autoclaved in a laboratory medium, discussed in the context of sub-lethally processed meats. J. Fd Technol. 2: 377–397.
Perigo, J. A. & T. A. Roberts, 1968. Inhibition of clostridia by nitrite. J. Fd Technol. 3: 91–94.
Roberts, T. A., 1971. The inhibition of bacteria in canned pasteurized hams by sodium nitrite. Proc. 17th Europ. Meet. Meat Res. Workers, Bristol, England, paper C9.
Vanin, A. F., 1967. Identification of divalent iron complexes with cysteine in biological systems by the EPR method. Translated from Biokhimiya 32: 277–282.
Woolum, J. C., E. Tiezzi & B. Commoner, 1968. Electron spin resonance of iron-nitric oxide complexes with amino acids, peptides and proteins. Biochim. biophys. Acta 160: 311–320.

Discussion

Additional remarks

The author added that NO can react at room temperature with isolated myofibrillar protein or actomyosin preparations if enough Fe (II) has been added. This yielded a yellow protein, whilst the aqueous phase remained colourless. This finding may confirm that Fe (II) and NO are able to form type II complexes with cysteyl residues of the protein matrix (Vanin, 1967; Woolum et al., 1968). When free cysteine is also present during the reaction with NO, both the protein and aqueous fraction are yellow. By heating the protein fraction with the complex II bound to the cysteyl residues, after complete removal of free NO by evacuation, with added cysteine for 1 h at 90°C, the aqueous phase becomes yellow too. There appears to be competition between free cysteine and cysteyl residues of the protein matrix during the formation of complex II.

The bacteriological significance of these findings is now being tested.

Toxicological data about the black Roussin salt are in the publication of Dobry & Boyer (1945).

Grever added that the Perigo factor also may be formed in a very simple nutrient medium containing only tryptone and agar.

Heating 3% tryptone agar with 20 mg/kg nitrite for 20 min at 110°C, resulted in a medium that slightly inhibited a *Cl. sporogenes* strain. If 10% tryptone agar was heated with 20 mg/kg nitrite the medium completely inhibited this strain.

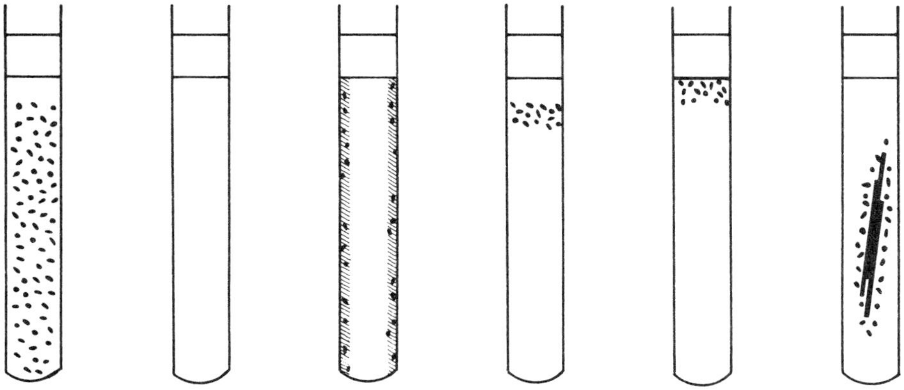

Growth phenomena in 1 1/2 % tryptone, 1% yeast extract agar in which the Perigo factor has been formed, inoculated with 1 ml of a spore suspension of *Cl. sporogenes* (10^7 spores per ml), in Miller-Prickett tubes.

1. Normal growth. At the top no growth due to the presence of oxygen.
2. Complete inhibition.
3. Inhibition, except in or near the liquid, caused by syneresis of the agar.
4. Inhibition, except in a zone at the top of the tube, due to inactivation of the inhibitor by the oxygen present.
5. Inhibition, except directly under the fat layer, due to some inactivation of the inhibitor by the fat. (There is no inactivation when the fat is replaced by parafin oil).
6. Inhibition except in the environment of a cooked-meat fibre.

The inhibiting effect could also be obtained by heating 3% tryptone – 0,05% cysteine agar with the nitrite.

When the formation of the inhibitor was studied by heating cysteine and nitrite alone it was found that the presence of iron was necessary.

After this was established it appeared that the inhibitor could be formed in 1.5% trypton, 1% yeast extract agar provided that iron was present (0.04% Mohr's salt).

The Perigo factor could also be formed in the mentioned media by heating at temperatures much lower than 110°C (60°C); 0.1% ascorbate must be added.

From observations of post-growth phenomena in parts of some inhibiting media various properties of the inhibitor could be derived: inactivation by oxygen, by fat and by a cooked meat-fibre.

Also it was observed that the Perigo factor is absent in the liquid formed by syneresis of the inhibiting agar medium.

Walters said that the results presented by van Roon and Grever correspond to those which are formed at BFMIRA, Leatherhead.

Acid hydrolysed casein has been used as an effective precursor to a Perigo type inhibitor and, after heating with nitrite, thioglycollate or ascorbate, one or more inhibitors have been extracted into lipid solvents. The black inhibitory extract contains iron which added to casein hydrolysed with acid stimulates inhibitor production.

Inhibitory compounds are also obtained on heating nitrite iron salts and ascorbate with cysteine or reduced glutathione; these have been purified and are being studied.

The anti-microbial spectrum of action of the Perigo type inhibitor from acid hydrolysed casein is very similar to that of Roussin black salt. So far as toxicity is concerned, neither substance has been found to nitrosate amines such as morpholine or N-methylanilline at pH values relevant to meat products but some nitrosation was detected at pH 3.

It has also been demonstrated that some detergents inactivate the inhibitor.

Possible influence of oxygen on the Perigo factor

Roberts (1971) could show inhibition of clostridia by the Perigo factor for months. But after opening, removal of the spores and reinocculation of the medium no inhibitory effect lasted. This may be due to sensitivity to oxygen.

Influence of Fe

Addition of $FeSO_4$ in amounts of 100 and 200 mg/l to the Perigo medium produced no increase in inhibition. This may be due to other limiting factors like the content of thioglycollate and nitrite. Furthermore a very small concentration of the inhibitor will suffice. However, at the moment no information is available about the minimum amount of Fe (II) needed.

Stability of the inhibitor

The inhibiting compound according to formula II shows slow decomposition during storage under nitrogen gas at 6°C and in the dark. There is no information available at present about its properties when the complex is bound to the protein matrix.

Other inhibitors

It is not known which inhibiting substance is present in the cans used in the experiments of Pivnick. Johnston et al. (1969) could not find inactivation of the inhibitor with fat.

Conclusions and recommendations of the microbiological session, 12th September 1973

1. Outbreaks of botulism (5 per year) have been reported involving unheated cured pork. There have been none to our knowledge where levels of nitrite used in present commercial practices have been employed.

There is much experimental evidence that *Clostridium botulinum* can develop and form toxin in cured meat or model systems without nitrite. The presence of nitrite retards or prevents such toxin formation, depending on the amount used.

In controlled experiments with cured meat on the laboratory or commercial scale, initial nitrite concentrations of orders exceeding 100 mg per kg (ppm) have been necessary to retard or prevent development of *Cl. botulinum*. These experiments include trials with realistic numbers of inoculated cells or spores.

Experiments involving large inocula are more realistic than might appear from the small numbers actually found in meat, because it is usually necessary to include resistant cells which are only a small proportion of those which occur.

A few similar observations have been made with *Clostridium perfringens* and enterotoxinogenic staphylococci.

2. The necessary amount of nitrite is strongly dependent on related factors e.g. salt.

It is likely that the level of nitrite required will vary with many factors including the product, the geographical location, local practices and ambient temperature.

Experiments to simulate the situation in meat, besides adequately describing the changes in nitrite, should always been done under realistic and as far as possible, specified conditions in respect of water activity, pH, P_{O_2}, E_h, additives, number of cells involved, inactivating effect, of any processing treatment and storage temperature.

3. Cured meats include several different categories of foods, with distinctive microbiological and toxicological hazards. Each type of product should therefore be considered on its own merits.

4. We cannot for the foreseeable future wholly rely on current commercial refrigeration and demestic food handling practice to substitute for nitrite without the risk of botulism and possibly also of food poisoning from *Cl. perfringens* and staphylococci.

5. Any search for substitutes for nitrite should be guided by knowledge of its inhibitory action on non-sporing as well as on spore-forming organisms.

6. Experiments in ordinary microbiological culture media at present seem to have little direct significance in trying to assess the role of nitrite in meat products.

Chemical and technological session

Reporter: P. C. Moerman

Proc. int. Symp. Nitrite Meat Prod., Zeist, 1973. Pudoc, Wageningen.

About the mechanism of nitrite loss during storage of cooked meat products

W. J. Olsman

Central Institute for Nutrition and Food Research TNO, Utrechtseweg 48, Zeist, the Netherlands
Dept: Netherlands Centre for Meat Technology

Abstract

The effects of pH, temperature and addition of an alkylating reagent and ethylenediamine tetraacetate (EDTA) on the rate of loss of nitrite are discussed. The heat of activation of the depletion process is 13 to 14 kcal per mole of nitrite, which is equal to the value found by Fox & Ackermann (1968) for the nitrosomyoglobin formation from metmyoglobin in the presence of ascorbate. It is suggested that the small amount of endogenous ascorbic acid in meat plays a key role in the mechanism of the nitrite depletion by acting as an electron carrier in the reduction of nitrite by the sulphydryl group of the meat proteins. The pH dependence and the effect of alkylation can be explained on the basis of such a mechanism.

Introduction

The amount of nitrite used in manufacturing meat products generally exceeds that, needed for the formation of nitrosomyoglobin or nitrosomyochrome, by a factor 10 or more. This surplus of nitrite is considerd to have an important function by inhibiting the outgrowth of *Clostridium* spores. During cooking and, subsequently, during the storage at ambient temperatures most of this nitrite is lost, however. Oxidation to nitrate only occurs at the higher oxidation-reduction potentials before heating (Möhler, 1971). Nitrogen oxides, such as nitric oxide and nitrous oxide are formed as decomposition products, but cannot account for the complete loss of nitrite. The chemical mechanism of this nitrite depletion has hardly been studied, although it is generally accepted that it disappears by way of a reductive process in which the sulphydryl groups of the meat proteins play a role. This paper discusses some of the results of studies on the effects of pH, temperature, EDTA and SH group blocking reagents. These investigations are confined to the depletion during storage at ambient temperatures.

Experimental

The nitrite loss was studied in model comminuted beef products containing about 70% lean meat. As homogeneity of the system was considered to be an important prerequisite for a successful study of the reaction kinetics, jelly formation was prevented completely or reduced to a neglegible extent by the addition of

0.5% polyphosphates, 4% starch and 2% sodium caseinate. The added amount of sodium nitrite was 100 or 200 mg/kg. No ascorbate was added. The emulsions were filled into lacquered cans, 75 × 57 mm, and heated to such an extent that, on the one hand the keepability of the product was guaranteed for the duration of the experiment and, on the other, the residual concentration of nitrite immediately after cooking was high enough to allow its accurate measurement over a range of at least one log cycle.

The methods of analysis have been described elsewhere (Olsman & Krol, 1972).

Results and discussion

Depletion model

In a rough approximation the depletion rate may be considered as being proportional to the nitrite concentration:

$$\frac{d\,[NaNO_2]}{dt} = k\,[NaNO_2]$$

or:

$$\frac{d \log [NaNO_2]}{dt} = k \qquad \text{Eqn 1}$$

The relation between the logarithm of the nitrite content and the storage time t is, therefore, reasonably well represented by a straight line:

$$\log [NaNO_2] = a + kt$$

where k is the depletion rate constant and a is another constant. This relationship has also been found by Nordin (1969). If, however, the nitrite loss is followed over a period long enough for the concentration to drop by a factor 10 or more, significant deviations from this first order model may occur, especially with products of lower pH-values. The gross reaction kinetics is then between first and second order with respect to the nitrite concentration, and may be represented satisfactorily by the equation:

$$\log [NaNO_2] = a + bt + ct^2 \qquad \text{Eqn 2}$$

in which a,b and c are constants. Naturally, this parabolic model is theoretically incorrect. It is a working model only and applies to the range of nitrite concentrations, analysed during the period of storage.

Influence of pH

The dependence of the nitrite depletion on pH value was determined by

Table 1. Relation between $[k]_{40}$ and pH.

Experiment nr.	$\log [k]^{1}_{40}$
1	8.50–1.37 pH
2	7.58–1.17 pH
3	6.81–1.06 pH
4	6.91–1.07 pH
5[2]	6.60–1.04 pH

1. $[k]_{40}$ in μM/min.
2. Pork products.

analysing a series of products of equal composition. They were prepared from one homogenized batch of meat, but different amounts of glucono-δ-lactone were added to obtain a range of pH values. The results of five experiments are summarized in Table 1. The $[k]_{40}$ is the time gradient of log $[NaNO_2]$ at a sodium nitrite concentration of 40 mg/kg and is calculated from the parabolic regression equation 2, fitting the results of analysis. Similar relations between the pH and log [k] values were found for [k] values at other nitrite concentrations. This pH-effect cannot be explained by differences in dissociation of HNO_2. Substituting the equilibrium equation

$$K_{HNO_2} = \frac{[H^+][NO_2^-]}{[HNO_2]}$$

in Equation 1 gives:

$$\frac{d\,\{\log [HNO_2] + \log K_{HNO_2} + pH\}}{dt} = k$$

and, as both the pH and the equilibrium constant K_{HNO_2} are independent of the storage time t:

$$\frac{d \log [NaNO_2]}{dt} = \frac{d \log [HNO_2]}{dt} = k$$

One explanation for the pH dependence is that nitrite or HNO_2 reacts with a compound which itself is subject to an acid-base equilibrium. However, the protein sulphydryl group, being the most obvious reactant, is largely or completely in its acid form in the range of pH values (6.4 – 5.4) studied (Jocelyn, 1972). Reactions with amino groups hardly take place under these circumstances at ambient temperatures and, moreover, they proceed by way of N_2O_3, the anhydric form of HNO_2, and therefore should be of second order with respect to nitrite (Hughes et al.,

1958). Another possibility is, that, by way of a reversible reaction of nitrite with a meat constituent, some intermediate is formed whose decomposition rate is dependent on pH.

SH-alkylation and EDTA

In an earlier study (Olsman & Krol, 1972) it was shown that blocking of the SH groups results in a decrease of the nitrite loss in meat products. However, at the pH value concerned (6.27) the effect was unexpectedly small. A more extensive investigation was carried out covering the pH range of 6.4 to 5.4. The SH group was blocked by alkylation with 4-vinylpyridine (VP) (Friedman et al., 1970). This reagent was used in preference to *N*-ethylmaleimide because the latter was suspected to be nitrosated to a nitrosamine at lower pH values. The effect of adding EDTA was also studied, as well as the combined effect of VP + EDTA. In Table 2 the k or $[k]_{40}$ values are given for the various products. The effect of the added compounds on the depletion rate is shown by expressing k as a percentage of that of the control.

Typical depletion curves for the products with pH= 5.91 are shown in Fig. 1. The results in the table demonstrate the role of the SH groups in the mechanism of the nitrite loss. Alkylation retards the process considerably. The effect is greater when the pH is lower. Nevertheless, the inhibition was far less pronounced than had been anticipated. On the basis of the statement of Friedman et al.(1970), that alkylation with VP readily proceeds at room temperatures at pH values of 5 and upwards, it must be accepted that the alkylation of the SH groups (at an excess of reagent of 1.6) is virtually complete. The fact that the inhibition is nevertheless lower than 50%, except at pH= 5.4, would suggest that the SH group is not the only nitrite consuming reactant. The non-alkylating counterpart of VP, ethylpyridine, appeared to have no effect at all on the nitrite loss. EDTA does affect the nitrite depletion. It manifests its action by strongly bent depletion curves (see Fig. 1). The depletion kinetics deviated significantly from the first order model at all pH values. At three of them the loss rate during the first days or weeks of storage exceeded that of the corresponding control product. After some time this situation was reversed. The data given in Table 2 for the EDTA products are, therefore, strongly

Table 2. Effects of addition of VP, EDTA and VP+EDTA on the nitrite loss during storage, in meat products with different pH values.

pH	Control	VP		EDTA		VP + EDTA	
	$k_c \times 10^4$	$k \times 10^4$	$(k/k_c) \times 100$	$k \times 10^4$	$(k/k_c) \times 100$	$k \times 10^4$	$(k/k_c) \times 100$
6.38	108	82	76	103[1]	95	72	67
6.14	217[1]	153	71	182[1]	84	106[1]	49
5.91	363	202	56	410[1]	113	174[1]	48
5.63	496	276	56	667[1]	134	244[1]	49
5.41	1327[1]	446[1]	34	928[1]	70	444[1]	33

1. $[k]_{40}$ values calculated from the parabolic regression equation (2); in all other cases the depletion met the first order kinetic model satisfactorily.

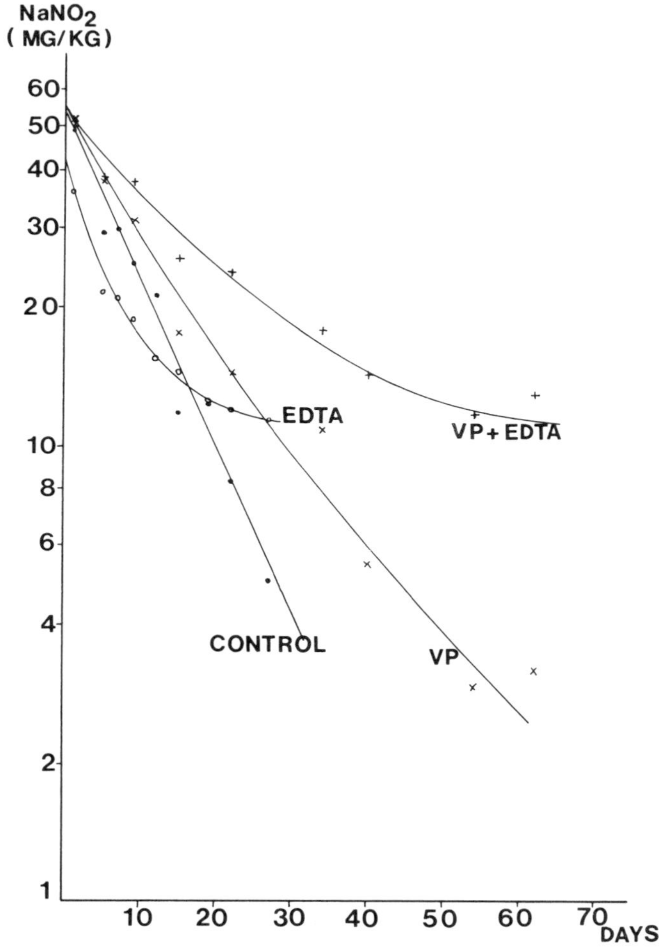

Fig. 1. Nitrite loss in products of pH 5.91 with VP and/or EDTA.

dependent on the particular nitrite concentration at which the [k] value was calculated and would be different for other values than $[k]_{40}$. This is also demonstrated in Fig. 1. These phenomena indicate that the nitrite loss is affected by metal traces which can be bound by EDTA (e.g. Fe, Cu and Sn, the latter originating from the tinplate of the can). As the nitrite breakdown is a reductive process, these materials obviously behave as electron carriers.

The products containing VP as well as EDTA behaved as expected from the separate effects of both compounds. The nitrite depletion in these products was even somewhat more retarded than in the ones with VP alone.

Heat of activation

In two experiments with different products the temperature dependence of the rate constant k of the nitrite depletion was determined. By means of the Arrhenius equation:

$$2.303 \log k = (\Delta H/RT) + C$$

(where R is the gas constant, T is the storage temperature in $^\circ$K and C is a constant) the following values for the heat of activation ΔH were calculated:

Table 3. Heat of activation of the nitrite depletion.

Experiment nr.	Product pH	Δ H in kcal/mole	
		first order model	parabolic working model
1	6.30	13.9±0.3	14.0±0.4
2	6.31	13.3±0.3	13.5±0.3

The ΔH values calculated on the basis of the parabolic model are also given. Here the k of the Arrhenius equation was substituted by the slope d log [$NaNO_2$]/dt at a nitrite concentration of 40 mg/kg. The Arrhenius plots for the k values calculated on the basis of the first order model are shown in Fig. 2.

The heat of activation ΔH is equal to that found by Fox & Ackerman, (1968) for the rate limiting step in the formation of nitrosomyoglobin (NO-Mb) from metmyoglobin (MetMb) and nitrite in the presence of ascorbate as a reducing agent. By extrapolating their data, calculated for pH-values between 4.5 and 6.0, to the pH of the meat products of our study a value of ΔH= 13.6 kcal/mole was found. Fox & Ackerman postulated a chain of reactions, the slowest of which was:

$RNO \rightarrow R\cdot + NO$ (R· being the ascorbic acid radical)

This could imply that ascorbic acid which is naturally present in small amounts in muscle tissue acts as an electron carrier in the reduction of nitrite. Very different concentrations of endogenous ascorbic acid in beef are reported in the literature: Grau (1968) 10 mg/kg; Niinivaara & Antila (1972) 20 mg/kg; Souci et al. (1968) 8 mg/kg for lean beef, but 126 mg/kg for the Musculus semimembranosus. Recently Davidek et al. (1971) reported figures from 30 to 70 mg/kg dehydroascorbic acid in minced pork. Whether these discrepancies should be attributed to individual variabilities or to lack of reliability of the methods of determination is uncertain.

The ultimate electron donor, which reduces the radical and so keeps the nitrite reduction going, might be the SH group of the meat proteins, which is available in excess. Support for such an electron transfer chain operating in cooked meat products is given by the results of Klein & Davidek (1972). It is further known that dehydroascorbic acid can be easily reduced by cysteine and other SH-compounds

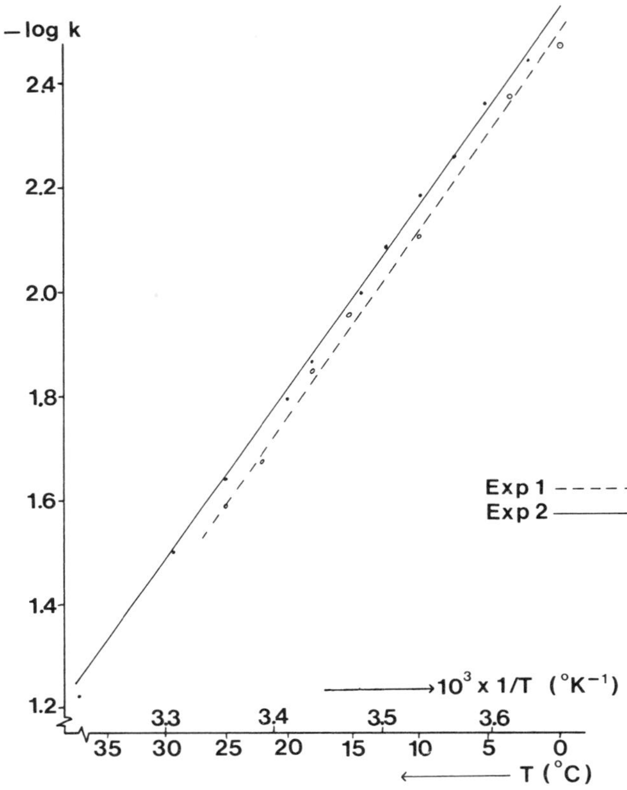

Fig. 2. Relation between reaction rate and temperature.

(Schmandke & Gohlke, 1966). As a consequence of the supposed mechanism there should be a proportionality between the rate of loss of nitrite in a meat product and its content of ascorbic acid. This is in agreement with our earlier finding that the depletion rate is proportional to the meat content divided by the water content of the product (Olsman & Krol, 1972). However, Fox & Thomson (1963) found the NO-Mb formation to be proportional to the square root of the ascorbic acid concentration. This discrepancy could be because ascorbic acid was the only reductant in their model systems, whereas in our meat products an excess of SH groups was available for reaction with the ascorbic acid radical. Tinbergen (1973) found a proportional relationship between the rate of NO-Mb formation and the degree of dilution of the dialysable fraction of a meat extract that provided the reducing system. In view of the ΔH of the nitrite depletion being 13 – 14 kcal/mole, it now seems improbable that nitrite is reduced directly by SH groups, because the ΔH of the reduction by cysteine is as high as 19.2 kcal/mole, and is dependent of the pH (Fox & Ackerman, 1968). This puts the effects of the alkylation into another light. The unexpectedly low inhibition of the nitrite loss could now be explained by the disappearance of the excess of the electron donating SH-groups so that the rate limiting step is the reduction of the ascorbic acid radical rather than the cleavage of R-NO. The effect of the pH on the rate constant of the nitrite depletion could also be explained by the ascorbate-mediated mechanism, assuming that the cleavage of the NO-complex is pH-dependent, as is suggested by the findings of Fox & Ackerman (1968).

The first experiments for verifying this hypothesis, however, yielded rather negative results. Addition of 10, 20, 30 and 40 mg/kg ascorbic acid to a meat product which contained 12 mg/kg endogeneous ascorbic acid, did not accelerate the nitrite depletion significantly. This does not refute the hypothesis. An explanation for the finding could be that the rate-limiting reaction

$$RNO \rightarrow R\cdot + NO$$

does not proceed spontaneously, but only by mediation of a catalyst which could be a trace metal ion. It has been shown before that metal ions have an effect on the nitrite depletion.

Much work will be needed to elucidate the mechanism of the nitrite depletion.

References

Davidek, J., J. Velisek & M. Nezbedova, 1971. Determination of dehydroascorbic acid in meat and meat products. Sb. Vys. Sk. chem.-technol. Praze, Potraviny 30 :17–22.

Fox, J. B. & J. S. Thomson, 1963. Formation of Bovine Nitrosylmyoglobin. I. pH 4.5–6.5. Biochemistry 2: 465–470.

Fox, J. B. & S. A. Ackerman, 1968. Formation of Nitric Oxide Myoglobin. Mechanism of the reaction with various reductants. J. Fd Sci. 33: 364–370.

Friedman, M., L. H. Krull & J. F. Cavins, 1970. Reactions of aminoacids, peptides and proteins with $\alpha\beta$-unsaturated compounds. XIV. Chromatographic determination of cystine and cysteine residues as S-β-(4-pyridylethyl)- cysteine. J. biol. Chem. 245: 3868–3871.

Grau, R., 1968. Fleisch und Fleischwaren. 2e Aufl. p. 61. Verlag Paul Parey, Berlin – Hamburg.

Hughes, E. D., C. K. Ingold & J. H. Ridd, 1958. Nitrosation, diazotation and deamination. Part I. Principles, backgrounds and method for the kinetic study of diazotation. J. chem. Soc. (London) 1958: 58–65.

Jocelyn, P. C., 1972. The biochemistry of the SH-group. p.54. Academic Press, London – New York.

Klein, S. & J. Davidek, 1972. Sodium ascorbate in meat products. Sb. vys. Sk. chem.-technol. Praze, Potraviny 34: 39–45.

Möhler, K., 1971. Bilanz der Bildung des Pökelfarbstoffs im Muskelfleisch. II. Hitzekatalysierte Bildung des Stickoxydkomplexes. Z. Lebensmittel-untersuch. u.-Forsch. 147: 123–127.

Niinivaara, F. P. & P. Antila, 1972. Der Nährwert des Fleisches. p. 118. Fleischforschung und Praxis Schriftenreihe Heft 8. Verlag der Rheinhessischen Druckwerkstätte, Alzey.

Nordin, H. R., 1969. The depletion of added sodium nitrite in ham. J. Inst. Technol. Aliment. 2: 79–85.

Olsman, W. J. & B. Krol, 1972. Depletion of nitrite in heated meat products during storage. 18th Meet. Meat Res. Workers, Guelph, Canada, Vol. II, p. 409–415.

Schmandke, H. & H. Gohlke, 1966. Dünnschichtchromatografische Bestimmung der L-Ascorbinsäure und Dehydroascorbinsäure in Lebensmitteln. Z. Lebensmitteluntersuch. u.-Forsch. 132: 4–12.

Souci, S. W, W. Fachmann & H. Kraut, 1968. Die Zusammensetzung der Lebensmittel. Band I; F-III, 1 d and F–III, 11.

Tinbergen, B. J., 1973. Low-molecular meat fractions active in nitrite reduction. These Proceedings, p. 29–36.

Discussion

Effect of pH on k

In cases where the nitrite depletion did not comply with 1st order kinetics, the $[k]_{40}$ was calculated. This is the change of log $[NaNO_2]$ with storage time at a sodium nitrite concentration of 40 mg/kg, which is a normal level in a cooked product. The effect of pH on the k value decreases as the nitrite concentration, for which the k is calculated, decreases. The data of Table 1 are only meant to demonstrate the presence of an effect of pH on the k value.

The alkylation effect

Many SH-groups can only be alkylated after they have been made available for reaction by heat denaturation. Hence, part of the alkylation takes place during cooking. At elevated temperatures the specificity of the alkylation decreases. Amino functions react competitively. Therefore, it may be possible that some of the thiol groups remain free, even with an excess of alkylating reagent with respect to thiol groups. The extent of inhibition found suggests that so few SH-groups are left, that not the cleavage of RNO but the reduction of the ascorbate radical is rate limiting, thus resulting in an inhibition of the nitrite depletion.

Ascorbic acid

The nitrosated ascorbate intermediate may be represented as:

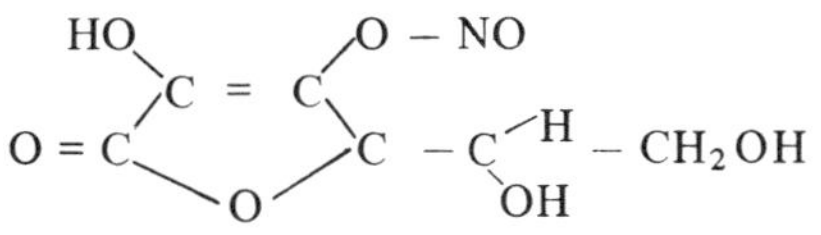

Other reductones are supposed to behave like ascorbic acid. The widely different values for the ascorbic acid content of muscle tissue reported in the literature seem to indicate that the current assay methods are not sufficiently specific for meat and meat products. Presumably the need for a specific method was never felt because of the unimportance of meat as a vitamin C source. On the other hand, it is not ruled out that appreciable variation may exist between ascorbic acid levels of muscles of different animals.

Storage conditions

The storage conditions in this study can be considered as anaerobic, because the redox potential of canned meat products is known to drop drastically during cooking.

Proc. int. Symp. Nitrite Meat Prod., Zeist, 1973. Pudoc, Wageningen

Fate of added nitrite

J. G. Sebranek, R. G. Cassens & W. G. Hoekstra

University of Wisconsin, Madison, WI 53706, USA

Abstract

The stable isotope of nitrogen was used to study the fate and distribution of nitrite in a cured meat product. The meat product was fractionated into water soluble, salt soluble (protein) and insoluble forms in order to conduct quantitative analysis for ^{15}N as a function of storage time and processing temperature. Residual nitrite, determined two days after processing, accounted for less than half of the label added in frozen samples and in samples processed at 71 °C, and the amount of label as nitrite decreased during storage. Samples processed at 107 °C were initially very low in residual nitrite compared to samples from other heat treatments. As residual nitrite decreased, the amount of label found in the non-nitrite water-soluble fraction and in the protein fraction (both soluble and insoluble) increased. About 5% of the label was lost as a gas during processing and 9–12% was present in the pigment fraction. Total recovery of label ranged from 72–86%.

Introduction

Sodium nitrite is used as a curing agent in meat. It causes the production of the typical cured meat color and distinctive flavor; it also specifically acts to inhibit outgrowth of *Botulinum* spores in the event of contamination and mishandling of certain products (Urbain, 1971; Greenberg, 1972). Nitrite is converted rapidly to forms undetectable as nitrite during processing and continues to be converted with time after processing until a fairly constant, low level is reached (Greenberg, 1972). This conversion depends on such factors as pH and processing and storage temperatures (Nordin, 1969). Some of the added nitrite forms nitrosomyoglobin and/or nitrosylhemochrome (Woolford et al., 1972; Bard & Townsend, 1971). Gaseous products such as NO and N_2O are formed (Woolford et al. 1972). The nitric oxide is believed to be involved in the nitrosomyoglobin and nitrosylhemochrome formation (Bard & Townsend, 1971). Some of the nitrite may also become bound to proteins (Mirna & Hoffman 1969, Olsman & Krol 1972).

A determination of the fate of the nitrite ion in cured meat is of importance in decisions concerning the addition of nitrite to meat products. We used the stable isotope of nitrogen (^{15}N) as a label for sodium nitrite in order to establish quantitatively the amount of nitrite nitrogen in various fractions of a typical cured meat as a function of processing temperature and storage time.

Materials and methods

A comminuted lucheon meat type product was prepared under conditions that simulated commercial production. The meat used was 80% lean picnic and included 3% water, 3% salt (sodium chloride), 0.25% dextrose and 0.25% sucrose. Sodium nitrite labeled with ^{15}N (96.1% enrichment from Prochem., Lincoln Park, New Jersey) was added at 156 mg/kg. Ascorbate was not used in the formulation because it accelerates breakdown of nitrite (Mirvish et al., 1972). Even though this may be desirable, ascorbate was purposely avoided in order to simplify our system.

The meat was ground and mixed with salt and sugar for 8 min. The nitrite and water were then added and mixing continued under vacuum for eight more minutes; product temperature after mixing was 2°C. The vacuum line to the mixer included two liquid nitrogen gas traps, connected in series to retain volatiles which might be produced during mixing.

The meat was stuffed into cans (11 ounce) which were closed under vacuum within 15 min of filling. Some of the cans were frozen at −55 °C immediately and the rest were held at 0 °C for 24 h. After 24 h half of the remaining cans were

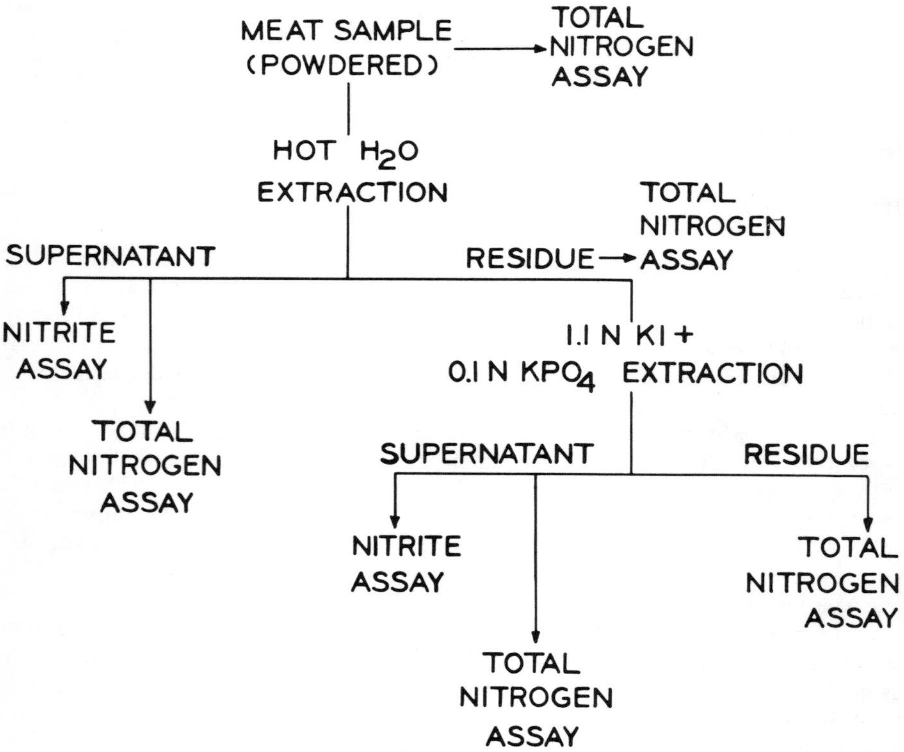

Fig. 1. Diagram for assay of meat samples.

cooked for 60 min in 71 °C water and half were retorted for 80 min at 107 °C. Following heat processing, the cans were chilled in 15 °C water. The frozen samples were stored in a freezer −18 °C. The 77 °C samples were stored at 5 °C and the 107 °C samples were stored at room temperature (22 °C).

The samples were fractionated as illustrated in Fig. 1; analysis for ^{15}N was conducted on each fraction by converting sample nitrogen to $(NH_4)_2SO_4$ which was in turn converted to N_2 on the mass spectrometer (Burris & Wilson, 1957; Bremner & Keeney, 1964; Davisson & Parsons, 1919). Sample analysis were conducted at 2, 23, 40, 49 and 65 days after processing.

The samples were extracted with water according to the AOAC procedure for colorimetric determination of nitrite. Nitrite assay was conducted on the water supernatant by the Griess reagent method (AOAC, 1970). Total nitrogen was determined on the water supernatant as a combination of the Devarda and Kjeldahl methods (Davisson & Parsons, 1919). The residue from the water extraction was analyzed by the total nitrogen method and then subjected to extraction with 1.1 *N*

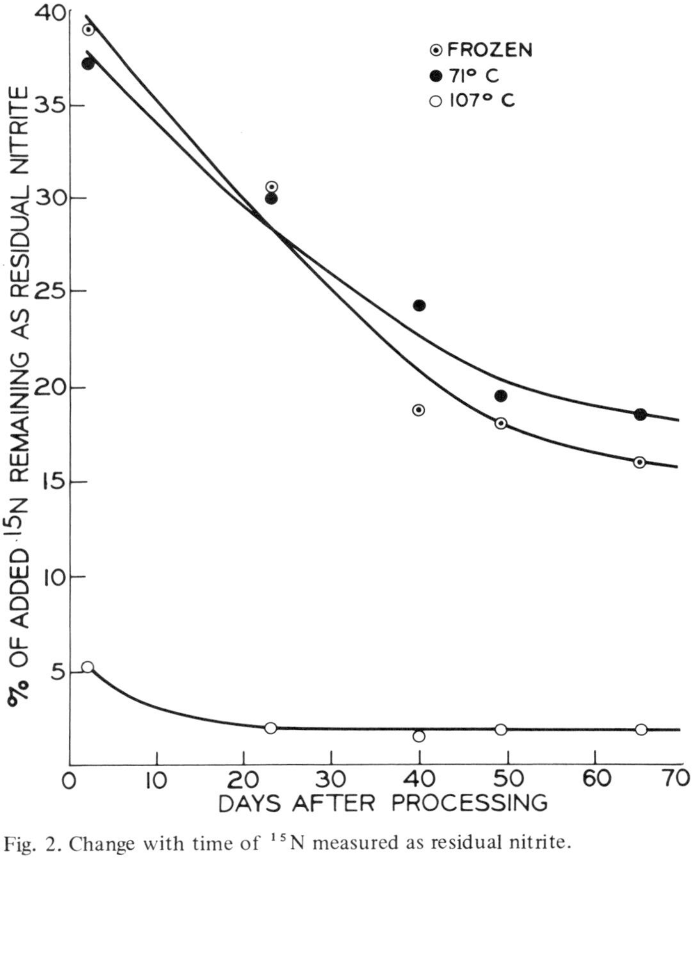

Fig. 2. Change with time of ^{15}N measured as residual nitrite.

KI and 0.1N KH_2PO_4 (Helander, 1957). The supernatant of the salt extraction was checked for residual nitrite and also subjected to the total nitrogen assay. The residue was also examined for total nitrogen.

The amount of label from nitrite that was associated with the pigment fraction was estimated by extraction with a solvent of 40 parts acetone and 3 parts water (Hornsey, 1956), followed by a simple Devarda reduction (Bremner and Keeney, 1964) which did not include the Kjeldahl digestion step.

Head space gases were analyzed by puncturing the can through an attached rubber septum with a Vacutainer test tube under vacuum. The collected gases were then injected directly into the mass spectrometer and analyzed for various nitrogen containing gases. Processing gas or volatiles generated during vacuum mixing were trapped as previously described and analyzed for various nitrogen containing gases.

Results and discussion

The first analysis was conducted two days after processing, and less than one-half of the added ^{15}N was identified as residual nitrite (Fig. 2). Residual nitrite

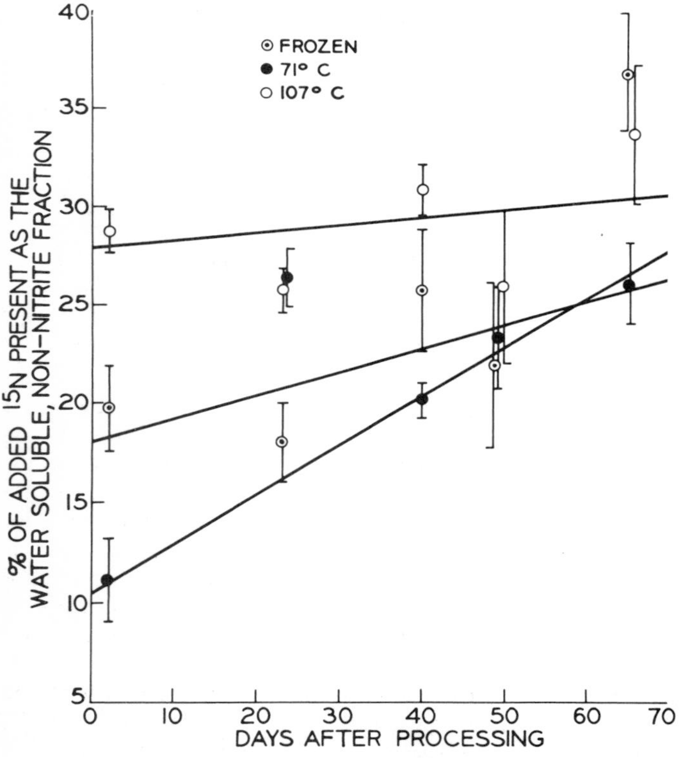

Fig. 3. Change with time of ^{15}N in non-nitrite water soluble compounds.

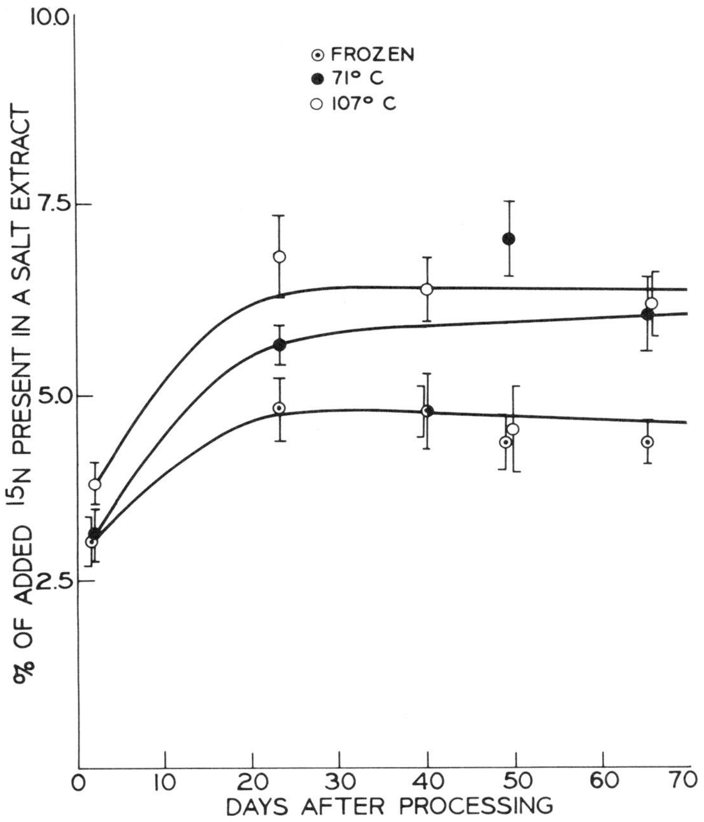

Fig. 4. Change with time of ^{15}N in compounds of a salt soluble extract.

continued to decrease with time as described by Greenberg (1972). The residual nitrite content and change during storage for both the immediately frozen (stored at $-18\ ^\circ C$) and $71\ ^\circ C$ processed samples (stored at $5\ ^\circ C$) paralleled each other closely and did not differ significantly. The residual nitrite content of the sample retorted at $107\ ^\circ C$ and stored at $22\ ^\circ C$, however, was already very low two days after processing. Nordin (1969) found nitrite conversion closely related to temperature. This was confirmed by comparison of the $71\ ^\circ C$ and $107\ ^\circ C$ samples. The frozen sample, in our experiment, however, showed an unexpected rapid conversion of nitrite; the cause for this is not known but an increased concentration of solute due to freezing may have played a role.

The total nitrogen assay on the supernatant from the water extraction showed more ^{15}N than could be accounted for by the residual nitrite. This water soluble, non-nitrite fraction accounted for a substantial portion of the total recovered ^{15}N and showed a significant increase over time for the frozen and $71\ ^\circ C$ sample (Fig. 3).

The salt extraction failed to solubilize much protein. This low protein solubility was expected since the samples were subjected to rather severe heat treatment, both in processing and in the hot water extraction. The protein that was extracted, however, showed a small but significant increase with time in the amount of ^{15}N (Fig. 4). The colorimetric test for nitrite in this supernatant was negative.

The residue from the salt extraction contained a small amount of ^{15}N initially which increased markedly and very significantly during the first 23 days of storage after which it plateaued (Fig. 5). The ^{15}N in this fraction of the frozen and 71 °C samples paralleled each other, while in the 107 °C sample it was significantly higher at 2 and 65 days of processing. The ^{15}N contained in this fraction, as well as that in the salt soluble extract, probably represents protein bound nitrogen. It has been suggested that nitrite binds to proteins through thionitroso bonding (Mirna & Hofmann, 1969; Olsman & Krol, 1972) as well as other means.

The amount of ^{15}N associated with the pigment fraction apeared at first to be surprisingly large. However, it was necessary to perform this extraction on a freshly

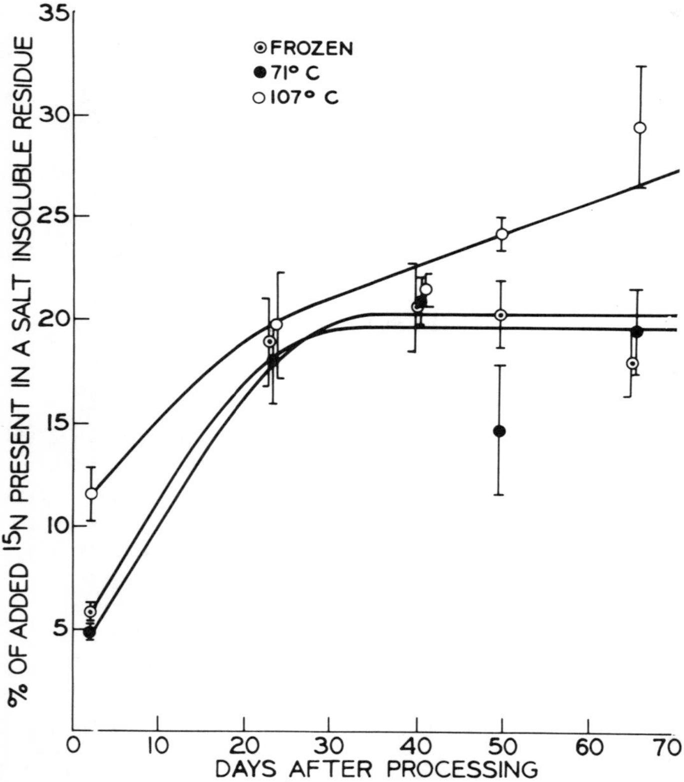

Fig. 5. Change with time of ^{15}N in compounds of a salt insoluble residue.

powdered sample and subsequent investigation showed that residual nitrite was carried along with the pigment in the small amount of water present. The amount of ^{15}N associated with the pigments was therefore corrected (for the 107 °C sample) by subtracting the ^{15}N of residual nitrite from the total ^{15}N in the uncorrected acetone-water extract and crediting the difference to pigment bound ^{15}N (Fig. 6). This correction may over-compensate, since nitrite may not be quantitatively recovered in acetone-water, but the error should be below 2% because less than 2% of the nitrite remained in the 107 °C sample beyond the first two days. This indirect method for pigment bound ^{15}N accounted for 8–9% of the ^{15}N in the 107 °C sample. The frozen sample should have little or no cured pigment formed, therefore all of the label in the uncorrected acetone water extract was presumed to be due to residual nitrite. Since residual nitrite levels were

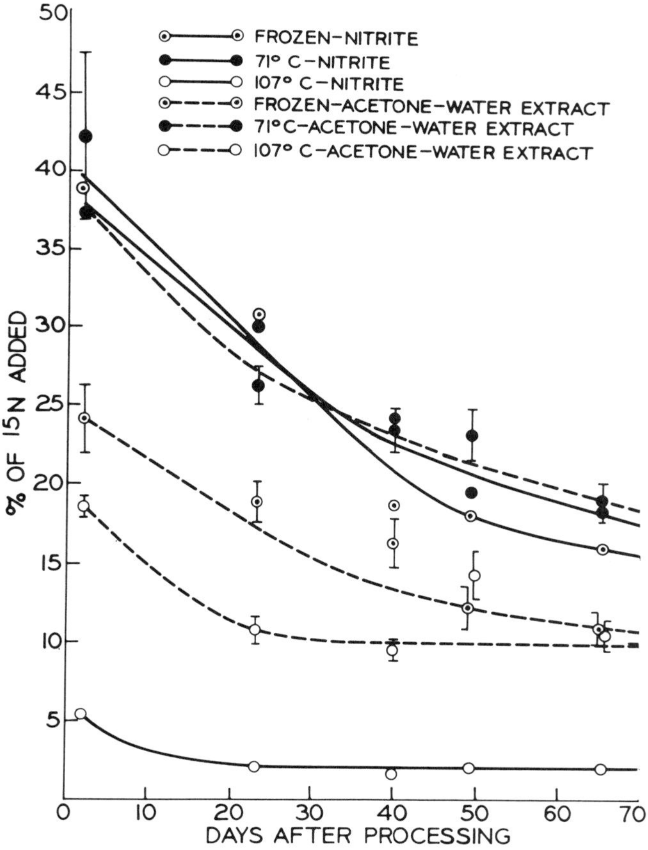

Fig. 6. Change with time of ^{15}N measured as residual nitrite and the change with time of ^{15}N in compounds in an uncorrected acetone water extract.

approximately the same for the frozen and 71 °C samples, the difference in ^{15}N content of the uncorrected acetone-water extract of two samples should be attributable to the pigment bound ^{15}N contained in the 71 °C sample. The difference was 9–12% which gave an estimation of the amount of pigment bound ^{15}N in the 71 °C sample.

Assuming that myoglobin comprises about 0.36% by wet weight of muscle (Lawrie, 1966) and the molecular weight of myglobin is 17,000, calculation reveals that on a mole of ^{15}N nitrogen per mole of myoglobin basis about 10% of the added label should be bound to myoglobin. Our results ranged between 9 and 12%. Meat will, however, contain varying amounts of myoglobin as well as some other heme pigments such as hemoglobin, cytochromes and vitamin B_{12}. Also, if cooked pigment has the capacity to bind two moles of ^{15}N nitrite as has been suggested (Tarladgis, 1962), then a maximum of 20% of the ^{15}N may be bound to pigments. Thus, our assumptions in correcting the pigment fraction seem reasonable.

Samples were also examined for ^{15}N in head space gases. Only a small amount of N_2 could be detected, which accounted for about 1% of the ^{15}N in the heat processed samples and none for the frozen sample.

Analysis of gas trapped during mixing showed a very small amount of N_2 and a somewhat larger amount of NO. Exact quantitation was not possible for the gas samples because of sampling limitations but estimates were made by comparison of peak heights with a reference of known concentration. The ^{15}N content in these two gases together was approximately 5% of the total ^{15}N added.

The sample processed at 107 °C also contained some exuded water and gelatin in the cans. This was examined by the total nitrogen assay and found to contain 2–3% of the total ^{15}N.

The total recovery of ^{15}N as determined by summing the individual fractions ranged from 72–86%.

Conclusions

It was confirmed that nitrite added to meat product is rapidly changed to other compounds during and after processing and that the rate of change slows until a rather constant low level of residual nitrite is reached. As the residual nitrite as such was changed we found an increase of ^{15}N occurred in two fractions: an apparent protein bound fraction and a non-nitrite, water-soluble fraction. The amount of ^{15}N in the pigment fraction was relatively constant during storage, did not vary greatly due to amount of heat processing and agreed with calculated predictions. Despite considerable effort, total recoveries could not be improved above the range 72–86%. However, because commercial conditions were being simulated, a low amount (156 mg/kg) of $NaNO_2$ was used. Use of larger amounts of nitrite might produce better recovery but result in artefacts and conclusions unrealistic for a commercial process. We have not attempted to quantitate nitrate which might be formed (Möhler, 1970) in the product since nitrate would be included in our total nitrogen assay.

References

AOAC, 1970. 'Official Methods of Analysis.' 11th edn Ass. Off. analyt. Chem., Washington, D. C.

Bard, J. & W. E. Townsend, 1971. Meat Curing. In: 'The Science of Meat and Meat Products', Price & Schweigert (Ed.), W. H. Freeman & Co., San Francisco, California.

Bremner, J. M. & D. R. Keeney, 1965. Steam distillation methods for determination of ammonium, nitrate and nitrite. Analytica Chim. Acta 32:485.

Burris, R. H. & P. W. Wilson, 1957. Methods for measurement of nitrogen fixation. In: 'Methods in enzymology IV' S. P. Colowick & N. O. Kaplan (Ed.), Academic Press Inc., New York.

Davisson, J. W. & J. Parsons, 1919. The determination of total nitrogen including nitric nitrogen. J. ind. Engng Chem. 11: 306.

Greenberg, R. A., 1972. Nitrite in the control of *Clostridium botulinum*, p. 25. Proc. Meat Ind. Res. Conf., Chicago, Illinois.

Helander, E., 1957. On quantitative muscle protein determination. Acta physiol. scand. 41 supp. 141

Hornsey, H. C., 1956. The color of cooked cured pork I. Estimation of the nitric oxide-haem pigments. J. Sci. Fd Agric. 7: 534.

Lawrie, R. A., 1966. 'Meat Science', Pergamon Press, New York.

Mirna, A. & K. Hoffmann, 1969. Über den Vergleib von Nitrit in Fleischwaren. Fleischwirtschaft 10: 1361.

Mirvish, S. S., L. Wallcave, M. Eagen & P. Shubik, 1972. Ascorbate-nitrite reaction; Possible means of blocking the formation of carcinogenic N-nitroso compounds. Science 177: 65.

Möhler, K., 1970. Bilanz der Bildung des Pökelfarbstoffs im Muskelfleisch. Z. Lebensmittel-unters. U. -Forsch. 142: 169.

Nordin, H. R., 1969. The depletion of added sodium nitrite in ham. Can. Inst. Fd Technol. J. 2: 79.

Olsman, W. J. & B. Krol, 1972. Depletion of nitrite in heated meat products during storage. p. 409. Proc. 18th Meet. Meat Res. Workers, Guelph, Ontario, Canada.

Perigo, J. A. & T. A. Roberts, 1968. Inhibition of clostridia by nitrite. J. Fd Technol. 3: 91.

Perigo, J. A., E. Whiting & T. E. Bashford, 1967. Observations on the inhibition of vegetative cells of *Clostridium sporogenes* by nitrite which has been autoclaved in a laboratory medium, discussed in the context of sublethally processed cured meats. J. Fd Technol. 2: 377.

Tarladgis, B. G., 1962. Interpretation of the spectra of meat pigments. II. Cured meats. The mechanism of colour fading. J. Sci. Fd Agric. 13: 485.

Urbain, W. M., 1971. Meat preservation. In: 'The Science of Meat and Meat Products.' Price & Schweighert (Ed.), W. H. Freeman Co., San Francisco, Calif.

Wasserman, A. E. & C. N. Huhtanen, 1972. Nitrosamines and the inhibition of clostridia in medium heated with sodium nitrite. J. Fd Sci. 37: 785.

Woolford, G., R. J. Casselden & C. L. Walters, 1972 Gaseous products of the interaction of sodium nitrite with porcine skeletal muscle. Biochem. J. 130:82P.

Discussion

Recovery of ^{15}N

It was remarked that the loss of ^{15}N was relatively high. Cassens answered that two years ago the recovery was only 30%. There is quite a lot of handling so that losses can be large. Walters confirmed that the losses in this kind of experiment are high. There was no indication that the type of heating or storage influenced the recovery: the data show too great a scatter. This can be explained since a small amount of ^{15}N had to be detected in a large amount of N-containing protein.

Loss of labelled ^{15}N at freezing

The suggestion that the unexpected rapid loss of nitrite ^{15}N during storage at $-18\ ^{\circ}C$ was due to concentration effects, was based on results presented by Tannenbaum and Fan (Proc. Meat Ind. Res. Conf. 1973, pp. 1–10) who obtained similar results at -6 to $-46\ ^{\circ}C$.

Nature of the ^{15}N compounds

Are there any indications about the nature of the ^{15}N-compounds in the water-soluble and protein-bound fraction? The author answered that his laboratory is initiating work with Sephadex column chromatography in an attempt to separate ^{15}N containing compounds in the water soluble fraction. Very preliminary results have shown that the ^{15}N is associated with compounds of rather low molecular weight.

Walters said that in similar experiments, involving model curing experiments at low temperature, low-resolution mass spectrometric peaks were observed only in the presence of nitrite which could have been derived from either ethane or ethylene. Both gas chromatography and high-resolution mass spectrometry showed that the compound present was ethylene. No route of information can be put forward for its production in very small amounts, but ethyl nitrite has been recognized in extracts of bacon and can break down loss of nitrous acid.

Nitrate and nitrite in animal diets

Additional information was given by Cassens concerning the work of Wang and Hoekstra (Dept. Biochemistry, Univ. Wisconsin). They have conducted experiments on excretion and retention of ^{15}N labelled sodium nitrate and sodium nitrite following ingestion by rats. The rats (6 in each group) received a single meal that contained either 2% of the diet as sodium nitrate or 1.6% as sodium nitrate. The animals had been previously trained to consume such a meal. Urine and faeces were collected at 6-h intervals for 72 h and the animals were killed to terminate the experiment. Urine, faeces and whole body were analysed for ^{15}N and the results were expressed as a percentage of the original amount of N.

The bulk of the dose was excreted within 48 h.

In future experiments organs will be examined separately.

Distribution of ^{15}N, 72 hours after feeding.

Labelled component	^{15}N in urine (%)	^{15}N in faeces (%)	^{15}N in body (%)	recovery of ^{15}N (%)
$^{15}NO_2$	68.4 ± 7.9	12.4 ± 0.7	10.2 ± 1.4	91.0 ± 8.2
$^{15}NO_3$	58.0 ± 10.9	6.3 ± 1.6	10.7 ± 2.4	75.0 ± 10.5

Proc. int. Symp. Nitrite Meat Prod., Zeist, 1973. Pudoc, Wageningen

Some compounds influencing colour formation

Norihide Ando

Laboratory of Chemistry and Technology of Animal Products, Faculty of Agriculture, Kyushu University, Fukuoka, Japan
Present address: Takamiya 2–14–20, Minami-ku, 815 Fukuoka-shi, Japan

Abstract

The effects of some food additives and some metal ions on the behaviour of nitrite and those of some endogenous low-molecular compounds in sarcoplasm from porcine skeletal muscle on the behaviour of nitrite and the formation of cooked cured meat colour were studied.

Glutamate, succinate, nicotinic acid and nicotinamide enhanced the decomposition of nitrite in the presence of ascorbate; nicotinamide also favoured the formation of nitric oxide from nitrite in the presence of ascorbate.

Of the five metal ions: Mg^{2+}, Ca^{2+}, Zn^{2+}, Fe^{2+} and Fe^{3+} tested, only Fe^{2+} significantly decomposed nitrite in the absence of ascorbate. With ascorbate, however, Mg^{2+}, Ca^{2+} and Zn^{2+} enhanced the decomposition of nitrite to some extent, whereas the effects of Fe^{2+} and Fe^{3+} were considerable.

The subfraction most active in forming cooked cured meat colour was collected from the low-molecular fraction of sarcoplasm by gel filtration. The compounds in this subfraction were identified by thin-layer chromatography.

The results indicated that the low-molecular substances IMP, ATP, reduced glutathione, glutamate, Fe^{2+} and possibly ribose are probably at least partly responsible for the colour formation.

Introduction

In recent studies on the formation of cured meat colour the following compounds have been reported to favour colour formation: various nitroso-intermediates of such reductants as ascorbic acid, cysteine, reduced nicotinamide-adenine dinucleotide (NADH) and hydroquinone (Fox & Thomson, 1963; Fox & Ackerman, 1968), nitroso-intermediate of ferricytochrome c, mitochondria and NADH (Walters & Taylor, 1963; Taylor & Walters, 1967), nitrosothiol groups formed in meat proteins (Mirna & Hofmann, 1969), arginine and lysine (Miyake et al., 1969), NADH, reduced nicotinamide-adenine dinucleotide phosphate (NADPH), flavin mononucleotide (FMN), flavin-adenine dinucleotide (FAD) and riboflavin (Koizumi & Brown, 1971), glucono-δ-lactone (Sair, 1963 and 1965; Meester, 1965), sodium ascorbate, sodium erythorbate, glucono-δ-lactone and sodium acid pyrophosphate (Sair, 1971).

In previous studies at our laboratory, it was observed that monosodium and monopotassium orthophosphates (Ando et al., 1961), disodium pyrophosphate and sodium hexametaphosphate (Ando et al., 1963), and sorbic acid (Ando & Nagata, 1969) enhanced the formation of cooked cured meat colour. In the investigations of the rapid curing process, it has been shown that of the four fractions separated from porcine skeletal muscle, sarcoplasm, myofibrils, mitochondria and microsomes fractions, the sarcoplasm fraction was the most active in the decomposition of nitrite and colour formation (Ando & Nagata, 1970). When the low-molecular and high-molecular fractions were separated from sarcoplasm by dialysis, the low-molecular fraction was clearly more active in the decomposition of nitrite and in the production of colour (Ando et al., 1971). As in the rapid curing process of cooked sausage many food additives are generally used in Japan, the effects of various food additives on the behaviour of nitrite were also investigated with aqueous model sytems (Nagata & Ando, 1971 and 1972).

Recently, the possibility of the formation of carcinogenic nitrosamines from nitrite has become a serious problem in processing meat products. In Norway the use of nitrate and nitrite in processing meat products has been banned since January 1, 1973, with some exceptions. At the present stage, however, nitrite is still widely used in most countries.

In this paper, therefore, with special reference to the rapid curing process, the results of investigation about (1) the effects of some food additives and some metal ions on the behaviour of nitrite, (2) the effects of some endogenous compounds on the behaviour of nitrite and colour formation with aqueous model systems will be reported.

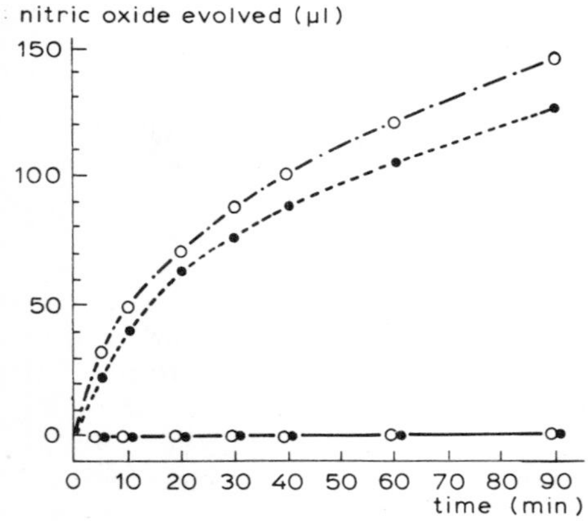

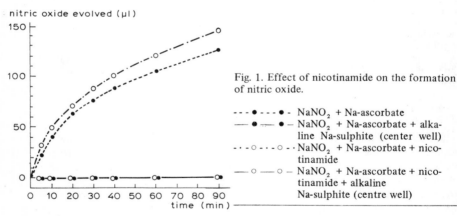

Fig. 1. Effect of nicotinamide on the formation of nitric oxide.

- - - ● - - - ● - $NaNO_2$ + Na-ascorbate
—— ● —— ● – $NaNO_2$ + Na-ascorbate + alkaline Na-sulphite (center well)
- · - ○ - · - ○ - · $NaNO_2$ + Na-ascorbate + nicotinamide
—— ○ —— ○ – $NaNO_2$ + Na-ascorbate + nicotinamide + alkaline Na-sulphite (centre well)

Warburg flask

Main compartment:	2 ml of 0.075% $NaNO_2$ solution in veronal buffer of pH 5.0 (0.05%)*
Side arm:	0.5 ml of 3% Na-ascorbate solution in veronal buffer of pH 5.0 (0.5%)*
	0.5 ml of 3% nicotinamide solution in veronal buffer of pH 5.0 (0.5%)*
Centre well:	0.2 ml of 10% Na-sulphite solution in 0.5N NaOH
Total volume:	3 ml
Atmosphere:	nitrogen
Incubating condition:	pH 5.0, 37°C, 90 min

( )*: the final concentration of additive used.

Materials and methods

Test solutions, prepared as described by Ando & Nagata (1969), contained sodium nitrite and one of the following: 0.3% monosodium glumate, 0.1% disodium succinate, 0.1% nicotinic acid, 0.1% nicotinamide, 0.02% $MgCl_2$, 0.02% $CaCl_2$, 0.02% $ZnCl_2$, 0.02% $FeCl_2$, 0.02% $FeCl_3$. Two samples from each solution were cooked at 75°C for 1 h, one immediately after preparation and one after standing for 72 h at 4°C. Before and after cooking, nitrite in the solutions was estimated by the method of Ando & Nagata (1969).

The effect of nicotinamide on the formation of nitric oxide was observed by a manometric technique under the experimental conditions specified in Fig. 1.

Experiments on some endogenous compouds

In the previous study (Ando et al., 1971), the low-molecular fraction of sarcoplasm obtained from porcine skeletal muscle was separated into 50 fractions of 5 ml each by Sephadex G–50 gel filtration; fractions 33 up to 43 (Region A in Fig. 2) were most active in decomposing nitrite, and had a high CFA (colour formation ability) and RA (reducing ability). They were also rich in ninhydrinpositive substance(s), carbohydrate(s) and unknown substance(s) with an absorption maximum at 248 nm.

In this work, therefore, Region A in Fig. 2 was collected first from the low-molecular fraction of sarcoplasm prepared from porcine skeletal muscle (Musculus adductores) just after slaughter, according to the procedures given in the previous

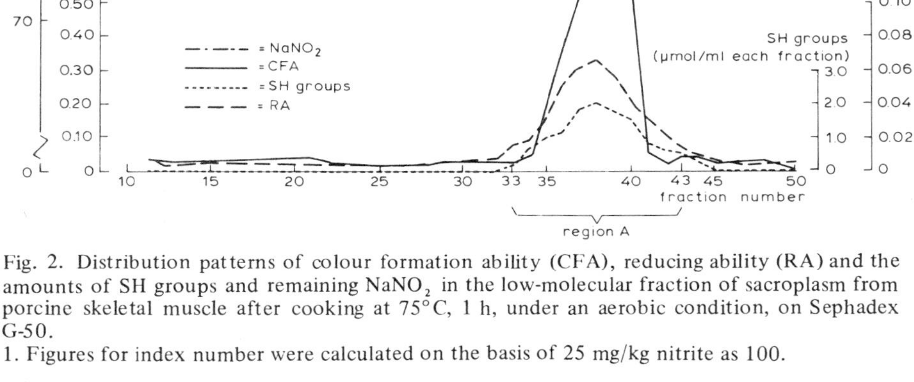

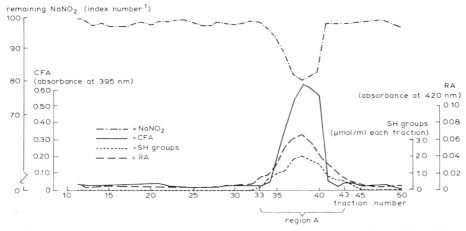

Fig. 2. Distribution patterns of colour formation ability (CFA), reducing ability (RA) and the amounts of SH groups and remaining $NaNO_2$ in the low-molecular fraction of sacroplasm from porcine skeletal muscle after cooking at 75°C, 1 h, under an aerobic condition, on Sephadex G-50.

1. Figures for index number were calculated on the basis of 25 mg/kg nitrite as 100.

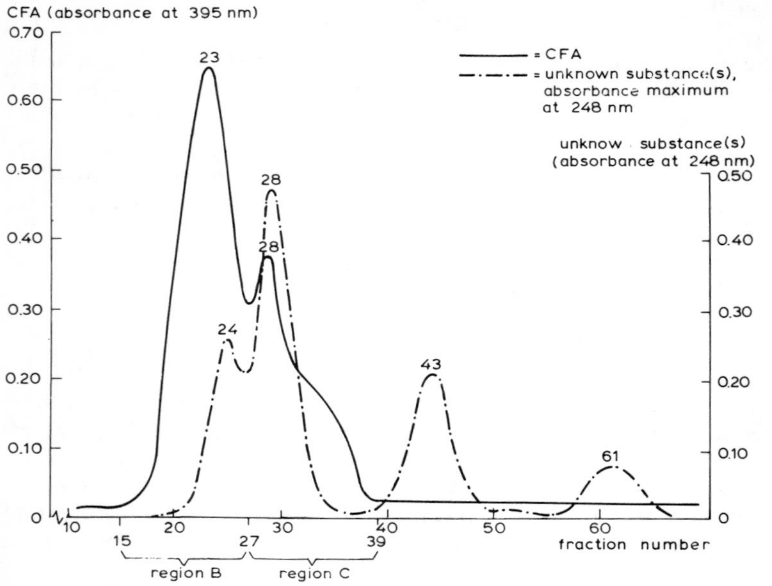

Fig. 3. Distribution patterns of colour formation ability (CFA) and unknown substances with an absorbance maximum at 248 nm in Region A in Fig. 2 on Sephadex G-15.

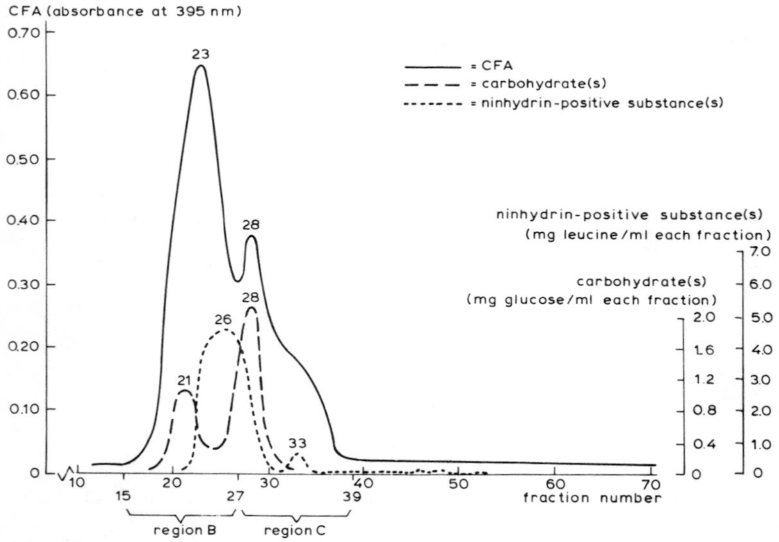

Fig. 4. Distribution patterns of colour formation ability (CFA), ninhydrin-positive substance(s) and carbohydrate(s) in Region A in Fig. 2 on Sephadex G-15.

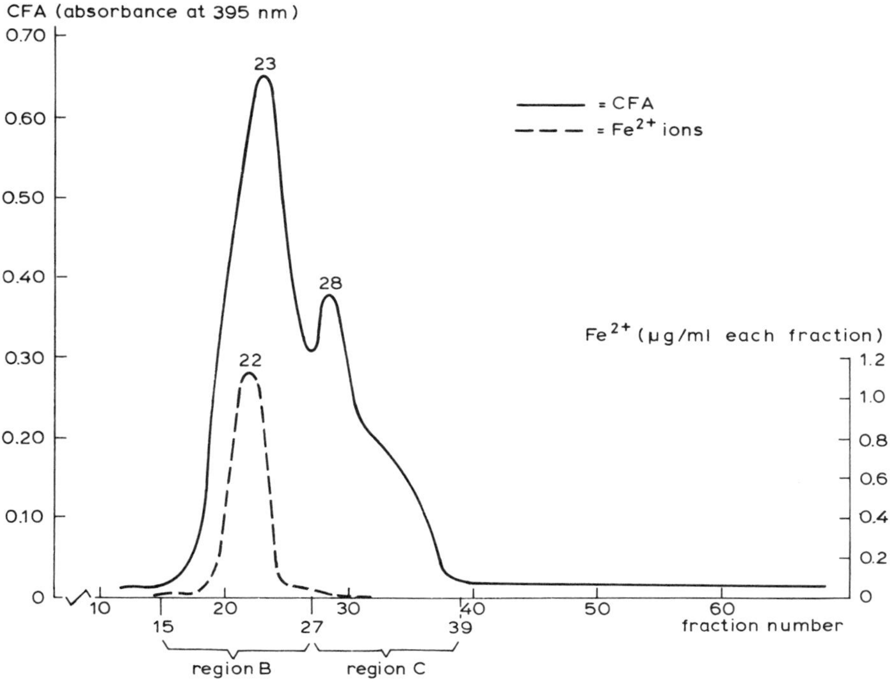

Fig. 5. Distribution patterns of colour formation ability (CFA) and Fe^{2+} in Region A in Fig. 2 on Sephadex G-15.

paper (Ando et al., 1971). Then the pooled fractions of Region A were again separated into 70 fractions of 5 ml each by Sephadex G–15 gel filtration. The distribution patterns of CFA, ninhydrin-positive substance(s), carbohydrate(s) and unknown substance(s) with an absorption maximum at 248 nm were determined by the method of Ando et al. (1971). The distribution of Fe^{2+} was determined with the o-phenanthroline-NH_2OH reagent (Jikken Nogei Kagaku, 1961).

As shown in Figs 3, 4 and 5, the fractions 15 up to 39 were found to be very active in colour formation. Two CFA peaks at the fractions 23 and 28 were observed. Consequently the fractions 15 up to 39 were divided into two regions, one comprising fractions 15 up to 27 (Region B in Figs 3, 4 and 5) and the other fractions 27 up to 39 (Region C in Figs 3, 4 and 5)

Since IMP has its absorption maximum at 248 nm, Region B and Region C were collected separately and then ninhydrin-positive substance(s), carbohydrate(s) and nucleotide(s) were identified by thin-layer chromatography (Randerath, 1963) on silica-gel plates (Merck), with the following developing solvent systems: phenol-water (75:25 w/w) for ninhydrin-positive substance(s), n-butanol-acetic acid-water (3:1:1 w/w) for carbohydrate(s), and isobutylalcohol-octylalcohol-ethylene glycol monoethyl ether-NH_3-water (30:10:70:15:55 v/v) for nucleotide(s). The ninhydrin-positive substance(s) were detected with ninhydrin-copper nitrate reagent, the carbodydrate(s) with anisaldehyde reagent, and the nucleotide(s) by looking at the spots on the silica-gel plate under ultraviolet light at 257 nm.

Results and discussion

Food additives

The results given in Tables 1 and 2 indicate that the food additives tested had more effect on the behaviour of nitrite in test solutions at pH 5.0 than at pH 6.0. Nitrite was not decomposed in any of the test solutions without ascorbate.

In the presence of ascorbate, however, as is shown in the test solutions at pH 5.0, the addition of glutamate, succinate, nicotinic acid and nicotinamide appreciably enhanced the decomposition of nitrite during cooking. In each test solution the amount of nitrite decomposed just after preparation was always larger than that after it had been kept for 72 h at 4 °C. This may have resulted from a decrease in the ascorbic acid due to autoxidation during the storage period.

The above fact is of importance when ascorbate is used in processing cooked sausage.

Glutamate and succinate are widely used as seasonings for processed meat

Table 1. Relative amount of $NaNO_2$ (%) after the influence of sodium glutamate and sodium succinate.

Test solution			Just after preparation at 0°C		After being kept for 72h at 4°C	
pH[1]	additive (%)		before cooking	after cooking[2]	before cooking	after cooking[2]
5.0	$NaNO_2$	0.01	100.0	100.0	100.0	100.0
5.0	$NaNO_2$ Na glutamate	0.01 0.3	100.0	100.0	100.0	100.0
5.0	$NaNO_2$ Na succinate	0.01 0.1	100.0	100.0	100.0	100.0
5.0	$NaNO_2$ Na ascorbate	0.01 0.1	99.8	69.0	92.0	73.7
5.0	$NaNO_2$ Na ascorbate Na glutamate	0.01 0.1 0.3	99.8	61.7	90.9	68.0
5.0	$NaNO_2$ Na ascorbate Na succinate	0.01 0.1 0.1	99.8	61.4	89.6	71.0
6.0	$NaNO_2$ Na glutamate	0.01 0.3	100.0	100.0	100.0	100.0
6.0	$NaNO_2$ Na succinate	0.01 0.1	100.0	100.0	100.0	100.0
6.0	$NaNO_2$ Na ascorbate	0.01 0.1	100.0	94.1	98.7	95.0
6.0	$NaNO_2$ Na ascorbate Na glutamate	0.01 0.1 0.3	100.0	94.1	98.6	94.8
6.0	$NaNO_2$ Na ascorbate Na succinate	0.01 0.1 0.1	100.0	93.9	98.6	95.0

1. Adjusted with veronal buffer.
2. Cooked at 75°C for one hour.

Table 2. Relative amount of $NaNO_2$ (%) after the influence of nicotinic acid (NA) and nicotinamide (NAM).

Test solution			Just after preparation at 0°C		After being kept for 72 h at 4°C	
pH[1]	additive (%		before cooking	after cooking[2]	before cooking	after cooking[2]
5.0	$NaNO_2$	0.01	100.0	100.0	100.0	100.0
5.0	$NaNO_2$ NA	0.01 0.1	100.0	100.0	100.0	100.0
5.0	$NaNO_2$ NAM	0.01 0.1	100.0	100.0	100.0	100.0
5.0	$NaNO_2$ Na ascorbate	0.01 0.1	99.8	68.7	91.5	75.5
5.0	$NaNO_2$ Na ascorbate NA	0.01 0.1 0.1	99.8	58.4	90.4	65.7
5.0	$NaNO_2$ Na ascorbate NAM	0.01 0.1 0.1	99.8	64.0	91.1	72.6
6.0	$NaNO_2$ Na	0.01 0.1	100.0	100.0	100.0	100.0
6.0	$NaNO_2$ NAM	0.01 0.1	100.0	100.0	100.0	100.0
6.0	$NaNO_2$ Na ascorbate	0.01 0.1	100.0	93.8	98.7	95.5
6.0	$NaNO_2$ Na ascorbate NA	0.01 0.1 0.1	100.0	91.6	97.9	93.1
6.0	$NaNO_2$ Na ascorbate NAM	0.01 0.1 0.1	100.0	93.2	97.6	94.5

1. Adjusted with veronal buffer.
2. Cooked at 75°C for one hour.

products in Japan. Saleh & Watts (1968) reported that the addition of NAD or NADP and various substrates from the glycolytic and respiratory pathway promoted the enzymatic reduction. Of all the substrates they tested only the use of glutamate was economically possible in processing meat products. The presence of succinate in meat seems possible as an intermediate of the tricarbonytic cycle. In the present experiments with non-enzymatic model systems both glutamate and succinate enhanced the decomposition of nitrite in the presence of a reductant.

It has been observed that nicotinic acid (NA) and nicotinamide (NAM) form red hemochrone (Coleman et al., 1949 and 1951; Olcott & Lukton, 1961; Kendrick & Watts, 1969) and NAM, but not NA, forms pink hemochromes in canned tuna (Brown & Tappel, 1957).

NAM is now widely used in processing meat products in Japan together with nitrite and ascorbate or isoascorbate to enhance colour formation. However the role of NAM in colour formation is still unclear. It has been suggested that NAM may protect NAD in the tissues from destruction by nucleosidase (Severin et al., 1963; Bailey et al., 1964; Kendrick & Watts, 1969) so that it may play a role in colour formation.

As shown in Table 2, NAM enhanced the decomposition of nitrite in the presence of ascorbate. The result of the manometric experiment, described in Fig. 1, indicated that NAM also forms nitric oxide in the presence of ascorbate, because no gas-evolution was observed in the presence of sodium sulphite, an absorbent for nitric oxide. Only nitric oxide may have been evolved from nitrite in this reaction system. These facts seem partly responsible for the better colour formation.

Metal ions

According to Weiss et al. (1953), Fe^{2+}, Fe^{3+}, Cu^{2+} and Zn^{2+} enhanced the formation of cured meat colour in the presence of ascorbate. The favourable effect of Fe^{2+} was also observed by Siedler & Schweigert (1959) and Reith & Szakály (1967).

From Tables 3 and 4 it can be seen that Mg^{2+}, Ca^{2+}, Zn^{2+} and Fe^{3+} did not decompose nitrite in the absence of ascorbate, while Fe^{2+} alone significantly

Table 3. Relative amount of $NaNO_2$ (%) after the influence of Mg^{2+}, Ca^{2+} and Zn^{2+}. (All the metals were added as chlorides).

Test solution			Just after preparation at 0°C		After being kept for 72 h at 4°C	
pH[1]	additive (%)		before cooking	after cooking[2]	before cooking	after cooking[2]
5.0	$NaNO_2$	0.01	100.0	100.0	100.0	100.0
5.0	$NaNO_2$ Mg^{2+}	0.01 0.02	100.0	100.0	100.0	100.0
5.0	$NaNO_2$ Ca^{2+}	0.01 0.02	100.0	100.0	100.0	100.0
5.0	$NaNO_2$ Zn^{2+}	0.01 0.02	100.0	100.0	100.0	100.0
5.0	$NaNO_2$ Na ascorbate	0.01 0.1	99.8	68.9	90.8	72.2
5.0	$NaNO_2$ Na ascorbate Mg^{2+}	0.01 0.1 0.02	99.8	65.6	90.8	69.0
5.0	$NaNO_2$ Na ascorbate Ca^{2+}	0.01 0.1 0.02	99.8	66.9	90.6	70.3
5.0	$NaNO_2$ Na ascorbate Zn^{2+}	0.01 0.1 0.02	99.8	65.5	90.4	68.0
6.0	$NaNO_2$ Mg^{2+}	0.01 0.02	100.0	100.0	100.0	100.0
6.0	$NaNO_2$ Ca^{2+}	0.01 0.02	100.0	100.0	100.0	100.0
6.0	$NaNO_2$ Zn^{2+}	0.01 0.02	100.0	100.0	100.0	100.0
6.0	$NaNO_2$ Na ascorbate	0.01 0.1	100.0	97.9	98.7	97.0
6.0	$NaNO_2$ Na ascorbate Mg^{2+}	0.01 0.1 0.02	100.0	97.4	98.6	97.2
6.0	$NaNO_2$ Na ascrobate Ca^{2+}	0.01 0.1 0.02	100.0	97.9	98.7	97.0
6.0	$NaNO_2$ Na ascorbate Zn^{2+}	0.01 0.1 0.02	100.0	97.9	98.9	97.0

1. Adjusted with veronal buffer.
2. Cooked at 75°C for one hour.

Table 4. Relative amounts of $NaNO_2$ (%) after the influence of Fe^{2+} and Fe^{3+}. (Both iron ions were added as chlorides).

Test solution			Just after preparation at 0°C		After being kept for 72 h at 4°C	
pH[1]	additive (%)		before cooking	after cooking[2]	before cooking	after cooking[2]
5.0	$NaNO_2$	0.01	100.0	100.0	100.0	100.0
5.0	$NaNO_2$ Fe^{2+}	0.01 0.02	99.8	67.0	100.0	71.7
5.0	$NaNO_2$ Fe^{3+}	0.01 0.02	99.8	100.0	100.2	100.0
5.0	$NaNO_2$ Na ascorbate	0.01 0.1	99.8	68.9	90.8	72.2
5.0	$NaNO_2$ Na ascorbate Fe^{2+}	0.01 0.1 0.02	99.6	16.1	82.3	18.1
5.0	$NaNO_2$ Na ascorbate Fe^{3+}	0,01 0.1 0.02	99.8	26.2	63.9	23.7
6.0	$NaNO_2$ Fe^{2+}	0.01 0.02	100.0	59.9	100.0	87.3
6.0	$NaNO_2$ Fe^{3+}	0.01 0.02	100.0	100.0	100.0	100.0
6.0	$NaNO_2$ Na ascorbate	0.01 0.1	100.0	97.9	98.7	97.0
6.0	$NaNO_2$ Na ascorbate Fe^{2+}	0.01 0.1 0.02	100.0	47.2	98.9	40.0
6.0	$NaNO_2$ Na ascorbate Fe^{3+}	0.01 0.1 0.02	100.0	73.5	95.0	73.3

1. Adjusted with veronal buffer.
2. Cooked at 75° C for one hour.

decomposed nitrite in the absence of ascorbate at pH 6.0 as well as at pH 5.0. The enhancement of decomposition by the addition of Fe^{2+} seems to be in good agreement with the observations of Olsman & Krol (1972).

In the presence of ascorbate, however, Mg^{2+}, Ca^{2+} and Zn^{2+} showed similar tendencies to those of the food additives in Tables 1 and 2.

We observed in our previous work (Ando et al., 1973) that the addition of sodium chloride increased the content of free Mg^{2+}, Ca^{2+} and Zn^{2+} and as sodium chloride is usually added to meat in processing meat products, the increased content of free bivalent metal ions may conceivably enhance colour formation in the presence of the reductants in meat products. This may presumably be one explanation for the observations of Weiss et al. (1953) and Olsman & Krol (1972) that the addition of EDTA exerts an unfavourable effect on the formation of cured meat colour.

Fe^{2+} and Fe^{3+} clearly enhanced the decomposition of nitrite in the presence of ascorbate at pH 6.0 and at pH 5.0, as shown in Table 4. The Fe^{3+} is likely to be active only after it has been reduced to Fe^{2+} by the ascorbic acid. Therefore the combined effect of Fe^{3+} and ascorbate on the decomposition of nitrite, as shown in Table 4, is puzzling. In comparison with the solution containing Fe^{2+} and ascorbate, the loss of nitrite during the 72 hours is unexpectedly high. There is no obvious explanation for these findings. The explanation may be continuous conversion of oxidized Fe^{2+} to maintain its presence and activity.

Endogenous compounds

According to Figs 3, 4 and 5, Region B contained nucleotide(s), ninhydrin-positive substance(s), and a fair amount of Fe^{2+}, while Region C was rich in nucleotide(s) and carbohydrate(s).

Thin-layer chromatography showed that Region B contained ATP, IMP, reduced glutathione (GSH), alanine, glutamate, histidine and Fe^{2+}, and Region C contained IMP, alanine, glutamate, glycine, histidine, tyrosine and ribose.

The results given in Table 5 indicate that GSH, in the system we used, enhances the decomposition of nitrite and colour formation. Neither ATP nor IMP enhanced the decomposition of nitrite and colour formation in the absence of GSH. In the presence of GSH, IMP seemed to enhance the decomposition of nitrite, whereas ATP seemed to enhance colour formation.

From the above observations, it may be suggested that in the fractions of Region B, colour formation may have been enhanced by the combined action of ATP, IMP, GSH, glutamate and Fe^{2+}, and in the fractions of Region C by the combined action of IMP, glutamate and possibly ribose, because it was observed in our previous studies that glucose and sucrose enhanced the decomposition of nitrite in the presence of ascorbate (Nagata & Ando, 1971).

Quite recently, Hamm & Bünnig (1972) reported that a fairly large amount of iron in porcine and bovine muscle was neither hemoglobin nor myoglobin bound. This nonheme-iron was as high as 68% of the total iron in porcine muscle and 29% in bovine muscle. As they had previously observed that no free iron ions existed in muscle tissue, they suggested that the above nonheme-iron may be bound to other components in muscle tissue. Olsman & Krol (1972) suggested that endogenous iron may exist in some co-ordination compounds.

As porcine skeletal muscle just after slaughter was used in the present experiments, there may have been a fairly large amount of nonheme-iron in the meat samples. Probably the Fe^{2+} found in Region B did not exist in the free state, but in some low-molecular co-ordination compounds in sarcoplasm.

Further detailed investigations are necessary to clarify the mechanism of the

Table 5. Effects of reduced glutathione (GSH), IMP and ATP on the behaviour of nitrite and the formation of cured meat colour.

Test solution[1]	Compounds added(%)	Relative amount of $NaNO_2$ decomposed after cooking[2] (%)	Absorbance at 395 nm
Control (C)	myoglobin 0.125 $NaNO_2$ 0.0025	3.2	0.065
(1)	(C) + GSH 0.1	14.4	0.365
(2)	(C) + IMP 0.2	4.4	0.065
(3)	(C) + ATP 0.2	3.2	0.098
(4)	(1) + IMP 0.2	19.6	0.281
(5)	(1) + ATP 0.2	14.0	0.416

1. The pH value of each test solution was adjusted to 5.5 with veronal buffer.
2. Cooked for one hour at 75°C.

promoting effects of the endogenous low-molecular compounds in sarcoplasm on the formation of cooked cured meat colour in the rapid curing process.

References

Ando, N., Y. Kako & Y. Nagata, 1961. Effects of orthophosphates on the color of cooked sausage. Bull. Meat Meat prod. 1: 1.

Ando, N., Y. Kako, Y. Nagata, T. Ohashi, Y. Hirakata, N. Suematsu & E. Katamoto, 1963. Effects of pyrophosphates on the color of cooked sausage. Bull. Meat Meat Prod. 2: 1.

Ando, N. & Y. Nagata, 1969. Effects of some food additives on the behavior of nitrite in processing meat products. Proc. 15th Eur. Meet. Meat Res. Workers, D 15.

Ando, N. & Y. Nagata, 1970. Effects of some fractions of porcine muscle on the behavior of nitrite and the formation of cooked cured meat color. Proc. 16th Eur. Meet. Meat Res. Workers, D 5.

Ando, N., Y. Nagata & T. Okayama, 1971. Effects of lowmolecular fraction of sarcoplasm from porcine skeletal muscle on the behavior of nitrite and the formation of cooked cured meat color in rapid curing process. Proc. 17th Eur. Meet Meat Res. Workers, C 8.

Ando, N., T. Ohashi & T. Ito, 1973. Some observations on the water-holding capacity of meat. Proc. 19th Eur. Meet Meat Res. Workers. Paris.

Bailey, M. E., R. W. Frame & H. D. Naumann, Studies of the photoxidation of nitrosomyoglobin. J. Agr. Fd Chem. 12: 89.

Brown, W. D. & A. L. Tappel, 1957. Identification of the pink pigment of canned tuna. Fd. Res. 22: 214.

Coleman, H. M., A. H. Steffen & E. W. Hopkins, 1949. Process for treating animal materials. Patent No. 2,491,646. December 20.

Coleman, H. M., A. H. Steffen & E. W. Hopkins, 1951. Process for treating animal materials. Patent No. 2,541,572. February 13.

Fox, J. B. Jr. & J. S. Thomson, 1963. Formation of bovine nitrosylmyoglobin. I. pH 4.5–6.5. Biochemistry 2: 465.

Fox, J. B. Jr. & S. A. Ackerman, 1968. Formation of nitric oxide myoglobin: Mechanism of the reaction with various reductants. J. Fd Sci. 33: 364.

Hamm, R. & K. Bünnig, 1972. Myoglobin, hemoglobin and iron in bovine and porcine muscle. Proc. 18th Meet. Meat Res. Workers. Session III.

Jikken Nogei Kagaku. Bekkan, 1961. Tokyo.

Kendrick, J. L. & B. M. Watts, 1969. Nicotinamide and nicotinic acid in color preservation of fresh meat. J. Fd Sci. 34: 292

Koizumi, C. & W. D. Brown, 1971. Formation of nitric oxide myoglobin by nicotinamide adenine dinucleotide and Flavins. J. Fd Sci. 36: 1105.

Meester, J., 1965. The application of glucono delta lactone in meat products. 11th Eur. Meet Meat Res. Workers. Beograd.

Mirna, A. & K. Hofmann, 1969. Über den Verbleib von Nitrit in Fleischwaren. I. Umsetzung von Nitrit mit Sulfhydrylverbindungen. Fleischwirtschaft 49: 1361.

Miyake, M., A. Tanaka & K. Kawakami, 1969. Studies on utilization of amino acids for foodstuffs. IV. Amino acid as a meat emulsifying agent. Rep. Fac. Fisheries, Prefectural University of Mie 6: 165.

Nagata, Y. & N. Ando, 1971. Effects of curing ingredients, seasonings, phosphates and preservatives on the behavior of nitrite. Eiyo To Shokuryo 24: 489.

Nagata, Y. & N. Ando, 1972. Effects of some food additives promoting the formation of cooked cured meat color and some metal ions on the behavior of nitrite. Eiyo To Shokuryo 25: 28.

Olcott, H. S. & A. Lukton, 1961. Hemichrome and hemochrome formation with anserine, carnosine, and related compounds. Archs Biochem. Biophys. 93: 666.

Olsman, W. J. & B. Krol, 1972. Depletion of nitrite in heated meat products during storage. Proc. 18th Meet Meat Res. Workers. Session VII.

Randerath, K., 1963. Thin layer chromatography. Academic Press. New York.

Reith, J. F. & M. Szakaly, 1967. Formation and stability of nitric oxide myoglobin. I. Studies with model systems. J. Fd Sci. 32: 188.
Sair, L., 1963. How to make cured meat red. Part II. Meat. 29: 36.
Sair, L., 1965. Sausage-A $2 billion market. IX. Cure acceleration. Meat 31: 44.
Sair, L., 1971. Cure accelerators. Proc. 17th Eur. Meet Meat Res. Workers, C 4.
Saleh, B. & B. M. Watts, 1968. Substrates and intermediates in the enzymatic reduction of metmyoglobin in ground beef. J. Fd Sci. 33: 353.
Severin S. E., L. N. Tsetlin & T. N. Druzhinina, 1963. Enzymic breakdown of diphosphopyridine nucleotide in cardiac and skeletal muscle. Biochemistry 28: 112.
Siedler, A. J. & B. J. Schweigert, 1959. Effects of heat, nitrite level, iron salts, and reducing agents on formation of denatured nitrosomyoglobin. J. Agr. Fd Chem. 7: 271.
Tayler, A.McM. & C. L. Walters, 1967. Biochemical properties of pork muscle in relation to curing. Part 2. J. Fd Sci. 32: 261.
Walters, C. L. & A.McM. Taylor, 1963. Biochemical properties of pork muscle in relation to curing. Fd Technol. 17: 354.
Weiss, T. J., R. Green & B. M. Watts, 1953. Effect of metal ions on the formation of nitric oxide hemoglobin. Fd Res. 18: 11.

Discussion

Succinate as a seasoning

In Japan succinate is used for seasoning; it enhances the decomposition of nitrite in the presence of ascorbate as well as glutamate.

Pre-rigor or post-rigor meat

Pre-rigor meat instead of post-rigor or aged meat was used because in aged meat the amount of low-molecular compounds is higher; so the situation is more complicated then.

Biochemical aspects

Some of the participants thought it likely that nitrite depletion could well be related to the biochemical process involved. However, Ando did not examine the effect of succinate and glutamate on the respiration of fresh meat. Only nitrite depletion and colour formation have been studied.

Proc. int. Symp. Nitrite Meat Prod., Zeist, 1973. Pudoc, Wageningen.

Nitrite and the flavour of cured meat I

D. S. Mottram and D. N. Rhodes.

ARC Meat Research Institute, Langford, Bristol, BS18 7DY, UK

Abstract

The effect of varying the concentration of sodium nitrite used in curing pork upon the flavour of bacon has been investigated. A taste panel was used to identify the various flavour characteristics and to examine products cured under different conditions. As the nitrite concentration was increased from zero to 1000 mg/kg an almost linear increase in the intensity of bacon flavour was found but above 1500 mg/kg further increase in flavour was small. Salt was shown to make a major contribution to bacon flavour but sodium nitrite has no detectable taste at concentrations similar to those found in bacon. The differentiation between salt pork and bacon in blind comparisons by flavour or odour was remarkably uncertain.

Volatile odorous compounds, isolated from pork and bacon, were analysed by gas chromatography and gas chromatography – mass spectrometry; no compound directly responsible for bacon flavour was isolated.

Introduction

Curing has an age-old history but not until the end of the last century was it established that nitrite rather than nitrate is the agent responsible for cured colour formation and probably for cured flavour. The relationship of nitrite to flavour was first described by Brooks et al. (1940) who compared taste panel preferences for pork cured in brines of differing nitrite and nitrate composition and concluded that 'the characteristic cured flavour of bacon is due primarily to the action of nitrite on the flesh and a satisfactory bacon can be made with only sodium chloride and sodium nitrite.' No taste panel data were given, although it was observed that the differences in samples from different treatments were small. At the same time parallel work at the Research Association of the British Food Manufacturers (Macara, 1939) reached similar conclusions.

The first rigorous test of the essential part played by nitrite in cured flavour formation was made by Cho & Bratzler (1970) who compared porcine musculus longissimus dorsi cured experimentally in brines with and without nitrite. They established significant differences in blindfold triangular tests, and in a suusequent two sample test the nitrite-containing sample was judged to possess a more intense cured flavour. Nitrite-containing and nitrite-free samples could also be distinguished when the samples were smoked and when sodium chloride was omited from the cure. However, the results also revealed that on 116 occasions comprising 288 individual tastings, no differentiation was made between salt pork and cured meat.

A similar conclusion regarding the role of nitrite in flavour formation in frankfurters was obtained by Wasserman & Talley (1972), although in a two-way scoring test the panel recorded similar scores for the flavour of smoked nitrite-free and nitrite-containing samples. Sausages prepared with and without nitrite were examined by Skjelkvale et al. (1973) who obtained significant differences between the samples in triangular tests, the differences being more easily distinguished in smoked than in unsmoked sausages.

In none of these studies was any relationship established between the amount of nitrite in the cure and the amount of cured flavour. Barnett et al. (1965) examined this among other factors which might affect the flavour of cured hams; his taste panels showed no preference for hams cured with nitrite levels over the range 100 to 1500 mg/kg, although 4500 mg/kg caused a bitter flavour which decreased the acceptability. No significant differences were observed between frankfurters cured with 150 or 75 mg/kg of nitrite (Wasserman & Talley, 1972) or sausages with 80 or 40 mg/kg (Skjelkvale, 1973).

Saltiness plays some part in the overall acceptability of bacon. The sensory appreciation of the saltiness of bacon is much less intense than that of aqueous solutions of equal strength, an effect due, probably, to fluid binding by the protein or to its slow release during mastication (Ingram, 1949).

A comprehensive study of the acceptability of Canadian and Danish bacon was carried out (in England) by the Canadian National Research Council in 1938–'39. They showed that the high salt content of Canadian bacon (7.5%) compared with Danish bacon (6.2%) was the major source of dissatisfaction (Winkler & Cook, 1941). These observations were followed by a laboratory taste panel study in which bacon containing salt in the range 4.5–9% was evaluated (Hopkins, 1947). A preference for 4.75% salt was observed with nitrate-free bacon, while with bacon containing 0.25% nitrate a 4% salt level was preferred. Nitrite at levels of 50–50 mg/kg did not apparently contribute to saltiness. Rhodes (1971, unpublished results) has also found a preference level of 3–4% salt in experimentally cured bacon, although 2% and 5% were only slightly marked down.

There are few reports concerning the actual constituents of cured meat responsible for flavour. Piotrowski et al. (1970) showed that precursors of basic meaty aroma are water extractable whereas components or precursors of the cured aroma are soluble in non-polar solvents. Gas chromatography of volatiles developed on heating these extracts showed that no single component had a meaty or cured flavour, but variations were observed in the pattern of volatiles among 6 types of ham studied. A number of volatile odorous compounds have been isolated from cured hams and sausages (Lillard & Ayres, 1969; Ockerman et al., 1964; Langner, 1969; Langner et al., 1970) most of which have been found in pork and other red meats. Cross & Ziegler (1965) also examined volatiles from cured hams and made a critical comparison with uncured pork. *n*-Hexanal and *n*-pentanal were found to be present in pork but almost absent from the cured product. It was suggested that the absence of these aldehydes was responsible for the flavour difference between ham and pork and that it was brought about by the modification of the course of the autoxidation of fat in the presence of nitrite.

The present work was undertaken to determine the effect of nitrite concentration on the flavour of bacon, and to investigate the nature of the volatiles contributing to flavour produced in pork and bacon.

Experimental

Taste panel studies on cured pork

Pilot-scale was carried out on pork middles with brines containing sodium chloride (20% w/v) and sodium nitrite at levels of 0, 250, 500, 1000 and 2000 mg/1. Pork middles (23–27 kg) were pumpted by repeated single needle injection to a 7 to 8% increase in weight, soaked in cover brines for 4 days at 4–5 °C and matured at this temperature for a further 6 days. Samples of twenty 3mm slices were vacuum packed and stored at −20°C until required. Salt and nitrite analyses were performed on random samples from each middle.

A standard cooking procedure was used in which the rashers were suspended from an aluminium frame in a covered pyrex casserole, and heated at 175 °C for 35 min. The samples were presented, one at a time to an experienced bacon tasting panel. Four samples were tasted at each sitting, the first on each occasion being salted pork (i.e. no nitrite in the cure) and the second a sample from the 2000 mg/kg nitrite cure. The components of the flavour of bacon and pork were decided on by the panel during preliminary tastings and discussions; they were pork flavour, bacon flavour, metallic flavour, saltiness, odour intensity and taste intensity when eating. While assessing taste intensity the panelists pinched the nose to eliminate the odour sensation. Samples were scored by marking the appropriate position on a 10-cm line calibrated at equally spaced intervals: nil, slight, moderate, strong, extreme. Numerical scores were subsequently obtained by measurement along the line (nil = 0, extreme = 10).

The colour difference between meat cured with and without nitrite made it essential to present the samples in such a way as to mask visual differences. Blindfolding of judges might have resulted in impaired ability to assess flavour. The use of green illumination in the panel booths eliminated the red colour of the meat, but bacon always appeared slightly darker when presented alongside pork samples. However, the samples were served separately, and the green illumination adequately prevented recognition of the identity of the meats.

Taste panel studies on nitrite solutions

Solutions containing different concentrations of sodium chloride and sodium nitrite were made up in freshly distilled water. The triangle taste test was used with the trained bacon tasting panel to determine if a difference could be detected between nitrite-free and nitrite-containing solutions. For each pair of solutions compared, tests were given with both the control and the nitrite solution as the odd sample. Samples were positioned randomly to prevent sample or positional bias.

The triangle test was also carried out using hot mashed potato as a carrier for nitrite and chloride. This potato was prepared from an 'instant' potato mix, reconstituted in the appropriate solutions.

Aroma panel studies on pork and bacon

Samples of pork and bacon (500 g of each) from the same animal were cooked in vacuum packs by immersing in water at 90 °C for 25 min. The cooked meat was minced and homogenised with 500 ml water and a few drops of an aqueous solution of a red food dye (Amaranth S) added to give similar colours to both homogenates. A two-way triangle test was used to determine if a difference could be detected in the aroma of the pork and bacon. Samples were served hot in pyrex beakers and were positioned randomly. Green illumination was employed in well ventilated taste panel booths.

The triangle odour test was also carried out with samples of grilled pork and bacon, presented minced in beakers covered with perforated aluminium foil to prevent selection being made on a visual basis.

G.l.c. analysis of volatile substances

One side of pork was cured by a normal commercial Wiltshire process and the other used as control; back fat and lean joints (approximately 500 g) were vacuum packed sealed in plastic bags and stored at −20 °C until required. The joints were cooked, by immersing the bags after thawing in water at 90 °C for 25 min and the whole content of the bag was minced. 500 g sample was homogenised with 1 l distilled water, 200 g sodium chloride added in a 3 l flask, and volatile compounds extracted for 12 h using a steam distillation-continuous extraction apparatus as described by Likens & Nickerson (1964). The solvent arm of the apparatus contained diethyl ether (50 ml) which had been re-distilled over a 350 X 15 mm column packed with glass helices. Any volatile material escaping from the apparatus was condensed in two cold traps connected in series to the condenser vent of the apparatus; the first trap was cooled with solid carbon dioxide/methanol, the second with liquid nitrogen. A slow stream of nitrogen was bubbled through the aqueous meat homogenate throughout the distillation.

The ethereal extract was dried over anhydrous sodium sulphate and concentrated to about 2 ml by distillation over a 150 × 10 mm column packed with glass helices, then by passing a slow stream of pure nitrogen over the surface of the chilled solution, to give a final volume of 250 μl. The concentrated extract was sealed in glass vials under an atmosphere of nitrogen and stored at −20 °C. Extracts were prepared from six bacon joints and combined for gas chromatographic analysis; similarly, the combined extract from six matching pork joints was prepared.

Gas chromatography of flavour extracts was carried out on a Pye 104 instrument equipped with a flame ionisation detector, using a 1.5 m × 6 mm glass column packed with 5% SE 30 on Chromosorb G (60–80 mesh) with an argon flow of 40 ml/min. The column temperature was maintained at 75 °C for 10 min, and then was programmed at 4 °C/min to 190 °C. The column effluent stream could be split before passage into the detector to permit evaluation of odours of components eluted from the column. The effluent splitter was also employed to allow the simultaneous use of the flame ionisation detector and a nitrogen-specific Coulson electrolytic detector.

Mass spectrometric analysis of the extracts was carried out with an LKB 9000 gas chromatograph – mass spectrometer, (electron energy, 70eV; ion source temperature, 250 °C; separator temperature, 200 °C). The gas chromatographic column (1.5 m × 6 mm, glass) was packed with 5% SE 30 on Chromosorb G (60–80 mesh) and operated with a helium flow of 20 ml/min. The column temperature was maintained at 75 °C for 5 min, and then was programmed at 4 °C/min to 190 °C.

Results and discussion

Taste panel data

No significant trends in the relationship between the taste panel assessments of metallic flavour, or of odour or taste intensity and the amount of nitrite used in curing were found (Table 1). The panel found a bacon flavour in salted pork (mean 2.54) and this increased with nitrite concentration almost linearly to a level

Table 1a. Taste panel evaluation of the flavour of cured pork prepared with brines containing different nitrite concentrations. Four samples served at each panel. Comparison of products from cures with 0, 500, 1000, 2000 mg/kg sodium nitrite (each figure is the mean of nine tasters and four replicates).

Characteristic	Treatment Means[1]				Analysis of variance: f-ratios		
	0	500	1000	2000	treatments	tasters	interaction
Bacon flavour	2.54	3.08	4.18	4.74	5.14**	2.53n.s.	2.73***
Pork flavour	3.03	2.35	1.33	1.00	5.23**	1.89n.s.	4.31***
Salt flavour	2.73	3.15	4.15	4.34	2.66n.s.	8.36***	1.86*
Metallic flavour	1.24	1.57	1.54	1.63	0.82n.s.	13.75***	0.58n.s.
Odour intensity	4.77	4.11	3.95	4.56	1.83n.s.	7.28***	1.85*
Taste intensity	5.24	4.78	4.98	5.37	1.46n.s.	9.80***	1.42n.s.

Table 1b. Comparison of products from cures with 0, 250, 1000, 2000 mg/kg sodium nitrite (each figure is the mean of nine tasters and two replicates).

Characteristic	Treatment means[1]				Analysis of variance: f-ratios		
	0	250	1000	2000	treatments	tasters	interaction
Bacon flavour	2.42	3.00	3.97	4.33	5.63**	2.69**	1.84n.s.
Pork flavour	3.35	2.36	1.46	1.24	9.41***	2.16n.s.	1.63n.s.
Salt flavour	2.57	3.39	3.80	4.00	2.27n.s.	4.45**	0.63n.s.
Metallic flavour	1.37	1.32	1.68	1.32	0.36n.s.	4.08**	0.77n.s.
Odour intensity	5.19	4.03	4.72	4.59	2.39n.s.	5.72***	1.39n.s.
Taste intensity	5.12	5.12	5.58	5.14	0.43n.s.	1.36n.s.	0.47n.s.

1. Means of scores of tasters derived from scale: 0 = nil; 2.5 = slight; 5 = moderate; 7.5 = strong; 10 = extreme.
*** significant at p = 0.001;
** significant at p = 0.01;
* significant at p = 0.05;
n.s. not significant.

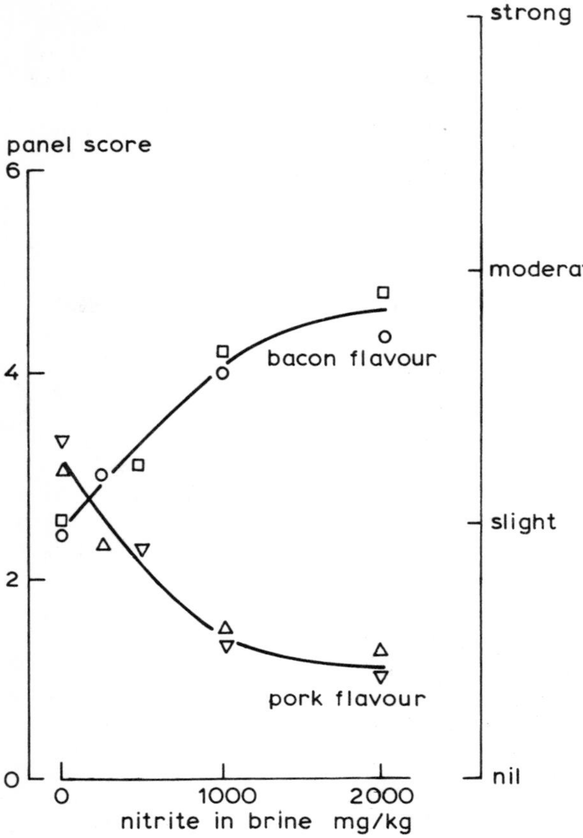

Fig. 1. The relationship between nitrite concentration in curing brines and panel assessment of bacon and pork flavours in the cooked bacon. (□ and ○ = bacon flavour, ▽ and △ = pork flavour from Tables 1a and 1b, respectively).

slightly above 1000 mg/kg; above this the increase in flavour fell off (Fig. 1). Thus a level of about 1500 mg/kg of nitrite was required to attain a near maximum effect on flavour: such a level contrasts with the much lower amount of nitrite needed to fully convert myoglobin to the nitroso-pigment characteristic of cured meat. This result does not imply that the 1500 mg/kg level is essential to produce satisfactory bacon; such a conclusion would require consumer marketing studies, but it establishes that a flavour-producing reaction between meat and nitrite is continuing well beyond the levels where colour formation is completed.

The panel mean scores for pork flavour displayed an inverse progression to those for bacon flavour, the relationship being extremely close (Fig. 2). The totals of the scores for pork flavour plus bacon flavour were, therefore, constant for all levels of nitrite with a mean of 5.5: this figure may be compared with the panel's assess-

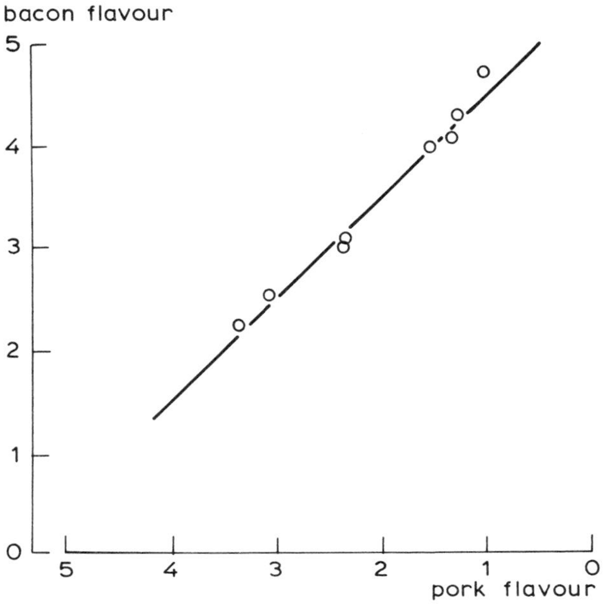

Fig. 2. The relationship between taste panel scores for pork and bacon flavours in bacon cured with nitrite levels from 0 to 2000 mg/kg.

ments of taste intensity and odour intensity, neither of which varied with nitrite level (Table 1 a and b) and gave means of 5.2 and 4.5, respectively. Analysis of variance of the two sets of results (Table 1a and b) showed frequent interaction between testers and treatments indicating that a wide divergence of opinion existed amongst the panelists in their differentiation between bacon and pork flavour when making judgements without prior knowledge of the identity of the meat in the samples submitted.

The panel means for odour intensity showed the cured samples to be slightly lacking in odour compared with the salted pork control and this was tested by

Table 2. Comparison of aromas of pork and bacon by two-way triangle test.

Samples compared	Odd sample	Number of correct answers / Number of judges
Boiled bacon and boiled pork	pork	10/13**
	bacon	9/13**
Grilled bacon and grilled pork	pork	20/29***
	bacon	21/29***

**significant at p = 0.01;
***significant at p = 0.001.

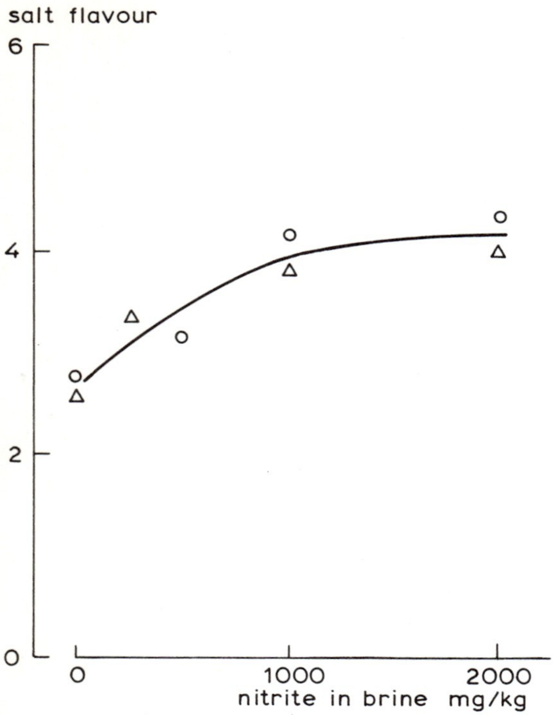

Fig. 3. Taste panel judgements of saltiness of bacon cured at different levels of nitrite. (○ and △ = results from Tables 1a and 1b, respectively).

triangle tests between pork and bacon, boiled and fried. An odour difference was then significantly established in the direct comparison; however, even under these critical circumstances, 30% of the judgements made were incorrect (Table 2).

Taste panel judgements of the saltiness of the samples were consistently higher in the cured meats than in the salted pork and showed a progressive increase with nitrite level (Fig. 3). The salt content of the samples varied between 3.9 and 6.1% but in a random way (Table 3) which could not account for the panel results. Nitrite itself has no detectable taste at 100 or 200 mg/kg in aqueous or 0.5% sodium chloride solution nor could it be detected when submitted in a neutral solid carbohydrate medium (mashed potato powder) (Table 4). Bacon is already known to exhibit a masking effect on saltiness (Ingram, 1949)and these results suggest that salt is bound into the cured meat structure more firmly than in untreated pork meat, though whether by a chemical or physical mechanism is not known. Unsalted pork was clearly recognised by the taste panel which gave it a mean mark of 0.32 for bacon flavour; whereas the presence of 4% salt led to the mark of 2.5 (Table 1).

These results underline the major part played by the presence of salt in the consumer's appreciation of bacon flavour; firstly, about one half of an expert panel's marking is accounted for by salt alone and secondly, about one third of

Table 3. Concentrations of curing salts in samples of cured pork served to taste panels.

$NaNO_2$ in brine (mg/kg)	Residual $NaNO_2$ (mg/kg)	Residual NaCl (%)
0	1	4.5
250	18	6.1
500	31	3.9
1000	72	5.0
2000	145	4.9

Table 4. Comparison of solutions with and without sodium nitrite by two-way triangle test.

Solution in:	$NaNO_2$ concentrations compared (mg/kg)	Odd sample	Number of correct answers number of judges
distilled water	0 and 100	0	7/14n.s.
		100	4/14n.s.
0.5% NaCl	0 and 100	0	6/14n.s.
		100	5/14n.s.
distilled water	0 and 200	0	7/14n.s.
		200	8/14n.s.
0.5% NaCl	0 and 200	0	5/14n.s.
		200	5/14n.s.
potato powder	0 and 200	0	5/12n.s.
		200	5/12n.s.
potato powder + 1% NaCl	0 and 200	0	6/12n.s.
		200	4/12n.s.

n.s. not significant

blind judgements on the identity of pork or bacon are in error. They show too that the moderate levels of nitrite in common usage for bacon production are enough to produce close to the maximum amounts of cured flavour attainable; reduction of these levels will, therefore, immediately result in loss of cured flavour.

Chemical analyses of volatiles

The aromas of pork and bacon are, by no means, distinguishable with certainty even by experienced judges (30% failures in our results; cf. 40% failures in taste testing by Cho & Bratzler (1970). Extracts of volatiles have been made and analysed and more than 40 peaks with retention times between 5 and 70 min have been found in gas chromatograms of both pork and bacon. Many of the peaks were complex and contained more than one substance and although observations of the odours of the more volatile peaks have shown some minor differences between cured and uncured patterns, no clear differences have been indentified. Five peaks, in both samples, contained nitrogen, one of which was thialdine a compound formed from hydrogen sulphide, ammonia and acetaldehyde in boiled beef (Brinkman et al., 1972). No light has been thrown by the results so far obtained on the theory of Cross & Ziegler (1965) that autoxidation of linoleic acid in pork during cooking is inhibited by nitrite present in cured meats.

References

Barnett, H. S., H. R. Nordin, H. D. Bird & L. J. Rubin, 1965. A study of the factors affecting the flavour of cured ham. 11th Eur. Meet. Meat Res. Workers, Belgrade.

Brinkman H. W., H. Copier, J. J. M. de Leuw & S. B. Tjan, 1972. Components contributing to beef flavour. Analysis of the headspace volatiles of beef broth. J. Agric. Fd Chem. 20: 177–181.

Brooks. J., R. B. Haines, T. Moran & J. Pace, 1940. The function of nitrate, nitrite and bacteria in the curing of bacon and hams. Fd Invest. Bd Spec. Rept. No. 49, HMSO, London.

Cho, I. C. & L. J. Bratzler, 1970. Effect of sodium nitrite on flavour of cured pork. J. Fd Sci. 35: 668–670.

Cross, C. K. & P. Ziegler, 1965. A comparison of the volatile fractions from cured and uncured meat. J. Fd Sci. 30: 610–614.

Hopkins, J. W., 1947. Precision of assessment of palatability of foodstuffs by laboratory panels II. Saltiness of bacon. Can. J. Res. F25: 29–33.

Ingram, M., 1949. Salt flavour in bacon. J. Soc. Chem. Ind. 68: 356–359.

Langner, H. J., 1969. Zur Bildung von freien Aminosäuren, fluchtigen Fettsäuren und fluchtigen Carbonylen in reifender Rohwurst. Fleischwirtschaft 49: 1475–1479.

Langner, H. J., V. Heckel & E. Malek, 1970. Aromastoffe in der reifenden Rohwurst. Fleischwirtschaft. 50: 1193–1199.

Likens, S. T. & G. B. Nikerson, 1964. Detection of certain hop oil constituents in brewing products. Proc. Am. Soc. Brew. Chem. 5–13.

Lillard, D. A. & J. C. Ayres, 1969. Flavour compounds in country cured hams. Fd Technol. 23: 251–254.

Macara, T., 1939. The development of the bacon flavour. Fd Res. Rep. No. 35, BFMIRA.

Ockerman, H. W., T. N. Blumer & H. B. Craig, 1964. Volatile chemical compounds in dry-cured hams. J. Fd Sci. 29: 123–129.

Piotrowski, E. G., L. L. Zaika & A. E. Wasserman, 1970. Studies on aroma of cured ham, J. Fd Sci. 35: 321–325.

Skjelkvale, R., M. Valland & H. Russwurm, 1973. Effekt av natriumnitritt på sensorisk kvalitet av kjøttvarer. Norsk Institutt for Naeringsmiddelforsking Informasjon 1973 No. 1: 1–15.

Wasserman, A. E. & F. Talley, 1972. The effect of sodium nitrite on the flavour of frankfurters. J. Fd Sci. 37: 536–538.

Winkler, C. A. & W. H. Cook, 1941. Canadian Wiltshire Bacon XIX. Comparative flavour tests on Canadian and Danish bacons. Can. J. Res. D19: 157–176.

Discussion

Tasting technique

It was stated that it was most unlikely that flavour components could have migrated from one sample to another during cooking or panelling.

Interpretation of results

Some participants thought it unjustified to draw further conclusions from the high number of panelists unable to distinguish between pork and bacon, because the number of times the difference was noticed was highly significant. The only conclusion can be that the difference between bacon and pork was only subtle.

Practical concentration of nitrite

In commercial Wiltshire bacon production when no nitrate is present, the usual level of nitrite in the brine is 1 500–2 000 mg/kg. In the presence of 0.1–0.5% nitrate, which is used in the majority of cases, the nitrite level is 500–1 000 mg/kg.

Proc. int. Symp. Nitrite Meat Prod., Zeist, 1973. Pudoc, Wageningen

Nitrite and the flavour of cured meat II

A. E. Wasserman

Eastern Regional Research Center, Agricultural Research Service, U.S. Department of Agriculture, Philadelphia, Pennsylvania 19 1 18, USA

Abstract

A review of the literature of the last 50 years shows very little interest in the effect of nitrite on flavor of cured meat products, particularly the relation of sodium nitrite to development of the flavor. Sensory evaluation studies indicate a need for the presence of nitrite; low concentrations may be sufficient to induce the flavor. Chemical analysis of head space vapors or extracts of cured hams has led to identification of up to approximately 50 compounds. Although many of them have been identified before in other meats or fowl, none has the characteristic cure aroma. Known reactions of nitrite and various meat components are discussed.

Introduction

Investigations into the role of nitrite in the cure process have been principally on the development of color and on the preservative, anti-clostridial effect. Authorization for the use of nitrite in cured meat as described in the U.S. Code of Federal Regulations (1971) is for the purpose of developing color. Recent reports, (Wolff & Wasserman, 1972), however, suggest a re-appraisal of the curing process because the nitrite may under some conditions react with amino compounds present to give small concentrations of nitrosamines. It is now being recognized what has been evident for many years: without the use of nitrite, the characteristic cured flavor is not developed in meat products. However, since 1940 very limited research has been carried out on this problem. The studies that have been reported can be classified either as processing and sensory evaluation products or chemical analysis of the reaction between nitrite and meat components. Insufficient information is available in either of these categories to permit a definitive discussion of the role of nitrite in the development of cure flavor, but this paper can serve to consolidate reported data and place them in a perspective that may encourage an increase in research activity.

Processing and sensory evaluation of meat products

The studies of Kerr et al. (1926), which established the basis for the use of nitrite in curing meat products, demonstrated that nitrite-cured meat was as acceptable as traditionally cured meat. Although flavor was not specifically mentioned in the report, it must be assumed that this was a consideration in the quality assessment of the products. Brooks et al. (1940) investigated the use of nitrite in the Wiltshire bacon curing process, particularly with respect to flavor formation. Although no sensory evaluation data were given and the responses appear vague, the authors concluded that the characteristic flavor of bacon and ham (as opposed to salt pork) is due to the reaction of nitrite with tissue constituents during curing or during cooking and that 10 mg NO_2^-/l cure was sufficient to give satisfactory flavor.

Barnett et al. (1965), in an extensive study of the development of ham flavor, found that hams pumped with pickle containing 0.1 g NO_2^-/l were as acceptable to a panel as those pumped with the normal pickle containing 1.5 g NO_2^-/l; the former value being equivalent to about 10 mg nitrite per kg. ham.

Unfortunately these authors did not compare hams without nitrite, but their data show that lower values can be used successfully. Tripling the normal concentration of nitrite in the pickle resulted in hams with bitter flavor.

Pork m. longissimus dorsi roasts, cured with 300 mg nitrite/kg but not smoked, could be distinguished in triangle taste tests from those cured without nitrite, and in paired comparison tests the panelists indicated the nitrite-treated products had more cure flavor (Cho & Bratzler, 1970). Smoking the cured roasts did not affect the outcome of the taste tests; nitrite-treated roasts were still differentiated in the triangle test and had more cure flavor.

More recently, Simon et al. (1972) found all meat frankfurters made with beef and pork were scored low in a hedonic taste panel evaluation when they contained no nitrite; 39 mg nitrite/kg, the lowest concentration tested, was sufficient to give an acceptable flavor. For some unexplained reason, however, all beef frankfurters were acceptable even in the absence of nitrite. It should be noted that in these studies the panelists were asked whether they liked or disliked the product – not how much cure flavor it contained.

In our laboratory, we (Wasserman & Talley, 1972) found the role of nitrite in frankfurter flavor to be complex, depending on the type of evaluation panel used. In triangle tests with both smoked and unsmoked frankfurters, there were significant differences between products prepared with or without 156 mg nitrite/kg (Table 1). However, in tests in which 'frankfurter' flavor was scored, the smoked, non-nitrite-treated sample was rated as highly as the cured sample. Analysis of variance showed an interaction between smoke and nitrite. Thus, while there is no question about the poor flavor of non-nitrite-, non-smoke-treated samples, cured franfurter flavor is also associated with smoke flavor.

Chemical studies

Ockerman et al. (1964) felt that since the aroma of dry cured ham was similar to its flavor, analysis of volatile components would identify the flavor factors. Vacuum distallates of ham cured with salt, sugar, and KNO_3 had a typical aroma and flavor; a number of carbonyls, acids, and bases were identified. While none were specifically associated with cured flavor, it was noted that the quantity of acids and carbonyls increased with the age of the ham.

Ockerman et al. (1964) felt that since the aroma of dry cured ham was similar to palmitoleic, oleic, and linoleic – decreased, first during curing and smoking, then

Table 1. Triangle test evaluation of the flavor of frankfurters prepared with cure in which sodium nitrite was either present or absent.

Experiment	Conditions	Number of correct answers number of judges
1	Cooked, no smoke; + NO_2 vs no NO_2	15/22***
2	Cooked, no smoke; + NO_2 vs no NO_2	28/36***
3	Cooked, no smoke; + NO_2 vs no NO_2	18/24***
4	Cooked, smoked; + NO_2 vs no NO_2	11/17**
5	Cooked, smoked; + NO_2 vs no NO_2	13/24*
6	Cooked, smoked; 50% NO_2 vs no NO_2	12/17**
7	Cooked, smoked; 50% NO_2 vs 100% NO_2	9/17n.s.

*p = .05;
**p = .01;
***p = .001;
n.s. = not significant

with aging. Flavor increased with age of ham. Flavor components in dry cured hams were also investigated by Lillard & Ayres (1969), who studied a number of hams of various ages from different parts of the United States. A total of 46 volatile compounds were identified. Although most of them had been reported previously in studies with chicken and other meats, a number of esters were unique to the aged hams. Large increases in amino acids were also noted, but variation due to age of the hams made it difficult to interpret these data. While chemical composition of uncured hams, unfortunately, was not compared to that of hams treated with nitrate, it was noted that the aroma and flavor of the country, or dry, cured hams was quite different from fresh pork. When the hams were cooked the flavor was intensified. The cured flavor was not water soluble but could be transferred by steam distillation at 60 °C under reduced pressure.

Volatile compounds of cured and uncured boiled hams were investigated by Cross & Ziegler (1965). Concentrating on the carbonyl compounds removed by stripping with nitrogen gas, they demonstrated the major difference between the two ham products was the absence of hexanal and valeraldehyde in the cured ham. Furthermore, after passage through a solution of 2,4-dinitrophenylhydrazine, the carbonyl-free volatiles of the uncured ham had a cured aroma, as did volatiles from cured and uncured chicken or beef. The cured aroma was removed by passage through solutions of mercuric chloride or mercuric cyanide. The authors conclude that cured ham flavor is the basic meat flavor derived from precursors other than triglycerides and that nitrite functions by inhibiting the oxidation of unsaturated lipids from which hexanal and valeraldehyde originate. Branched-chain aldehydes, derived from amino acids through the browning reaction, were found in equal concentrations in both cured and uncured hams, and thus, presumably, were not involved in the characteristic cured flavor.

Piotrowski et al. (1970) in our laboratory reported that a basic meaty aroma could be isolated in water extracts of cured ham similar to that from raw ham, beef, or veal. A fraction having the cured ham aroma was obtained by treatment of the cured ham residue of the water extraction with chloroform-methanol (2:1). This is the lipid portion of the meat. It is difficult to decide whether the cured aroma is derived from the lipids or whether it is lipid soluble, but unpublished data show that cure aroma can be transferred from the lipid fraction to a bland vegetable oil by bubbling N_2 through the former. Smoke components undoubtedly play an important role in defining cure flavor; their interaction has been demonstrated in the section on Sensory Evaluation. However, we have found that nitrite treatment enhances smoke aroma of ham (Wasserman, 1973). Uncured hams, smoked simultaneously with cure-pumped hams, had only a faint smoke aroma. Although there has been a question whether smoke and cure flavor may not be the same, it is possible to distinguish between the two. On heating aqueous extracts of cured, smoked ham, aromas recognizable as smoky are volatilized at the beginning, leaving saltier, more cured notes to evolve as heating progresses. Gas chromatographic profiles of pyrolyzates of water- and lipid-soluble fractions from cured and uncured hams were complex and, although no identifications were made, it was concluded that differences resulting from nitrite treatment of the meat were more quantitative than qualitative (Piotrowski et al., 1970). None of the compounds eluting from the gas chromatograph had a meaty or a cured aroma.

More recently, Baily and co-workers (1973) have also analyzed uncured and nitrite-treated hams for aroma constituents by gas chromatography of extracts, They have identified about 50 components in the volatile fraction, many of them previously reported in meat. Although several previously unreported compounds were identified, Bailey indicated neither the new nor the known components had cured or hammy aromas. He also showed that, at this time at least, the major differences between cured and uncured hams are quantitative and appear to be associated with lower concentrations of cabonyl compounds derived from the unsaturated lipids in the cured hams.

Two studies from Russian workers also link carbonyl compounds with ham aroma. Iskandarian (1970) reported that pyruvic acid in cured pork was an intermediate for the formation of volatile carbonyl compounds that impart a pronounced ham aroma to meat. Solovyov et al. (1970) found that cured ham differed from cooked pork in having a greater content of volatile carbonyls, particularly methylglyoxal (pyruvaldehyde).

The effect of nitrite on the amino acid composition of hams has been investigated. The free amino acids have been implicated in the development of the basic meaty aroma and primary amines are known to react with nitrite via the Van Slyke reaction. Definitive data, however, have not been reported. Solovyov et al. (1970) reported higher total amino acids in cured ham, with losses of some free amino acids. Piotrowski et al. (1970) investigated the amino acid composition of raw, cooked, and cooked-smoked hams that were either uncured or treated with nitrite-containing cure and found variations in the different preparations but couldn't identify a trend due to treatment.

What is cured flavor? If the aroma of smoke, and possibly the effect of spices, are discounted, is there actually a cured flavor? Cross & Ziegler (1965) and Bailey (1973) suggest that nitrite acts to reduce oxidation of unsaturated lipids which from odorous carbonyls. Bailey (1973) has claimed that the odor of cured and uncured meats are the same immediately after processing and that 'cured' aroma develops after storage. This would mean the difference between aromas of nitrite-treated and uncured products develops as oxidation compounds form in the uncured product. Thus 'cure' aroma actually would be a lack of certain odors that are associated with cooked pork.

On the other hand, it has been reported that the flavor of dry cured ham increases with aging (Lillard & Ayres, 1969); unfortunately these studies were not done with controls cured for the same period of time without nitrite. Oxidation under these conditions is extensive and the effect of nitrite could have been magnified.

Nitrite, however, reacts with other meat components as well. It combines with -SH compounds to form nitrosothiols. Cross & Ziegler (1965) found dimethyldisulfide in the mercuric salt trap that removed cure aroma from ham volatilies bubbled through it. They postulated this was derived from methanethiol. However, when Barnett et al. (1965) studied the effect of adding sulfur compounds to the meat prior to pickling, or curing meat from hogs fed sulfur compounds, they found, in general, there was no difference in preference between the treated and untreated ham. In some instances the sulfur compounds produced undesirable odors in the product even at low concentrations; methanethiol is an example.

Nitrite also reacts with primary amino acids, as described above; it is also known to react with secondary and tertiary amines. There are undoubtedly other meat components that react to some small extent with nitrite. It is not known whether these reactions affect the aroma.

The concluding remarks by Barnett et al. (1965) are still valid today. 'on a number of occasions we have compared hams prepared by present-day curing and processing practices with those generally employed 20 or more years ago, and almost invariably the 'modern' ham has been preferred. In spite of this, may old-timers recall the day when merely cutting a ham filled the room with tantalizing aroma, and the flavor on eating was equally as delightful'.

'However, perhaps hams never did taste like they used to.'

References

Bailey, M., 1973. Influence of nitrite on meat flavor. Presented at the Meat Ind. Res. Conf., Chicago, Ill., March, 1973.

Barnett, H. W., H. R. Nordin, H. D. Bird & L. J. Rubin, 1965. A study of factors affecting the flavour of cured ham. Proc. 11th Meeting Eur. Meat Res. Workers.

Brooks, J., R. B. Haines, T. Moran & J. Pace, 1940. The function of nitrate, nitrite and bacteria in the curing of bacon and hams. Dept. Sci. ind. Res. (Br.), Fd Invest., Special Rep. No. 49.

Cho, I. C & L. J. Bratzler, 1970. Effect of sodium nitrite on flavor of cured pork. J. Fd Sci. 35: 668.

Code of Federal Regulations. Title 9 (Revised as of January 1, 1971).

Cross, C. K. & P. Ziegler, 1965. A comparison of the volatile fractions from cured and uncured meat. J. Fd Sci. 30: 610.

Iskandaryan, A. K., 1970. Pyroracemic acid conversion into volatile carbonyls and the development of pronounced ham aroma. Abstr. Proc. 16th Meet. Eur. Meat Res. Workers, p. 119.

Kerr, R. H., C. T. N. Marsh, W. F. Schroeder & E. A. Boyer, 1926. The use of sodium nitrite in the curing of meat. J. agr. Res. 33: 541.

Lillard, D. A. & J. C. Ayres, 1969. Flavor compounds in country cured hams. Fd Technol. 23: 251.

Ockerman, H. W., 1962. Separation and identification of volatile chemical compounds in dry-cured hams. Ph. D. Thesis, North Carolina State College.

Ockerman, H. W., T. N. Blumer & H. B. Craig, 1964. Volatile chemical compounds in dry-cured hams. J. Fd Sci. 29: 123.
Piotrowski, E. G., L. L. Zaika & A. E. Wasserman, 1970. Studies on aroma of cured ham., J. Fd Sci. 35: 321.
Simon, S., D. E. Ellis, B. D. MacDonald, D. G. Miller, R. C. Waldman & D. O. Westerberg, 1972. Influence of nitrite on quality of packaged frankfurters. Proc. 18th Meet. Meat Res. Workers.
Solovyov, V. I., M. M. Tchirikova, O. P. Shtchyogleva, G. I. Edelman, V. D. Sinitsyna, V. A. Alexakhina & N. V. Kulishenko, 1970. Some physicochemical indices of specific ham aroma of cooked ham. Abstr. 16th Meet Eur. Meat Res. Workers, p. 176.
Wasserman, A. E. & F. Talley, 1972. The effect of sodium nitrite on the flavor of frankfurters. J. Fd Sci. 37: 536.
Wasserman, A. E., 1973. Unpublished data.
Wolff, I. A. & A. E. Wasserman, 1972. Nitrates, Nitrites and Nitrosamines. Science 177: 15–19.

Discussion

The study of nitrite and flavour formation

The experiments of Brooks were repeated by the British Food Manufacturing Industries Research Association with the same results. So already in 1940 it was clear that nitrite was necessary for the flavour. The gap between 1940 and 1965 can easily be explained by the lack of methods to study the nature of this flavour. Advanced techniques nowadays make a new approach possible.

Origin of cured meat flavour

Herring said that the flavour of cured meats varies according to product type. Cured flavour per se can only be measured on meat treated with nitrite and/or salt. He asked whether the flavour of all other products like cured primal cuts, wieners, dry sausages could be considered as a blend of cured flavour with that contributed by other factors such as bacteria, salt smoke and spices. A very typical example is the country style cured ham! Significant differences in flavour between frankfurters with and without nitrite were only obtained without smoke.

As long as the nature of the nitrite flavour is not known it is very difficult to look for other solutions. Further studies are necessary.

Cross and Siegler have inferred, and Bailey has indicated that cure flavour is due to the inhibition of oxidation of unsaturated fatty acids by nitrite. Cure flavour, therefore, would be a negative flavour.

NO_2 from smoke

Several speakers asked whether nitrogen oxides from the smoke plays a role. Wassermann stated that its role was proved by the identification of nitrosomyglobin in smoked frankfurters, prepared without nitrite. Furthermore it was stated that nitrite so formed may be a catalyst for a Maillard type reaction which is important both for smoke flavour and smoke colour.

Possible interaction of nitrite and fat in nitrosamine formation

Schram reported that the presence of lipid hydroperoxides was found to enhance the rate of nitrosamine formation in model systems. It is likely that the same would occur in bacon.

Proc. int. Symp. Nitrite Meat Prod., Zeist, 1973. Pudoc, Wageningen

Ascorbate and nitrosamine formation in cured meats

R. A. Greenberg

Swift & Company, Research and Development Center, Oak Brook, Illinois 60521, USA

Abstract

Nitrite is an essential component of cured meat manufacture. It is responsible for cured flavor, color, prevention of 'warmed-over' flavor, and retards botulinal toxin development, particularly under conditions of product mishandling.

Nitrosamines can be formed from the reaction of nitrite and secondary or tertiary amines under acid conditions. Nitrosamines have been observed occasionally in virtually all types of cured meats. Only in bacon, after preparation for the table, has a nitrosamine been found to be generally present. It is probably that a precursor of nitrosopyrrolidine (nitroso-hydroxy-proline?) is formed during curing and/or subsequent storage. The heat of cooking apparently catalyzes conversion of the precursor to nitrosopyrrolidine.

Ascorbate has been a curing adjunct for many years. Recent observations in several laboratories have demonstrated the ability of ascorbic acid and sodium ascorbate to block nitrosation.

Work is under way to determine whether increasing ascorbate levels in bacon curing formulations can reduce nitrosopyrrolidine levels in cooked bacon. Preliminary results are quite promising.

Introduction

The observation that nitrosamines can be found in various processed meats has resulted in one of the most serious problems ever encountered by the meat industry. In the United States, about two-thirds of all pork is consumed in the cured form. Growing shortages of beef, poultry, and other traditional protein foods have made it mandatory that consumer acceptable cured meat products not be removed from the food supply without sound reason.

While nitrosamines have been observed to be occasionally present in other cured meats, the only meat product in which the positive result cannot be traced to excessive levels of nitrite, improper processing conditions, or questionable analytical technique, is bacon. Bacon is a popular food. (Its uncured counterpart, salt belly, is not). Commercial bacon samples fried according to typical U.S. consumer procedures have been found by analysts from the U.S. Food and Drug Administration (USFDA) to contain up to 108 μg/kg of nitrosopyrrolidine (Table 1).

Table 1. *N*-Nitroso pyrrolidine levels found in commercial bacon (μg/kg).

Sample	Raw	Pan-fried	Fat cooked out
1	0	30	45
2	0	95	96
3	0	106	207
4	0	106	142
5	0	10	81
6	0	20	68
7	0	32	61
8	0	108	100

(Fazio et al., 1973)

Ascorbate as a curing adjuvant

Ascorbic acid, isoascorbic acid, sodium ascorbate, and sodium isoascorbate have been used for many years to improve color characteristics of cured meats. US Department of Agriculture regulations permit about 470 mg/kg of the acid or approximately 550 mg/kg of salt (Table 2). The permitted level is about twice that actually necessary. Thus, most American processors use about 200–250 mg/kg in their products. Watts & Lehmann (1952) reported that color fixation, odor, and flavor were greatly improved when ascorbic acid was included in the formulation of cooked 'nitrited' ground pork. Mills et al. (1958) noted that frankfurters containing sodium isoascorbate or sodium ascorbate had a more desirable internal color than did control lots. Ascorbate-free control product was less uniform in color and tended to lose its color more rapidly when held in a lighted display case. Wirth (1973), in his recently published paper in 'Fleischwirtschaft' on the possible consequences of reducing or prohibiting nitrite in cured meats, showed that nitrite formulation levels could be reduced somewhat in several products, without organoleptic difficulty, when sufficient ascorbate was utilized.

Ascorbate as a nitrosation preventive

Because many modern drugs are secondary and tertiary amines, the possibility of

Table 2 US Regulations on addition of ascorbic acid, isoascorbic acid, sodium ascorbate, and sodium isoascorbate to cured meats.

	Per 100 gal pickle at 10% pump level	Per 100 lb meat or meat byproduct
Ascorbic acid or isoascorbic acid	75 oz	3/4 oz
Sodium ascorbate or sodium isoascorbate	87.5 oz	7/8 oz

in vivo production of nitrosamines has been under consideration for some time by the pharmaceutical industry. Kamm et al. (1973) have recently demonstrated a protective effect of ascorbic acid on hepatotoxicity for rats fed a combination of sodium nitrite and aminopyrine. Articles such as from Lijinsky et al. (1973) have postulated that amine drugs (or even cigarette smoke) could combine with free nitrite from food and form nitrosamines in the human stomach. Tannenbaum (1972) has shown that nitrite is present universally in human saliva. A man typically ingests as much nitrite each day from his own saliva (6–12 mg) as he would receive from the free nitrite present in three typical meals in which a cured meat was the main component. Thus, ascorbate appears to be an interesting candidate as a nitrosation preventive, from several points of view.

Dimethylnitrosamine has been observed to be present in occasional samples of commercial frankfurters at the retail level (Wasserman et al., 1972). However, attempts to purposefully produce DMNA in frankfurters under controlled laboratory conditions have required concentrations of sodium nitrite up to about ten times the legal limit (Fiddler et al., 1972).

Mirvish et al. (1972) and Fiddler et al. (1972a) have postulated that the ascorbate-nitrite reaction could be a possible means of preventing nitrosation in cured meats. Fiddler et al. (1973), studying the effect of various meat curing ingredients on DMNA formation in a model system, found that ascorbic acid, sodium ascorbate, and sodium isoascorbate markedly inhibited DMNA formation in systems containing high concentrations of sodium nitrite. The results of their experiments with frankfurters formulated with 1500 mg/kg sodium nitrite are shown in Table 3 (Fiddler et al., 1973a). Frankfurters formulated with no ascorbate or isoascorbate developed DMNA during moderate thermal processing. Product formulated with 550 and 5500 mg/kg of these salts did not contain DMNA. Even when subjected to severe overprocessing, the effect of increasing concentration of ascorbate and isoascorbate was to reduce the DMNA level in the finished product.

Table 3. Effect of ascorbate and isoascorbate on dimethylnitrosamine formation in frankfurters formulated with 1500 μg/kg sodium nitrite.

	Processed 2 h	Processed 4 h
Ascorbate (μg/kg)	dimethylnitrosamine (μg/kg)	
0	11	22
550	0	7
5500	0	4
Isoascorbate (μg/kg)		
0	10	11
550	0	6
5500	0	0

(Fiddler et al., 1973)

Table 4. Effect of ascorbate on nitrosopyrrolidine formation in fried bacon. (170 μg/kg sodium nitrite added).

Ascorbate (μg/kg)	Storage before curing at 4 °C (days)	
	1	13
	Nitrosopyrrolidine (μg/kg)	
0	7	18
250	6	11
1000	0	0

(Herring, 1973)

Ascorbate as an answer to the bacon problem

If nitrosation can be effectively blocked by ascorbate under commercial bacon manufacturing conditions, we would have a simple way out of the nitrosopyrrolidine dilemma. Obviously, the use of 200–250 mg/kg sodium ascorbate – the current practice – does not control the development of nitrosopyrrolidine precursors in bacon. The market basket survey from the US Department of Agriculture presented in Table 1 makes this abundantly clear. Increasing the use of ascorbate up to and beyond presently acceptable levels may be at least a partial solution. With this in mind, a series of tests were begun by the American meat industry, the US Department of Agriculture, and USFDA. The results of experiments studying the effect of increasing ascorbate formulation levels on nitrosopyrrolidine formation in bacon containing typical sodium nitrite cures are shown in Tables 4, 5, and 6. The data in Table 4 represent bacon produced in a commercial plant, stored before curing for 1 and 13 days at 4 °C. Although nitrosopyrrolidine was present after frying in bacon formulated with 0 and 250 mg/kg sodium ascorbate, no nitrosopyrrolidine was found in products formulated with 1000 mg/kg sodium ascorbate. The data in Table 5 show the results of a companion experiment in which the

Table 5. Effect of ascorbate on nitrosopyrrolidine formation in fried bacon. (170 μg/kg sodium nitrite added).

Ascorbate (μg/kg)	Stored after curing at 7 °C (weeks)		
	0	1	6
	nitrosopyrrolidine (μg/kg)		
0	0–13	5–11	5–10
500	0–5	0	0
2000	0	0	0

(Herring, 1973)

Table 6. Effect of ascorbate on nitrosopyrrolidine formation in fried bacon formulated with 170 μg/kg sodium nitrite.

Ascorbate (μg/kg)	Nitrosopyrrolidine (μg/kg)
0	20 – 92
250	14 – 26
500	0 – 7
1000	10
2000	0

(Wasserman, 1973)

product was stored up to 6 weeks at 7 °C after curing. Again, increasing ascorbate level in the cure formulation reduced nitrosopyrrolidine levels in the fried bacon. The data in Table 6 show the results of still another commercial plant experiment and demonstrate a decrease in nitrosopyrrolidine with increasing ascorbate.

More definitive large-scale commercial tests are currently under way. As of the writing of this paper, the results are not yet available.

Other aspects of increasing ascorbate concentration in bacon

We have observed no organoleptic difficulties with bacon formulated with up to 2000 mg/kg ascorbate.

Increasing ascorbate intake in the human diet should have a salutary rather than a harmful nutritional effect (Yew, 1973).

One of the considerations to be resolved should increasing levels of ascorbate prove a practical means of handling the bacon nitrosopyrrolidine problem is the effect on nitrite's ability to prevent botulinal toxin formation. Nitrite's antibotulinal efficacy has been demonstrated in canned comminuted ham (Christiansen et al., 1973) and in frankfurters (Hustad et al., 1973), as well as in bacon (Greenberg, 1973). Will high levels remove sufficient nitrite from the product so as to result in a botulism hazard?

Table 7. Effect of ascorbic acid on putrefactive spoilage of pasteurized lunch meat.

Sodium nitrite (μg/kg)	Ascorbic acid (μg/kg)	Putrid spoilage (%)
0	468	100
117	0	100
117	78	18
117	156	26
117	312	20
117	468	30

(Schack & Taylor, 1966)

Work is under way to answer the question. A suggestion that this may not be a problem can be deduced from some work on luncheon meat published in the Schack & Taylor patent (1966). Luncheon meat containing 117 mg/kg sodium nitrite and inoculated with *Clostridium sporogenes* P.A. 3679 spores did not show statistically different putrefactive spoilage at ascorbic acid formulation concentrations from 78 to 468 mg/kg. Complete and rapid spoilage occurred in products containing no sodium nitrite and 468 mg/kg ascorbic acid and in products containing 117 mg/kg sodium nitrite and no ascorbic acid. The data are shown in Table 7.

References

Christiansen, L. N., R. W. Johnston, D. A. Kautter, J. W. Howard and W. J. Aunan, 1973. The effect of nitrite and nitrate on toxin production by *Clostridium botulinum* and on nitrosamine formation in perishable canned comminuted cured meat. Applied Microbiology 25: 357–362.

Fazio, T., R. H. White, L. R. Dusold and J. W. Howard, 1973. Nitrosopyrrolidine in cooked bacon. J. Ass. off. analyt. Chem. 56: 919–921.

Fiddler, W., E. G. Piotrowski, J. W. Pensabene, R. C. Doerr and A. E. Wasserman, 1972. Effect of sodium nitrite concentration on N-nitrosodimethylamine formation in frankfurters. J. Fd Science 37: 668–670.

Fiddler, W., E. G. Piotrowski, J. W. Pensabene, R. C. Doerr & A. E. Wasserman, 1972a. Some current observations on the occurrence and formation of N-nitrosamines. Proc. 18th Meet. Meat Res. Workers, Vol. II, p. 416, Guelph, Ontario, Canada.

Fiddler, W., J. W. Pensabene, I. Kushnir & E. G. Piotrowski, 1973a. Effect of frankfurter cure ingredients on N-nitrosodimethylamine formation in a model system. J. Fd Sci. 38: 714–715.

Fiddler, W., E. G. Piotrowski, J. W. Pensabene & A. E. Wasserman, 1973. Studies on nitrosamine formation in foods. 33rd Ann. Meet. Inst. Fd Technol., Miami, USA, 1973.

Greenberg, R. A., 1973. The effect of nitrite on botulinal toxin formation in bacon. Proc. Meat Ind. Res. Conf. Chicago, USA, 1973, p. 69–70.

Herring, H. K., 1973. Effect of nitrite and other factors on the physico-chemical characteristics and nitrosamine formation in bacon. Proc. Meat Ind. Res. Conf. Chicago, USA, 1973, p. 47–60.

Hustad, G. O., J. G. Cerveny, H. Trenk, R. H. Deibel, D. A. Kautter, T. Fazio, R. W. Johnston & O. E. Kolari, 1973. Effect of sodium nitrite and sodium nitrate on botulinal toxin production and nitrosamine formation in wieners. Appl. Microbiol. 26: 22–26.

Kamm, J. J., T. Dashman, A. H. Conney & J. J. Burns, 1973. Protective effect of ascorbic acid on hepatotoxicity caused by sodium nitrite plus aminopyrine. Proc. Nat. Acad. Sci. USA 70: 747–749.

Lijinsky, W., H. W. Taylor, C. Snyder & P. Nettesheim, 1973. Malignant tumours of liver and lung in rats fed aminopyrine or heptamethyleneimine together with nitrite. Nature 224: 176–178.

Mills, F., D. S. Ginsberg, B. Ginger, C. E. Weir & G. D. Wilson, 1958. The effect of sodium ascorbate and sodium iso-ascorbate on the quality of frankfurters. Fd Technol. 12: 311–314.

Mirna, A., 1973. What might be the results of forbidding or reducing the addition of nitrate and nitrite curing salts to meat products? From the chemical point of view. Fleischwirtschaft 53: 357–360.

Mirvish, S. S., L. Wallcave, M. Eagen & P. Shubik; 1972. Ascorbate-nitrite reaction: possible means of blocking the formation of carcinogenic N-nitroso compounds. Science 177: 65–68.

Schack, W. R. and R. E. Taylor, 1966. Extending shelf life of cured meats. U.S. Patent #3,258,345.

Tannenbaum, S. R., 1972. Nitrosamine content of food and current research. Proceedings of the 25th Annual Reciprocal Meat Conference of the American Meat Science Association, Ames, Iowa, USA, 1972, p. 96–106.
Wasserman, A. E., W. Fiddler, R. C. Doerr, S. F. Osman & C. J. Dooley, 1972. Dimethylnitrosamine in frankfurters. Fd Cosmet. Toxicol. 10: 681–684.
Wasserman, A. E., 1973. Personal communication.
Watts, B. M. & B. T. Lehmann, 1952. Ascorbic acid and meat color. Fd Technol. 6: 194–196.
Yew, M. S., 1973. 'Recommended Daily Allowances' for Vitamin C. Proc. Nat. Acad. Sci. USA 70: 969–972.

Discussion

Trials on increase of ascorbate levels

Addition by the author

A large scale commercial plan test was conducted to further test the practicality of controlling nitrosamine in bacon by increasing ascorbate levels in the cure. Production size 20 000 pound lots each were manufactured, using four bacon pickles to test two levels of sodium ascorbate (330 mg/kg and 1 000 mg/kg) with and without 0.4% sodium tripolyphosphate. The pumped bellies were cured for one day at 4 °C (40 °F), processed about 22 hours to 55 °C (128°–130 °F) internal temperature. Chilled bellies were formed 3 to 4 days after smoking and vacuum packed one day after forming. Some products were fried in electric pans (average frying yield 33.5%), frozen and shipped for nitrosamine analysis to the laboratories of the US Department of Agriculture-Agricultural Research Service and the US Food and Drug Administration, Other products were frozen after manufacture and shipped to those laboratories for subsequent frying.

The samples fried at the packer and analysed by the USFDA laboratory were all negative (for unknown reasons). Similar samples tested at USDA-ARS showed a nitrosopyrrolidine reduction in both phosphated and non-phosphated bacon in the higher ascorbate level product.

Analysis of products fried at USFDA also showed reduced nitrosopyrrolidine levels in the higher ascorbate level product than that formulated with 330 mg/kg.

Analyses by USDA – ARS. US commercial plant bacon study.

	NPP (in µg/kg) assays of fried bacon (170 mg/kg NO_2)			
	with 0.4% TPP		without TPP	
	fried by packer	fried by ARS	fried by packer	fried by ARS
330 mg/kg ascorbate	4	11	15	3
1 000 mg/kg ascorbate	3	0	3	0

Analyses (three samples of each variable) by USFDA (all products fried at USFDA). US commercial plant bacon study.

	with 0.4% TPP	without TPP
330 mg/kg ascorbate	20 – 38 –20	20 – 15 – 18
1 000 mg/kg ascorbate	11 – 9 – 6	6 – 7 – 6

Residual nitrite level in cured meats at time of consumption (3 oz. portions, before home preparation).

	$NaNO_2$ in original formulation	$NaNO_2$ as consumed
Bacon	120 mg/kg	3.0 mg
Frankfurthers	150 mg/kg	1.4 mg
Ham	150 mg/kg	7.6 mg
Total		12.0 mg

(Tannenbaum – Proc. of 25th Annual Reciprocal Meat Conf. AMSA, 1972 – reports daily NO_2 excretion in saliva for all human subjects studied to be 6–12 mg per day).

Addition by Dr. Schram

Trials have been made with vacuum packed raw bacon with 500 mg/kg nitrate, stored at 25 °C. The followed results were obtained:

Added nitrite (mg/kg)	Added ascorbate mg/kg)	After 14 days			After 28 days		
		NO_2 mg/kg	DMN μg/kg	NNP μg/kg	NO_2 mg/kg	DMN μg/kg	NNP μg/kg
100	0	3.5	0.5	12	1.9	0.5	4.9
100	1000	2.0	0.4	6	2.6	0.3	4.9
200	0	7.5	1.4	24	2.1	0.3	4.5
200	1000	4.5	0.4	2	3.6	0.5	4.6
0	0	22	1.5	20	1.6	0.4	4.3

The effect of ascorbate was a rapid decrease of the amount of nitrite, so less nitrosamine was formed. Nevertheless it has a positive effect against *Cl.botulinum.* In inoculated packs without added nitrite 9 out of 10 appeared to be toxic, with only 100 mg/kg nitrite 1 out of 10 was toxic, the remaining packs showed no toxin formation.

The level of ascorbate was fairly constant between 500 and 600 mg/kg. In the pouches without added nitrite the nitrite concentration showed a maximum after about two weeks.

Effect of ascorbate on DMNA formation

Work has been carried out at the Meat Research Institute in conjunction with the Laboratory of the Government Chemist to show the effect of ascorbate on the formation of dimethyl nitrosamine (DMNA). A reduction in nitrosamine formation in the presence of ascorbate has been found. Slices of pork eye-muscle, impregnated with dimethylamine (DMA) were cured in nitrite-containing brines, and the cured samples heated to 90 °C. In the absence of ascorbate at pH 5.8 a yield of about 0.1% of the theoretical amount of DMNA was found. At the same pH in the presence of ascorbate at 2.0, 1.0, 0.5 molar ratios to nitrite, this yield was reduced by about 80% i.e. a yield of 0.02%.

For pork middles similarily impregnated with DMA and cured, a similar effect of ascorbate was noticed in lean samples of the middle heated in a canning process, but in the case of fried samples the situation was more confused.

Minimum level of nitrite

Some results were given by Ingram concerning Wiltshire cure of backs with different amounts of nitrite, without nitrate. This was done in a factory under normal conditions. With 2000 and 1000 mg/kg nitrite in the brine there were no problems. With 500 mg/kg the colour was less good and also the flavour and keepability. With 250 mg/kg nitrite in the brine many problems occurred with respect to colour, flavour and keepability, there was even deterioration. These experiments were carried out with backs, which are more stable than shoulders, so it is expected that 100 mg/kg nitrite in the bacon (1000 mg/kg in the brine) will be the required level.

Role of ascorbate

Does ascorbate only act as a nitrite scavenger or does it interfere with the nitrosation reaction? No definite data are yet available, but probably ascorbate does not act only as a scavenger.

Alternatives for ascorbic acid

Alternatives are often themselves of questionable safety. It is known that several compounds like alloxane, dialluric acid and cyclohexanetrione, in very small concentrations can act like ascorbic acid. There is nothing known about their activity with respect to the nitrosation action, nor about their toxicity.

Effect of frying temperature on nitrosamine formation

It was stated that the temperature distribution during frying was extremely uneven. This has a considerable effect on the formation of nitrosamines. It was found that the concentration of nitrosamines in fat was much higher than in lean meat. Experiments have been carried out on this subject. Bacon was fried in a pan (starting with either a hot or cold pan), broiled on a grill, fried on a 'baconer' (direct heated with electric elements) or heated in a microwave oven. Little differences in the nitrosamine concentration were found, however lower levels were obtained in the bacon cooked in the microwave oven. It was commented that controlling the cooking of bacon is difficult. It is possible to set up controlled grilling and frying, but even then the quantity of nitrosamines, found in samples of bacon, varies over very wide ranges. Schramm reported that in the Unilever laboratory it was found that N-nitrosopyrrolidine is formed only in the fatty tissue and hardly at all in the lean. Nitrosamines have never been found in cured meat products that have been boiled nor in raw bacon.

Sawyer supported the Unilever finding, that no evidence of volatile nitrosamines had been obtained in the examination of boiled bacon at the laboratory of the Government Chemist, London.

Proc. int. Symp. Nitrite Meat Prod., Zeist, 1973. Pudoc, Wageningen.

Shelf stable cured ham with low nitrite-nitrate additions preserved by radappertization[1]

Eugen Wierbicki and Fred Heiligman

Irradiated Food Products Division, Food Laboratory, U.S. Army Natick Laboratories, Natick, Massachusetts, USA

Abstract

Radappertization (radiation sterilization) is effective in reducing the requirements for the curing agents, nitrites and nitrates, in cured ham. These agents benefit the organoleptic qualities of cured ham, and nitrite, among other of its functions, inhibits toxin production by *Clostridium botulinum*. The use of these agents, however, has been under reappraisal by the industry, the Food and Drug Administration, and the US Department of Agriculture. It has been found that under certain conditions nitrites may react with the free amines in food to form nitrosamines, which are carcinogenic.

The experiments reported in this paper were designed to learn whether or not the amounts of nitrites and nitrates, commonly used in combination with ascorbates/erythorbates in cured meats, could be reduced by use of radappertization. To this end a series of experiments was conducted to determine the minimum amounts of these additives needed to produce

1. The characteristic color of cured ham,
2. The cured ham flavor, and
3. The necessary control of *Cl. botulinum*.

Results showed that the levels of nitrites and nitrates commonly used by industry can be substantially reduced without impairment of any of the foregoing attributes.

An important factor in achieving this notable reduction is that radappertization destroys *Cl. botulinum*, thus eliminating the need for that amount of nitrite required for controlling *Cl. botulinum* in ham cured by non-radappertized processes.

Introduction

Background

Ham is the most important of cured meats from a commercial standpoint. In

1. Radappertization according to the definition given by Goresline et al. (Nature 204 (1964) 237–8) is: . . . . 'the application to foods of doses of ionizing radiation sufficient to reduce the number and/or activity of viable organisms to such an extent that very few, if any, are detectable in the treated food by any recognized method (viruses being excepted) while no spoilage or toxicity of microbial origin is detectable no matter how long or under what conditions the food is stored in the absence of recontamination'. . . The article continues: . . . 'the prefix rad discounts a heat treatment' . . . . .

The smokehouse processing as described in this article implies a heat treatment to inactivate enzymes and make the item ready-to-eat, but not shelf stable by the heat.

1972, over 584.7 million killograms (mil. kg) of ham was processed under Federal Inspection and 78.0 mil. kg of canned ham imported into the United States for a total of 662.7 mil. kg. Only 12.6 mil. kg (1.9%) of this total were canned shelf-stable hams packed in small cans (Anonymous, 1973b). Radappertization (radiation sterilization) processing of ham, to produce a shelf-stable product while preserving its organoleptic qualities, has been a part of the Food Radiation Program since its beginning (Wierbicki et al., 1965; Anonymous, 1968). It was one of the first radappertized meats of acceptable quality to be achieved even when processed with the relatively high radiation sterilizing dose of 4.5 – 5.6 Mrad (Anonymous, 1965; Wierbicki et al. 1965). Initially, the 12D dose was determined to be 2.9 Mrad when the product was in the non-frozen state (Anellis et al., 1967). With the advance of technology of radappertization of meats, it was shown that the quality of foods is improved by irradiating them in the frozen state (Wierbicki et al., 1965; Josephson et al., 1968). Continued investigation of irradiated foods at various product temperatures down to –196 °C has shown that as irradiation temperatures are lowered, higher irradiation doses are required to achieve the same biocidal effect under the 12D concept (Grecz et al., 1971). Moreover, the cost of processing increases as temperature is lowered below the limit of mechanical refrigeration. Therefore, the most favorable balance of quality, cost and effective irradiation dose appears to be –30±10 °C (Josephson et al., 1968; Josephson & Wierbicki, 1973). The minimum required irradation dose for sterilization (MRD) of ham (12D dose) at –30±10 °C is 3.7 Mrad. Table 1 lists the MRD for nine radappertized foods, including ham, under the 12D concept (Anellis et al., 1965, 1967, 1968, 1969 and 1972). Irradiation of ham in the frozen state is beneficial not only with regard to the organoleptic qualities of the product but also for significantly reducing the loss of vitamins by irradiation (Thomas & Josephson, 1970; Thomas & Wierbicki, 1971). Because

Table 1. Minimum required irradiation dose (MRD) for sterilization (Mrad)[1].

Food	Irradiation temperature (°C)	MRD
Bacon	5 to 25	2.9
Beef[2]	–30 ± 10	4.7
Beef[2]	–80 ± 10	5.7
Chicken[2]	–30 ± 10	4.5
Ham	5 to 25	2.9
Ham[2]	–30 ± 10	3.7
Pork	5 to 25	4.6
Pork[2]	–30 ± 10	5.1
Schrimp[2]	–30 ± 10	3.7
Codfosh cakes	–30 ± 10	3.2
Corned Beef	–30 ± 10	2.4
Pork sausage	–30 ± 10	2.7

1. 12D dose.
2. Preliminary data, Microbiol. Div., Fd Lab., NLABS.

Table 2. Chemical composition of non-irradiated smoked ham. Experiment 68/101: ground whole hams; average of duplicates

Lot No.[1]	Anal. Lab[2].	H_2O (%)	Protein (%)	Fat (%)	NaCl (%)	Ash (%)	Sugar (%)	Free nitrite mg/kg	Total P (%)	pH
I	Ind.	67.1	18.8	9.8	2.3	.	.	60	0.33	6.3
		66.8	20.1	9.8	2.0	.	.	80	0.35	6.5
	NLab.	67.1	19.2	7.9	2.2	4.6	0.78	60	0.35	6.6
		66.8	18.4	8.4	2.3	4.7	0.78	45	0.35	6.6
	Ave.	67.0	19.1	8.7	2.2	4.6	0.78	61	0.35	6.5
II	Ind	68.1	18.6	9.8	2.2	.	.	77	0.34	6.4
		65.4	21.0	9.9	2.2	.	.	33	0.35	6.0
	NLab.	64.9	18.9	11.6	1.9	4.0	0.24	3	0.34	6.4
		68.0	19.0	7.2	2.7	4.8	0.19	28	0.37	6.5
	Ave.	66.6	19.4	9.6	2.2	4.4	0.22	35	0.35	6.3
III	Ind.	69.6	17.4	7.8	2.6	.	.	10	0.37	6.2
		69.5	20.3	6.4	2.1	.	.	14	0.35	6.4
	NLab	65.4	19.2	11.4	1.8	3.9	0.22	3	0.33	6.4
		68.9	19.7	7.2	2.6	4.8	0.17	15	0.37	6.4
	Ave.	68.4	19.2	8.2	2.3	4.4	0.20	11	0.36	6.4

1. Lot I = Sugar (sucrose) cure; Lot II = No-sugar cure; Lot III = No-sugar + ascorbate.
2. Ind. = meat packing plant; NLab. = US Army Natick, Lab.

of the high organoleptic and nutritional quality of radappertized ham in the frozen state and because it does not require refrigeration during transportation and storage, the military expressed an interest as long ago as 1966 in using 7.1 mil. kg. radappertized ham per year, provided approvals from the U.S. Food and Drug Administration (FDA) and the US Department of Agriculture (USDA) could be obtained (Anonymous, 1966). Under the Food, Drug and Cosmetic Act of 1958, radappertized ham, like other irradiated foods, requires FDA's approval prior to umlimited human consumption. This requirement was the impetus for the intensive wholesomeness studies of radappertized meats.

In an initial study on the need of sugar in the cure, it was shown that sugar can be eliminated without affecting product quality of either irradiated or non-irradiated hams. In reviewing the data, it is interesting to note that our studies in 1968 showed that less residual nitrite is present in the ham cured without sugar than in the sugar-cured ham, and that further reduction of the residual nitrite can be obtained by adding sodium ascorbate (550 mg/kg) to the non-sugar cured ham (Table 2). The data reported in Table 2 were obtained on the ham with normally added nitrite (156 mg/kg) and nitrate (700 mg/kg).

In a succeeding study the objective was to investigate the addition of sodium chloride to ham during curing. Analysis of various commercial hams showed that the salt content ranges from 2 to over 6%. Deboned, rolled, smoked ham, with no 'added substance'[1] was procured from two meat packers, both using the same

1. 'Added substance' = (moisture + salt) – (protein × 3.79).

Table 3. Quality variability among and between smoked hams. Least significant difference test.

Experimental ham	Treatment	n	Ham No. I	II	III	IV	Radiation effect
68/98:	irradiated[1]	40	6.1[2]	5.9	6.1	6.4	6.1
Packer A,	non-Irrad	40	6.7	6.6	6.6	6.9	6.7[2]
2.5% salt	ham effect		6.4	6.2	6.3	6.6	
68/99:	irradiated[1]	40	5.5	5.2	5.5	.	5.4
Packer B,	non-Irrad.	40	6.3[3]	6.1[4]	5.3	.	5.9
5.5% salt	ham effect		5.9[4]	5.6	5.4		

1. 3.7–4.7 Mrad at −30 ± 10 °C.
2. Significantly preferred to irradiated ham.
3. Significantly preferred to samples rating <5.6.
4. Significantly preferred to samples rating <5.4.

additions of the curing ingredients except for salt. Both the irradiated and nonirradiated hams from the two procurements were evaluated by consumer-type panels and the data analyzed for acceptance. Table 3 shows the results. The responses confirm the fact that consumers prefer low-salt ham. As a result of this confirmation, we standardized the salt addition to radappertized ham at a level of 2.2 ± 0.3% in the finished product. The concentration of salt added to ham in relation to nitrite is an important factor in the control of the growth of *Cl. botulinum* (Duncan & Foster, 1968; Pivnick & Thacker, 1970) as well as for establishing the MRD required for radappertization (Rowley et al., 1968).

A series of acceptance tests were run on the non-sugar cured ham with the normal additions of nitrite (156 mg/kg), nitrate (500 to 700 mg/kg), and sodium chloride content of 2.2 ± 0.3%, radappertized in the close range of 3.7 – 4.7 Mrad

Table 4. Acceptance of 12D[1] radappertized ham. (Hams stored at room temp. for 1 to 12 months prior to serving; testing period: Oct. 1971-July 1973).

Experiment no.	Recipes	Number of men rating	Average acceptance rating
71/90	grilled	16	8.06
71/123	baked	16	7.46
71/123	baked	19	7.26
71/123	baked	22	7.40
72/80[2]	baked	17	7.71
72/80[2]	baked	45	7.06
72/80[2]	baked	17	7.85
72/101	baked	25	6.60
72/101	baked	24	7.08
73/17	baked	22	7.40

1. 3.7–4.3 Mrad at −30 ± 10 °C.
2. Apollo 17 ham.

at $-30 \pm 10^{\circ}$C. Very high acceptance ratings were given to this product. The 9-point hedonic scale devised by Peryam & Pilgrim (1957) was used in the evaluation. Table 4 presents representative results.

It is of more than passing interest that the ham from the Experiment No. 72/80 (Table 4) was used at the request of the National Aeronautics and Space Administration (NASA) by the astronauts of the Apollo 17 flight to the moon in December 1972. The ham slices, 12mm in thickness and weighing approximately 105 ± 5 g, were eaten at three meals in sandwiches made with radurized bread (50 000 rads) using radiation insect-disinfested rye flour (50 000 rads). The verdict of the astronauts was most encouraging. They reported: 'the juicy, chewy (irradiated ham and cheese on (irradiated) rye was one of the space culinary delights enjoyed by the Apollo 17 astronauts.' (Anonymous, 1973a).

The use of nitrate, nitrite and ascorbate/ erythorbate in cured meats

The use of nitrite, nitrate, ascorbate and erythorbate in cured meats is authorized by the US Meat Inspection Regulations (Anonymous, 1970b), the specific stipulation being: Nitrite, direct addition of 156 mg/kg, 'to fix color'; nitrate, as a 'source of nitrite'. When used in combination with nitrite, nitrate addition 'shall not result in more than 200 mg/kg nitrite in finished product'. Ascorbate and erythorbate, 'to accelerate color fixing or preserve color during storage.'

In 1969, the FDA, the meat industry and the USDA began to reappraise the use of nitrites and nitrates as the curing agents for meats, poultry and fish. The results of a number of investigations indicated that under certain conditions nitrites may react with free amines of foods or during digestion of cured meats in the stomach to form carcinogenic compounds known generally as nitrosamines (Lijinsky & Epstein, 1970; Wasserman et al., 1972; Wolff & Wasserman, 1972; Friedman, 1973). A government-industry working group was thereupon organized to study the safety of nitrites and nitrates in foods, the first meeting being held on January 21, 1970 (Friedman, 1973). Since that time, much information has been generated through coordinated government-industry research efforts.

In March 1972, Greenberg reported that sodium nitrite definitely inhibits toxin production by *Cl. botulinum* in perishable canned cured ham and that the level of nitrite at the time of product manufacture rather than the residual nitrite concentration is the key protective factor. In the same study it was shown that at a low inoculum of *Cl. botulinum* the allowable amount of sodium nitrite (200 mg/kg) is required to inhibit the toxin formation and that sodium nitrate did not demonstrate anti-botulinal activity (Greenberg, 1972; Christiansen et al., 1973). Similar results were reported in March, 1973, for wieners (Bard, 1973) and bacon (Herring, 1973).

Nitrosamines can be formed by the reaction of secondary or tertiary amines with free nitrite at a proper pH. The rate of the nitrosamine formation from secondary amines is directly related to the free nitrite concentration in the second power (Tannenbaum & Fan, 1973). This fact emphasizes the importance of reducing insofar as possible the free nitrite in cured meats. No nitrosamines have been found in canned cured ham or in wieners. Nitrosopyrrolidine was found in bacon after preparation for serving. Pan fried bacon showed the greatest amount and microwave-cooked bacon the smallest amount of this nitrosamine. The use of ascorbate in the bacon cure showed some promise for the reduction of the formation of nitrosopyrrolidine during frying of bacon (Herring, 1973). Fiddler et al. (1973) have shown that the commonly used reductants (ascorbic acid, sodium ascorbate and sodium erythorbate) were effective in reducing the formation of *N*-nitroso-dimethylamine in experimental (model) frankfurters; the inhibitory effect of the three reductants was practically the same. Studies have also shown that nitrite is important in the production of the characteristic flavor of cured meats (Bailey & Swain, 1973; Herring, 1973; Wasserman & Talley, 1972).

The amount of nitrite used in processing cured meat products exceeds the quantity needed for the formation of cured meats color nitrosomyoglobin, or its heat denatured form nitrosomyochrome. The surplus of nitrite is needed for the inhibition of the growth and toxin formulation by *Cl. botulinum*. The radappertization process destroys *Cl. botulinum* spores thus eliminating the need of nitrite in controlling botulism in radappertized cured meats. It is emphasized, however, that nitrite is needed for other purposes.

The series of experiments described in this paper was designed to determine the minimum amounts of nitrite and nitrate (in combination with ascorbate/erythorbate) needed for the characteristic color and flavor of the products while controlling *Cl. botulinum* by radappertization. The research was conducted on deboned, defatted, smoked ham.

In March 1972, samples of prototype radappertized hams were shown to meat industry personnel and to food distributors during a servey to learn whether the industry had a potential interest in food radappertization. Continuation of our efforts on research and development in this field (Josephson & Wierbicki, 1973) was encouraged.

Experimental

Raw material

Fresh, raw pork ham, shankless, weighing 6 to 8 kg was used. The fresh ham had not been frozen before processing. The salt was non-iodized, white refined sodium chloride, with or without anticaking agents. Food Grade sodium tripolyphosphate (TPP), sodium nitrite, sodium nitrate, sodium erythorbate and sodium ascorbate, conforming with the USDA requirements, were the curing agents.

Curing

Fresh, raw pork hams were skinned, deboned and all visible cartilage, ligaments, tendons, connective tissue, lymph glands and surface and internal fat were removed. The skinned, deboned, defatted hams were then sectioned into chunks 70 to 750 g and mixed in a food mixer with the curing brine. The brine was adjusted to a 15 percent level, i.e., 15 kg brine per 100 kg meat. The composition of the curing brines (pickles) for high and low nitrite/nitrate hams is given in Table 5. Deviations from this composition for individual experiments will be noted in the text. Meat and brine were mixed for 10 to 15 minutes to produce a tacky mixture. After

Table 5. Experimental curing pickles.

Composition	Curing pickles 1		2	Addition to ham[1] 1		2
Water	25	(kg)	25	0		0
NaCl	4	(kg)	4	2.4	(%)	2.4
TPP	750	(g)	750	0.45	(%)	0.45
Na-ascorbate	54	(g)	54	275	(mg/kg)	275
Na-erythorbate	54	(g)	54	275	(mg/kg)	275
$NaNO_2$	29.4	(g)	4.9	150	(mg/kg)	25
$NaNO_3$	117.6	(g)	9.8	600	(mg/kg)	50

1. Based on 15% pickle additions and 100% smokehouse yield-to-green ham.

mixing, the mixture was stuffed into drilled, prestuck, fibrous, easy-to-peel casings of a size to fit the primary container used for packaging the finished product. After stuffing, the casings were tied (or clipped) under pressure to remove entrapped air. The hams, stuffed into casings, were refrigerated for at least two (2) hours, but not more than twenty-four (24) hours, at 2 °C to 3 °C before smokehouse processing. No. 6–½ Union Carbide casings were used for making ham rolls to fit No. 404 × 700 (10.80 × 17.78 cm) or 404 × 202 (10.80 × 5.40 cm) metal cans, (ham experiments 70/86 and 72/111). No. 11 Union Carbide casings were used to make rectangular hams. The casings, loosely stuffed, were tightly packed into steel wire cages 9 × 13 × 75 cm in size. The rectangular hams were cut into 12 mm thick slices weighing 115 ± 5 g, for packaging into the flexible pouches.

Smokehouse processing

The cured raw product, stuffed into the casings, was processed in the smokehouse in accordance with the following schedule:

Time	Dry bulb temperature	Wet bulb temperature
1 hour without smoke	65 °C	49 °C
2 hours with smoke	65 °C	49 °C
5 hours with smoke	77 °C	57 °C
2 hours with smoke	82 °C	65 °C

Cooking continued without smoke, until the internal temperature of the ham was 68 °C, and then continued at a dry bulb temperature of 77 °C without steam until the weight of the finished product was in the range of 92 to 100 percent of the weight of the raw hams prior to curing. After cooking, the hams were chilled to 5 °C or less (internal temperature) within 12 hours, and then kept in a 1 to 5 °C refrigerator until cut and packaged.

Packaging (cutting)

Ham rolls were removed from their casings and cut into sections to fit the containers. The product was packaged either in beaded metals cans coated inside with enamel conforming to the provisions of the Federal Food Drug and Cosmetic Act, 21CFR121.2514, Resinous and Polymeric Coatings, specifically designated as the epoxy-phenolic type either V21 (American Can Co.) or 221 ALN (Continental Can Co.), or in flexible pouches with outside dimensions of 11.5 × 17.8 cm. Pouches were constructed from a multiple layer laminate material: the outside layer was 0.0025 cm polyiminocaproyl (Nylon 6), the middle layer 0.009 cm 1145 alloy aluminum foil, and the inside layer chemically bonded laminate of 0.001 cm polyethylene terephthalate and 0.006 cm medium density polyethylene, such as Scotchpak 9 or equivalent, with the polyethylene as the food contactant. Details regarding packaging of radappertized foods are available in a recent review paper by Killoran (1972). The filled cans or flexible pouches were sealed under a vacuum of not less than 25 inches gauge (not more than 125 mm Hg). Some samples (see 'Results and Discussion') were dipped (and immediately removed) in an aqueous solution containing 2.5 or 5% ascorbic and 2.5 or 5% citric acid prior to vacuum sealing in flexible pouches.

Irradiation and post-irradiation storage

Prior to irradiation, the ham samples were chilled to a temperature of −40 °C in the center of the hams. The irradiation was performed at the NLABS (U. S. Army Natick Laboratories) either in the 1 250 000 curie cobalt−60 isotope source (dose rate about 30 000 rad/min) or in the 10 Mev Linear accelerator (dose rate about 10^8 rads/sec). The product temperature during irradiation with cobalt−60 was −30 ± 10°C; during electron irradiation, the product temperature range was −40 to −10 °C. The minimum dose received was 3.7 Mrad. The dose range, depending on the experiment, was from 3.7 – 4.2 to 3.7 – 4.7 Mrad. Before evaluation, the irradiated samples were stored at ambient temperature (25 ± 3 °C) and the non-irradiated controls in the −29 °C freezer.

Evaluation

Consumer-type panel Consumer-type tests were performed by the Food Acceptance Group, Behavioral Sciences Division, Pioneering Research Laboratory, NLABS. Selected samples were tested for preference using the 9-point hedonic scale (Peryam and Pilgrim, 1957) in which 9 = *like extremely* and 1 = *dislike extremely.* For ham and other meat products, a score of 5 (neither like nor dislike) indicates marginal acceptability, ratings of 6 to 7 indicate an acceptable product and ratings above 7, a highly acceptable product. The test data were statistically analyzed using analysis of variance and least significant differences (LSD) between the means.

Technological panel Some samples were sensory tested by an 8 to 12 member trained panel for preference, using the 9−point hedonic scale, and for color, odor and flavor, by means of a 9−points quality scale (See Table 6). Most ham samples were

Table 6. Technological examination for color, odor, flavor and texture.

Scale	Description
9	excellent
8	very good
7	good
6	below good above fair
5	fair
4	below fair above poor
3	poor
2	very poor
1	extremely poor

tested cold. Frozen samples were defrosted in a 3 to 5 °C refrigerator prior to serving. All irradiated samples were tested for the absence of *Cl. botulinum* toxin prior to sensory testing. If to be served hot, 12-mm thick ham slices were held in a 163 °C preheated electric oven until an internal product temperature of 63 to 65 °C was reached.

Color evaluation Most of the samples were visually evaluated for cured meat color only. Evaluations were usually made on both cold and hot samples. The 12–mm thick ham slices, vacuum sealed in flexible pouches, were held for 15 minutes in a 150 °C preheated oven. After heating, the samples were held at room temperature for 15 minutes before opening the pouches; they were then placed on white plates for evaluation. Evaluation was made at '0-Time' (within 5 to 20 minutes after removal of the ham slices from the packaging) and after 2 hours' exposure to light (room temperature 23 to 28 °C). The ham slices were wrapped in 0.0006 cm polyvinyl film, transparant to light, to prevent discoloration of the ham slices due to surface dehydration during the 2–hour exposure. The polyvinyl wrap was removed for color rating of the ham samples. In this series of evaluations for cured meat color, reference ham samples were shown to the panelists prior to the color rating by experts. The reference samples were of excellent color with assigned ratings of 8 or 9 (Table 6) by experts. The data obtained were statistically analyzed by the NLABS Data Analysis Office using Analysis of Variance and the Student's Range test.

Chemical analyses Ham samples were chemically analyzed for:
1. moisture, fat, protein, salt (NaCl), ash and total phosphate (AOAC Official Methods, 1970a);
2. residual sodium nitrite and sodium nitrate using ether extracted defatted samples (AOAC Official Method, 1970a);
3. thiobarbituric acid (TBA) values were determined by the method of Tarladgis et al. (1960);

Nitrosamines determination Samples from two ham experiments (hams 72/111 and 73/09) were analyzed for the following six nitrosamines:

dimethylnitrosamine (DMNA)
methylethylnitrosamine (MENA)
diethylnitrosamine (DENA)
nitrosomorpholine (NO-Mor)
nitrosopyrrolidine (NO-Pyr)
nitrosopiperidine (NO-Pip)

The analyses were performed at the USDA Eastern Regional Research Laboratory in Philadelphia, Pennsylvania, through the cooperation of Dr Fiddler and Wasserman.

Results and discussion

Ham 70/86 experiment

Various levels of nitrite, ranging from 10 to 200 mg/kg, were investigated, the objective being to ascertain the acceptance of the end product by the consumer-type panel. Four determinations were made; the first, 10 days after the ham processing, and the fourth, after nine months' storage. Results of these tests are given in Table 7. In this experiment, 350 mg/kg sodium nitrate and the maximum allowable amounts (550 mg/kg) of ascorbate (as 1 : 1 mixture of sodium ascorbate/ sodium erythorbate) were used. As the data indicate, there are no significant differences in the preference ratings given the irradiated ham regardless of the nitrite added to the ham during curing. The non-irradiated samples received higher preference ratings, but significant differences for the same nitrite level (200 mg/kg were found only between the samples stored for one month before testing. Tests for residual nitrites (Table 8) revealed a rapid depletion in both irradiated and

Table 7. Ham 70/86: effect of the added nitrites on preference. (3.7–4.4 Mrad at −30 ± 10 °C). The numbers in brackets refer to the amount of $NaNO_2$ added to the hams (mg/kg).

Time at 21–25 °C	Number of panelists	Average preference scores[1]					LSD[3]
		B (10)	C (25)	D (50)	E (200)	F[2] (200)	
10 days	35	6.9	6.9	6.7	6.7	7.2	0.8
1 month	35	6.3	6.2	6.3	5.9	7.4	1.4
4 months	35	5.6	5.6	5.8	5.7	7.1	1.5
9 months	32	7.1	6.4	6.5	6.2	–	0.8

1. 9 = like extremely; 1 = dislike extremely
2. Non-irradiated control stored at −29 °C
3. Least significant difference.

Table 8. Ham 70/86: effect of the added nitrites on the residual nitrites. (3.7–4.4 Mrad at −30 ± 10 °C). The numbers in brackets refer to the amount of $NaNO_2$ added to the hams (mg/kg).

Time at 21°C	Residual nitrites in ham (mg/kg)					
	(0)	(10)	(25)	(50)	(200)	(200)[1]
10 days	4	5	4	8	12	34
1 month	3	3	2	2	3	17
4 months	–	2	2	2	5	–
9 months	–	3	4	4	6	–

1. Non-irradiated control stored at −29 °C.

non-irradiated samples. The ham sample without nitrite added showed 3–4 mg/kg residual nitrite (Table 8), the same amount as that found for the irradiated ham samples with added 10 to 200 mg/kg after one month of storage. This 3–4 mg/kg analytical nitrite is apparently the base (zero) line for the accuracy of the analytical methodology used. Even though the consumer-type panel found the ham samples with 10 mg/kg nitrite equally acceptable to 200 mg/kg, examination by expert technologists of the 10 mg/kg nitrite ham revealed some non-cured spots and a metallic flavor was detected.

Ham 72/108 experiment

In this experiment, various combinations of nitrite and nitrate were used to study their effects on the color of the ham. Ascorbate/erythorbate (550 mg/kg) level was kept constant. Table 9 gives the results. There were no significant differences in the color of the samples with the exceptions of: (a) non-irradiated

Table 9. Ham 72/108: effect of nitrite and nitrate on color[1]. (Technological panel: n = 10).

mg/kg added		Non-irradiated		Irradiated[2]	
$NaNO_2$	$NaNO_3$	cold	hot	cold	hot[3]
150	500	7.7 ± 0.7	7.1 ± 0.6	7.4 ± 0.5	7.4 ± 0.7
150	100	7.5 ± 1.0	7.3 ± 0.5	6.9 ± 0.9	6.6 ± 1.0
15	50	7.6 ± 0.8	7.3 ± 0.5	7.1 ± 0.8	6.1 ± 1.0
15	100	7.7 ± 0.7	7.0 ± 0.8	7.2 ± 0.6	6.6 ± 1.1
25	50	7.7 ± 0.7	6.7 ± 0.7	7.0 ± 0.9	6.1 ± 1.1
25	100	7.8 ± 0.6	6.9 ± 0.3	7.1 ± 1.0	5.2 ± 1.5
50	50	7.8 ± 0.6	7.5 ± 0.7	7.1 ± 0.7	6.8 ± 0.9
50	100	7.8 ± 0.8	7.2 ± 0.6	7.4 ± 0.7	6.8 ± 0.9

1. Determined 5 to 20 min after pouch opening.
2. 3.7–4.4 Mrad at −30 ± 10 °C.
3. Heated in sealed pouches for 15 min at 150 °C.

Table 10. Ham 72/108: Effect of nitrite and nitrate on color. (Irradiated: 3.7–4.4 Mrad at −30 ± 10 °C.

mg/kg added		n	Cold		Hot[1]	
$NaNO_2$	$NaNO_3$		0[2]	2[3]	0	2
25	100	3x11	7.2 ± 0.7	6.4 ± 1.2	7.2 ± 0.4	6.8 ± 0.9
25	50	3x11	7.1 ± 1.0	6.2 ± 1.2	7.2 ± 0.7	6.7 ± 0.7
15	50	2x11	6.7 ± 1.0	5.8 ± 1.1	7.2 ± 1.0	6.4 ± 0.7

1. Heated in sealed pouches for 15 min at 150 °C.
2. 0 = determined 5 to 20 min after pouch opening.
3. 2 = determined after 2 hours exposure to light at 23 to 28 °C.

cold vs. hot 25/50 (nitrite/nitrate) and 25/100 samples and, (b) the irradiated cold vs. hot 15/50, 25/50 and 25/100 samples. Therefore, these three irradiated samples were subjected to additional evaluations for color involving exposure to light (at '0'-time and after 2 hours). Table 10 gives the results. The data indicate that exposure to light has a significant effect on the color scores of these samples. The 15/50 nitrite/nitrate samples were scored significantly lower when evaluated cold. It was noted that the 15/50 samples had under-cured spots in some cases and a slight metallic flavor when analyzed closely by the expert panel. Based on this experiment, it was concluded that not less than 25 mg/kg should be used in curing radappertized ham.

Ham 73/29 experiment

In this experiment, attempts were made to eliminate the addition of nitrate to the low nitrite (25 mg/kg) ham (Cure 1) by varying the addition of ascorbate/erythorbate in the cure (Cures 2, 3, 4 and 5). The data (Table 11) were statistically analyzed and showed the following: (a) Cure 1 ham (with nitrate) received significantly higher color ratings than the other four cures (cures without nitrate); (b) no significant differences were found among the four no-nitrate samples which contained either no ascorbate/erythorbate or varying amounts from 250 to 1000 mg/kg; (c) the irradiation and the exposure to light under the extreme conditions used in this experiment (2 hours at 23 to 28 °C room temperature) significantly decreased the color of the ham samples, particularly in the cures without nitrate; (d) dipping the ham slices into an aqueous solution containing 5% ascorbic and 5% citric acid or evaluating the ham samples cold or hot had no significant effect on the ham color. It was particularly surprising not to find any effect on color when ham slices were dipped into the ascorbic/citric acid solution prior to vacuum packaging. USDA Meat Inspection Regulations, permitting the use of citric acid or sodium citrate for just this purpose, read: 'in cured products or in 10 percent solutions to spray surfaces of cured cuts prior to packaging to replace up to 50 percent of the ascorbic acid, erythorbic acid' or their sodium salts (Anonymous, 1970b). With this experiment, we have shown the need for nitrate, in association with nitrite, in the cures of radappertized ham in order to obtain the typical color of the product.

Table 11. Ham 73/29: Effect of ascorbate/erythorbate without nitrate on color. (Technological panel: n = 12).

	Cure no.				
	1 (25 NO_2 500 A/E 100 NO_3)[1]	2 (25 NO_2 0 A/E)	3 (25 NO_2 250 A/E)	4 (25 NO_2 500 A/E)	5 (25 NO_2 1000 A/E)
Irradiated[2]	6.5 ± 1.4	4.8 ± 1.5	5.1 ± 1.6	4.9 ± 1.5	5.2 ± 1.5
Non-irradiated	7.0 ± 1.1	6.7 ± 1.1	6.6 ± 1.2	6.9 ± 1.1	6.4 ± 1.1
Non-dipped	7.2 ± 1.0	5.6 ± 1.8	5.6 ± 1.6	5.8 ± 1.6	6.0 ± 1.5
Dipped[3]	6.4 ± 1.3	5.9 ± 1.4	6.1 ± 1.5	6.1 ± 1.6	5.6 ± 1.4
Evaluated cold	6.8 ± 1.4	5.8 ± 1.9	5.9 ± 1.8	5.9 ± 1.9	5.9 ± 1.6
Evaluated hot	6.8 ± 1.0	5.7 ± 1.4	5.8 ± 1.3	5.9 ± 1.3	5.8 ± 1.2
0-h exposure	7.2 ± 1.0	6.1 ± 1.5	6.2 ± 1.4	6.3 ± 1.5	6.0 ± 1.5
2-h exposure	6.3 ± 1.3	5.3 ± 1.7	5.4 ± 1.6	5.5 ± 1.6	5.7 ± 1.4
All treatments	6.8 ± 1.2	5.7 ± 1.6	5.8 ± 1.6	5.9 ± 1.6	5.8 ± 1.5

1. Additions (mg/kg) during curing of $NaNO_2$, ascorbate/erythorbate and $NaNO_3$.
2. 3.7–4.4 Mrad at -30 ± 10 °C.
3. See text.

Ham 73/39 experiment

In this experiment an attempt was made to show whether or not ascorbate/erythorbate can be eliminated from the nitrite/nitrate cures for radappertized ham. We used high (normal) nitrite (150 mg/kg)/nitrate (600 mg/kg) cure and low nitrite (25 mg/kg)/nitrate (50 mg/kg) cure as the reference (Table 12, Cures No. 1 and 2). The remaining three cures omitted either ascorbate/erythorbate or nitrate, or both.

Table 12. Ham 73/39: Effect of nitrite plus nitrate on color. (Technological Panel: n=12).

	Cure No.				
	1 (150 NO_2 600 NO_3 550 A/E)[1]	2 (25 NO_2 50 NO_3 550 A/E)	3 (25 NO_2 50 NO_3 0 A/E)	4 (25 NO_2 0 NO_3 550 A/E)	5 (25 NO_2 0 NO_3 0 A/E)
Irradiated[2]	7.5 ± 1.0	6.4 ± 1.3	5.6 ± 1.6	4.8 ± 1.8	4.3 ± 2.0
Non-irradiated	7.6 ± 1.1	6.9 ± 1.5	7.0 ± 1.3	7.5 ± 1.0	6.2 ± 1.8
Evaluated cold	7.9 ± 0.9	6.8 ± 1.5	6.3 ± 1.8	6.4 ± 2.1	5.9 ± 2.1
Evaluated hot	7.2 ± 1.1	6.6 ± 1.4	6.3 ± 1.4	5.9 ± 1.9	4.5 ± 1.9
0-h exposure	7.9 ± 1.0	7.2 ± 1.4	7.2 ± 1.2	6.8 ± 1.7	6.0 ± 2.0
2-h exposure	7.2 ± 1.0	6.2 ± 1.3	5.4 ± 1.5	5.5 ± 2.0	4.5 ± 2.0
All Treatments	7.6 ± 1.1	6.7 ± 1.4	6.3 ± 1.6	6.2 ± 2.0	5.2 ± 2.1

1. Additions (mg/kg) during curing of $NaNO_2$, ascorbate/erythorbate and $NaNO_3$.
2. 3.7–4.4 Mrad at -30 ± 10 °C.

The statistical evaluation of the data (Table 12) showed the following: (a) ham cured with Cure No. 1 had the best color; (b) the next best was Cure No. 2; (c) Cure No. 2 ham (containing both 50 mg/kg nitrate and 550 mg/kg ascorbate/ erythorbate) had significantly better color than the No. 4 or 5 cure hams, neither containing nitrate; (d) Cure No. 2 ham, containing both nitrate and ascorbate/ erythorbate suffered less discoloration by irradiation and the exposure to light than Cure No. 3 ham, which contained nitrate without ascorbate/erythorbate; (e) Cure No. 5 ham which omitted both nitrate and ascorbate had the poorest color. This experiment confirmed our previous overall observations that ascorbate/erythorbate is needed in association with nitrate and nitrite to achieve a radappertized ham of acceptable color.

Ham 73/36 experiment

In this experiment, five different cures of hams were evaluated by a 12-member panel not only for color but also for flavor, odor and preference. All hams received 25 mg/kg sodium nitrite during the curing process. The variables were: 100 and 50 mg/kg nitrate (Cures No. 1 and II); sodium citrate substituted for sodium erythorbate in the ascorbate/erythorbate mixture (Cures III and IV); and, sodium pyrophosphate and sodium acid orthophosphate substituted for TPP (Cure No. V) in equivalent amounts. Results of this investigation are given in Tables 13 and 14. In all instances, irradiated samples scored significantly lower than the nonirradiated controls. The effect of the individual cures differed, depending on the quality attribute of the ham. Color evaluations showed no significant difference between the Cures I and II (100 and 50 mg/kg nitrate). However, the color of the ham cured with Cure No. I significantly differed from the color of the hams cured with Cures

Table 13. Ham 73/36: Effect of various cures on flavor. (Constants: 25 mg/kg $NaNO_2$ added; technological panel: n=12). Served without previous exposure to light.

	Cure No.				
	I (P-1[1] 275 – Asc. 275 – Eryth. 100 – NO_3)[2]	II (P-1 275 – Asc. 275 – Eryth. 50 – NO_3)	III (P-1 275 – Asc. 275 – Cit. 50 – NO_3)	IV (P-1 275 – Asc. 275 – Cit. 0 – NO_3)	V (P-2[1] 275 – Asc. 275 – Eryth. 50 – NO_3)
Irradiated	6.4 ± 1.3	5.9 ± 1.5	5.8 ± 1.5	5.8 ± 1.5	5.9 ± 1.5
Non-irradiated	7.4 ± 0.8	7.3 ± 0.8	7.2 ± 0.8	7.2 ± 0.9	7.3 ± 1.1
Non-dipped	6.9 ± 1.2	6.4 ± 1.4	6.5 ± 1.4	6.4 ± 1.4	6.6 ± 1.5
Dipped	7.0 ± 1.1	6.8 ± 1.3	6.5 ± 1.4	6.6 ± 1.4	6.6 ± 1.5
Served cold	6.9 ± 1.4	6.3 ± 1.7	6.4 ± 1.7	6.5 ± 1.4	6.4 ± 1.5
Served hot	7.0 ± 1.0	6.9 ± 0.9	6.6 ± 1.0	6.5 ± 1.4	6.8 ± 1.5
All treatments	6.9 ± 1.2	6.6 ± 1.4	6.5 ± 1.4	6.5 ± 1.4	6.6 ± 1.5

1. P-1= 0.45% TPP added; P-2 = equivalent Na-Pyro P+NaH_2PO_4 added.
2. mg/kg added: Asc. = sodium ascorbate; Eryth. = sodium erythorbate; NO_3 = $NaNO_3$; Cit. = sodium citrate.

Table 14. Ham 73/36: Effect of various cures on preference. (Constants: 25 mg/kg $NaNO_2$ added; technological panel: n=12) Served without previous exposure to light.

	Cure No.				
	I (P-1[1] 275 – Asc. 275 – Eryth. 100 – NO_3)[2]	II (P-1 275 – Asc. 275 – Eryth. 50 – NO_3)	III (P-1 275 – Asc. 275 – Cit. 50 – NO_3)	IV (P-1 275 – Asc. 275 – Cit. 0 – NO_3)	V (P-2[1] 275 – Asc. 275 – Eryth. 50 – NO_3)
Irradiated	6.4 ± 1.2	5.9 ± 1.6	5.9 ± 1.6	5.7 ± 1.6	6.0 ± 1.4
Non-irradiated	7.2 ± 0.9	7.0 ± 1.0	7.1 ± 0.8	7.0 ± 1.2	7.1 ± 1.4
Non-dipped	6.8 ± 1.2	6.4 ± 1.5	6.6 ± 1.3	6.4 ± 1.6	6.5 ± 1.5
Dipped	6.9 ± 1.1	6.5 ± 1.3	6.4 ± 1.5	6.3 ± 1.5	6.6 ± 1.5
Served cold	6.7 ± 1.3	6.2 ± 1.7	6.3 ± 1.8	6.3 ± 1.6	6.4 ± 1.6
Served hot	6.9 ± 0.9	6.7 ± 1.0	6.7 ± 0.8	6.4 ± 1.6	6.7 ± 1.4
All Treatments	6.8 ± 1.1	6.5 ± 1.4	6.5 ± 1.4	6.3 ± 1.6	6.5 ± 1.5

1. P-1 = 0.45% TPP added; P-2 = equivalents Na-Pyro P + NaH_2PO_4 added.
2. mg/kg added: Asc. = sodium ascorbate; Eryth. = sodium erythorbate; NO_3 = $NaNO_3$; Cit. = sodium citrate.

No. III, IV and V. Previous observations were confirmed; namely: (a) color scores of the hams were significantly lower after 2-hour exposure to light; (b) there was no effect of dipping the ham slices into ascorbic/citric acid solution (this time 2.5% concentration of each acid was used) prior to packaging; (c) evaluating cold vs. hot did not affect the color scores; and (d) the ham cured without nitrate (Cure No. IV) suffered most in the loss of color by the irradiation and exposure to light.

There were no significant differences in flavor (Table 13), odor and preference (Table 14) of the hams due to differences in the curing pickles. However, in all quality attributes the hams cured with Cure No. I (25 mg/kg nitrite and 100 mg/kg nitrate) received slightly higher scores, and the effect of radiation on the scores was less pronounced than that of hams of the other four cures. The overall results have shown that: (a) the best low nitrite/nitrate cure for radappertized ham is Cure No. I; (b) there is no advantage in replacing erythorbate with citrate in the ascorbate/erythorbate mixture (Cure II vs III); and (c) there is no advantage in replacing TPP with sodium pyrophosphate and sodium acid phosphate (Cure II vs V).

Ham 73/30 experiment

The main objective of this experiment was to show the effect of nitrite and nitrate on the overall consumer preference for ham as well as the effect on flavor and odor characteristics of the product. There were five groups of ham samples, each processed with different curing pickles. All cures contained the same concentrations of salt, TPP, and ascorbate/erythorbate (Table 5). The cures varied in the nitrite/nitrate levels added to the ham during curing. They were as follows: 150/600, 25/100, 25/0, 0/100, and 0/0 mg/kg for cures, A, B, C, D, and E, respectively. There were four separate experiments for the five groups. In each

Table 15. Ham 73/30: Effect of nitrite and nitrate on preference. Consumer-type panel: n=35; samples (a)[1]

Cure	Variables (mg/kg)	Days of storage			
		1	7	14	21
A	150 NO_2 600 NO_3	6.8 ± 1.5	7.2 ± 1.3	7.0 ± 1.2	6.5 ± 1.5
B	25 NO_2 100 NO_3	6.1 ± 1.8	6.5 ± 1.8	6.2 ± 1.6	6.3 ± 1.8
C	25 NO_2 0 NO_3	6.1 ± 1.8	6.2 ± 1.8	6.0 ± 1.6	5.6 ± 1.9
D	0 NO_2 100 NO_3	4.5 ± 1.7	4.3 ± 2.1	4.2 ± 1.9	4.1 ± 2.5
E	0 NO_2 0 NO_3	4.2 ± 2.1	4.3 ± 2.0	4.1 ± 2.2	3.7 ± 2.1

1. Non-irradiated, no-vacuum, stored at 2 to 3°C.

experiment, hams were cut into 12 mm slices, each weighing 115 ± 5 g per slice, and packed in flexible pouches. The differences between the four experiments were as follows: (a) pouches unsealed (no vacuum), ham non-irradiated and stored at 2 to 3 °C for 1, 7, 14 and 21 days; (b): pouches unsealed, ham non-irradiated and frozen stored at −29 °C for 10, 30 and 90 days; (c): pouches vacuum sealed, ham irradiated (3.7 – 4.3 Mrad at −30 ± 10 °C and stored at 23 to 28 °C for 10 days and 3, 6 and 12 months; (d): pouches vacuum sealed and the ham frozen stored at −29 °C for 10 days and 3, 6 and 12 months (control samples for the sample used in

Table 16. Ham 73/30: Effect of nitrite and nitrate on preference. Consumer-type panel: n=35.

Cure	Variables (mg/kg)	Days of storage			
		10(b)[1]	30(b)[1]	10(c)[1]	10(d)[1]
A	150 NO_2 600 NO_3	7.3 ± 1.2	7.1 ± 1.0	6.4 ± 1.7	7.2 ± 1.2
B	25 NO_2 100 NO_3	7.2 ± 1.3	6.9 ± 1.1	6.3 ± 1.8	7.1 ± 1.6
C	25 NO_2 0 NO_3	6.8 ± 1.4	6.9 ± 1.4	6.0 ± 1.7	7.3 ± 1.6
D	0 NO_2 100 NO_3	5.5 ± 2.0	4.8 ± 1.8	5.8 ± 2.0	5.2 ± 2.0
E	0 NO_2 0 NO_3	5.2 ± 1.9	4.4 ± 1.7	5.5 ± 1.8	5.5 ± 1.5
LSD[2]:		0.7	0.7	0.7	0.7

1. (b): Non-irradiated, no-vacuum, stored at −29°C;
   (c): Irradiated, vacuum packed, stored at 23 to 28°C;
   (d): Non-irradiated, vacuum packed, stored at −29°C.
2. LSD: Least significant difference.

Table 17. Ham 73/39: Effect of nitrite and nitrate on odor and flavor. Technological panel: n=8; samples (a)[1].

Cure	Variables (mg/kg)	*Odor* days after processing				*Flavor* days after processing			
		1	7	14	21	1	7	14	21
A	150 NO_2 600 NO_3	7.4	7.3	6.6	7.5	7.6	7.3	6.5	7.4
B	25 NO_2 100 NO_3	7.0	6.6	6.3	6.8	7.1	6.4	6.0	6.0
C	25 NO_2 0 NO_3	6.6	6.5	6.8	5.4	6.6	6.1	7.0	4.9
D	0 NO_2 100 NO_3	5.3	5.3	4.5	4.9	4.6	3.6	3.6	4.0
E	0 NO_2 0 NO_3	5.4	4.3	5.1	4.8	5.1	3.5	3.8	3.9

1. Non-irradiated, no-vacuum, stored at 2 to 3°C.

(c) above). All evaluations were made on cold (3–5 °C) ham samples. As of this time, only samples (a) have been evaluated. The results are given in Tables 15–20. Consumer-type panel preference data are given in Tables 15 and 16. Similar data were obtained by the technological panel. Data clearly indicate the low quality of the hams cured without nitrite (Cures D and E). The differences were highly significant when compared with the ham with nitrite added (Cures A, B and C). In the samples (a) series (Table 15), the ham of Cure A was also significantly preferred to the hams of Cures B and C after 1, 7 and 14 days storage. After 21 days' storage,

Table 18. Ham 73/30: Effect of nitrite and nitrate on odor and flavor. Technological Panel: n=8; samples: (b), (c) and (d)[1].

Cure	Variables (mg/kg)	*Odor* days after processing				*Flavor* days after processing			
		10(b)	30(b)	10(c)	10(d)	10(b)	30(b)	10(c)	10(d)
A	150 NO_2 600 NO_3	7.3	7.3	6.6	7.0	7.4	7.4	6.5	6.6
B	25 NO_2 100 NO_3	7.5	7.1	6.4	6.8	7.5	7.1	6.4	6.6
C	25 NO_2 0 NO_3	6.6	7.4	5.9	7.0	6.9	7.1	5.3	6.8
D	0 NO_2 100 NO_3	5.1	5.9	6.0	6.1	4.6	4.9	5.9	4.6
E	0 NO_2 0 NO_3	5.1	5.3	5.6	5.6	4.9	6.1	5.0	4.6

1. (b): Non-irradiated, no-vacuum, stored at –29°C;
(c): Irradiated, vacuum packed, stored at 23 to 28°C;
(d): Non-irradiated, vacuum packed, stored at –29°C.

ham cured with 25 mg/kg nitrite without nitrate (Cure C) dropped significantly in preference. The first evaluations of samples (b), (c) and (d) are given in Table 16. Data confirm the need of nitrite to attain hams of high acceptance (Cures A, B and C vs D and E).

Tables 17 and 18 record the odor and flavor scores as determined by an 8-member technological panel using the scoring system given in Table 6. The data clearly indicate the need of nitrite for an acceptable odor and flavor (Cures A, B and C vs D and E). Lack of a typical odor and flavor as well as color were the main reasons given by the consumer-type panel for the low preference scores for hams of Cures D and E given in Tables 15 and 16. This response confirms the recently reported findings of Bailey and Swain (1973) that nitrite is needed for the flavor and taste of ham and other cured meats.

Odor and flavor scores for Cure C ham (25 mg/kg nitrite but no nitrate) were significantly lower when storage time increased from 14 to 21 days (Table 17). It appears that a small amount of nitrate (for example, 100 mg/kg, Cure B) might be needed for non-irradiated ham when the product is to be stored for an extended period of time. As shown previously, addition of 100 mg/kg nitrate to 25 mg/kg nitrite cure is needed to minimize the loss of the typical ham flavor and odor caused by irradiation (Table 18: samples (c) and (d), cures B and C).

Tables 19 and 20 show the effect of nitrite and nitrate on the TBA values (mg malonaldehyde per 1000 g). The significantly higher TBA values in the ham samples of Cures D and E stored in open pouches in refrigerator and the freezer (Table 19) clearly indicate the antioxidant effect of nitrite in ham; this is particularly observable in the samples stored for 7 days or longer. It is interesting that the ham cured with 25 mg/kg nitrite without nitrate (Cure C) showed a high TBA value after 21 days' storage (Table 19). This high value might be associated with the lower scores for odor and flavor (Table 17) and preference (Table 16) for the samples of the same group. Results suggest that a small amount of nitrate should remain in the cure for non-irradiated hams. Table 20 lists TBA values for irradiated and non-

Table 19. Ham 73/30: Effect of nitrite and nitrate on TBA. (a): non-irradiated, no-vacuum, stored at 2 to 3°C; (b): non-irradiated, no-vacuum, stored at −29°C.

Cure	Variables (mg/kg)	Days after processing				Days after processing	
		1	7	14	21	10	30
		samples (a)				*samples (b)*	
A	150-NO_2 600-NO_3	0.00	0.05	0.02	0.08	0.04	0.05
B	25-NO_2 100-NO_3	0.01	0.07	0.12	0.37	0.06	0.14
C	25-NO_2 0-NO_3	0.02	0.37	0.08	3.04	0.08	0.05
D	0-NO_2 100-NO_3	0.92	1.50	2.71	3.28	0.86	2.06
E	0-NO_2 0-NO_3	0.73	1.73	3.03	4.29	0.50	2.54

Table 20. Ham 73/30: Effect (duplicate samples) of nitrite and nitrate on TBA. (c): irradiated[1], vacuum packed, stored at 23 to 28°C for 10 days; (d): non-irradiated, vacuum packed, stored at −29°C for 10 days.

Cure	Variables (mg/kg)	(c)		(d)	
		1	2	1	2
A	150 NO_2 600 NO_3	0.26	0.26	0.06	0.05
B	25 NO_2 100 NO_3	0.23	0.23	0.12	0.07
C	25 NO_2 0 NO_3	0.21	0.21	0.06	0.09
D	0 NO_2 100 NO_3	0.23	0.25	0.29	0.31
E	0 NO_2 0 NO_3	0.38	0.40	0.33	0.30

1. 3.7–4.4 Mrad at −30 ± 10°C.

irradiated hams after 10 days of storage. Irradiation of hams with nitrite additions (Cures A, B and C) slightly increased the TBA values; however, the values are below the rancidity level. There was practically no effect of irradiation on the TBA values of the no nitrite hams (Cures D and E). These experiments are continuing and the final conclusion must await the end of the storage studies.

Ham 72/111 experiment

The main objective of this investigation was to determine the effect of adding nitrite and nitrate during curing on residual nitrite and nitrate in the finished product. The constant constituents of the cure for all hams were salt, TPP and ascorbate/erythorbate as given in Table 5. The variables were sodium nitrite and sodium nitrate. Five different combinations were used (Table 21). Proximate chemical analyses showed them to be in the range given for the hams in Table 2, except for pH, which was within the range of 6.0 to 6.2 for this experiment. The data in Table 21 indicate that no difference is apparent whether 25 or 150 mg/kg sodium nitrite are added to ham during curing since the residual nitrite approaches the analytical 'zero' line (0.9 to 1.9 mg/kg) after 10 days' storage. This finding confirms the results found for wieners (Bard, 1973). Non-irradiated frozen control ham cured with high nitrite/nitrate (Table 21, cure 01) had only 4.7 to 7.3 mg/kg free nitrite, thus indicating that the depletion of the added nitrite occurred mainly during the curing and smokehouse processing of the ham, probably by reacting with constituents of ham muscles other than myoglobin, as reported by Olsman and Krol (1972).

The amount of analytical sodium nitrate in the ham shows an interesting relationship to the addition of nitrite/nitrate to the hams during curing. Whereas the amount of the analytical (residual) nitrate in the high nitrate ham (Cure No. 01) is close to the amount added during curing in the low nitrite ham (Cure No. 02) and

Table 21. Ham 27/111: Added and residual nitrite and nitrate. (Averages of duplicate determinations).

Cure No.	mg/kg added		mg/kg residual			
	NO_2	NO_3	$NaNO_2$		$NaNO_3$	
			10	90	10	90[1]
	Irradiated[2]					
01[3]	150	600	2.4	2.2	477	515
02[3]	25	100	1.6	2.4	140	141
03	0	600	1.8	1.6	737	728
04	0	100	1.0	1.4	118	137
05	0	0	0.9	1.7	58	69
	Non-irradiated					
01[3]	150	600	4.1	7.3	572	666
02[3]	25	100	1.1	2.3	145	169
03	0	600	1.4	1.4	871	835
04	0	100	1.1	1.6	128	175
05	0	0	1.0	1.3	51	67

1. Days of storage: irradiated at 23 to 28°C, non-irradiated at −29°C.
2. 3.7−4.4 Mrad at −30 ± 10°C.
3. Samples used for nitrosamines determination.

no-nitrite hams (Cures No. 03 and 04), the amounts of the analytical nitrate is significantly greater than the amounts added during curing. Even the no-nitrate cured ham (Cure No. 05) contained 51 to 69 mg/kg analytical free nitrate. Bard (1973) reported similar results (3 to 47 mg/kg nitrate) in the wieners cured without nitrate. It might be an analytical error or it might be due to nitrate impurities in the water used for making the cures. This factor is being investigated.

Ham 73/09 experiments

The main objective of this experiment was to determine nitrosamines in high nitrite/nitrate (150/600 mg/kg) and low nitrite/nitrate (25/100 mg/kg) hams, irradiated by both the gamma rays from Cobalt−60 source and by electrons from the linear accelerator. The Cobalt−60 and electron irradiated samples were made from paired 12 mm ham slices, cut from the same ham rolls. The hams were of excellent quality as shown by the preference (acceptance) ratings in Table 22. It is of interest that in this experiment there were no significant differences between the preference ratings of high vs low nitrite/nitrate hams, but somewhat higher ratings were obtained for electron vs Cobalt−60 irradiated samples. There was no indication of the presence of the six nitrosamines in any of the samples of ham 72/111 and 73/09 analyzed.

Minimum radiation dose (MRD) for low nitrite/nitate ham

To complete the technology of low nitrite/nitrate ham the minimum radiation

Table 22. Ham 73/09: Acceptance of high and low nitrite irradiated Ham[1]. (Consumer-type panel: n=32).

Sample series No.	mg/kg added		Irradiation Source	Acceptance ratings	
	$NaNO_2$	$NaNO_3$		10 days	30 days
1	150[2]	600	Cobalt-60	6.8 ± 1.5	6.2 ± 1.7
2	25[2]	100	Cobalt-60	6.4 ± 1.5	6.0 ± 2.2
3	150	600	Electrons	6.5 ± 2.0	6.8 ± 1.4
4	25	100	Electrons	7.0 ± 1.2	6.8 ± 1.4
LSD (Least significant difference):				not significant	0.7

1. 3.7–4.4 Mrad at –30 ± 10°C; used also for nitrosamines determinations.
2. Paired sets of samples, Cobalt-60 vs electrons for both the low and high nitrite – nitrate hams.

sterilizing dose under the 12D concept (MRD) must be determined. The MRD for the high (normal) nitrite (156 mg/kg) /nitrate (700 mg/kg) ham at –30 ± 10 °C is 3.7 Mrad (Table 1). In the inoculated pack studies to determine the MRD for radappertized ham and other foods, enzyme-inactivated (pre-cooked) foods, diced into small pieces before inoculation and vacuum closing of the cans, were used (Anellis et al; 1965, 1967, 1969 and 1972). In the inoculated ham pack studies, *Cl. botulinum* spores are exposed only to residual levels of nitrite and nitrate, which are very low (Anellis et al., 1967; this report, Tables 8 and 21). The presence of salt (NaC1) in the inoculated ham pack studies has an effect on the surviving *Cl. botulinum* spores, whereas the residual levels of nitrite and nitrate have, or should have, little or no effect on the MRD for radappertized ham (Rowley et al., 1968). Since the salt content in the high and the low nitrite/nitrate hams is the same (2.2 ± 0.3%), little if any, deviation in the MRD from the high nitrite/nitrate ham (3.7 Mrad at –30 ± 10 °C) is expected. The inoculated pack study on the low nitrite/nitrate ham to determine the MRD has been in progress since December 1972.

Conclusions

1. Nitrite can be reduced from 156 mg/kg (the USDA maximum allowed and commonly used by the meat industry) to 25 mg/kg in radappertized ham.
2. A small amount of nitrate (100 mg/kg) and the USDA allowable amounts of ascorbate/erythorbate are needed for the characteristic flavor, odor and taste of cured meat color, formed by nitrite, in radappertized ham.
3. In addition to color, the use of 25 mg/kg nitrite, 50 to 100 mg/kg nitrate, and ascorbate/erythorbate are neede for the characteristic flavor, odor and taste of radappertized ham which determine the consumer acceptability of the product.
4. Small amounts of nitrate (50 to 100 mg/kg), in addition to nitrite and ascorbate/erythorbate, are needed apparently to prevent rancidity development in non-irradiated (and probably in irradiated) ham when stored in opened containers in a refrigerator. This factor should be investigated further.

5. Further reduction of nitrate from 100 mg/kg to 50 mg/kg in radappertized ham should be investigated in a paired comparison (50 vs 100 mg/kg $NaNO_3$) of irradiated and non-irradiated ham samples during a storage period of two years.
6. The mechanism of the prevention of discoloration of radappertized ham by nitrate and ascorbate/erythorbate is not known and should be investigated.
7. No nitrosamines were found in the radappertized, shelf-stable ham preserved by gamma rays from the cobalt–60 source or by the electrons from the linear accelerator. There was no indication, furthermore, of the presence of nitrosamines in the non-irradiated, frozen stored, control hams used in this investigation.

Acknowledgments

We acknowledge with great appreciation the technical assistance in generating certain technical data given in this report: J. S. Cohen for the data of the ham 73/30 experiment; Abe Anellis for allowing the use of some of the preliminary unpublished data in Table 1; Lucy J. Rice for conducting technological evaluations and monitoring the consumer tests; O. J. Stark for the chemical analyses; Mary A. Wall for the statistical treatments of the data; and W. Fiddler and A. E. Wasserman, USDA, for the determination of the nitrosamines in the ham 72/111 and 73/09 experiments. We are thankful also to M. S. Peterson, A. Brynjolfsson and E. S. Josephson for their critical reviews of the manuscript.

This paper reports research undertaken at the U.S. Army Natick (MA) Labs. and has been assigned No. TP–1414 in the series of papers approved for oral presentation and/or publication. The findings in this report are not to be construed as having the official endorsement of the Department of the Army.

References

Anonymous, 1965. The technical Basis for Legislation of Irradiated Foods. Report of a Joint FAO/IAEA/WHO Expert Committee, Rome, 21–28 April 1964. FAO Atomic Energy Series No. 6, The United Nations, Rome.

Anonymous, 1966. Review of the Food Irradiation Program. Hearing, Joint Committee on Atomic Energy, Congress of the United States (JCAE), September 12.

Anonymous, 1968. Status of the Food Irradiation Program. Hearings, JCAE, July 18 and 30.

Anonymous, 1970a. Official Methods of Analysis, AOAC, 11th Edition, Chapter 24: Meat and Meat Products.

Anonymous, 1970b. USDA. Meat Inspection Regulations, Revision Pursuant to Wholesome Meat Act. Fed. Register, Vol. 35 (193), Part II, October 3.

Anonymous, 1973a. AEC Authorizing Legislation, Fiscal Year 1974. Hearings, JCAE, March 20 and 22.

Anonymous, 1973b. Federal Meat and Poultry Inspection, Statistical Summary for 1972. U.S.D.A. Animal and Plant Health Inspection Service, Report No. MPI-l, Washington, D. C., May.

Anellis, A., N. Grecz, D. A. Huber, D. Berkowitz, M. D. Schneider & M. Simon, 1965. Radiation sterilization of bacon for military feeding. Appl. Microbiol. 13 (1) 37–42.

Anellis, A., D. Berkowitz, C. Jarboe & H. M. El-Bisi, 1967. Radiation sterilization of prototype military foods, II Cured ham. Appl. Microbiol. 15 (1) 166–177.

Anellis, A. & S. Wierkowski, 1968. Estimation of radiation resistance values of microorganisms in food products. Appl. Microbiol. 16 (9) 1300–1308.

Anellis, A., D. Berkowitz, C. Jarboe & H. M. El-Bisi, 1969. Radiation sterilization of prototype military foods, III Pork loin. Appl. Microbiol. 18 (4) 604–611.

Anellis, A., D. Berkowitz, W. Swantak & C. Strojan, 1972. Radiation sterilization of prototype military foods: Low-temperature irradiation of codfish cake, corned beef and pork sausage. Appl. Microbiol. 24 (3) 453–462.

Bailey, M. E. & J. W. Swain, 1973. Influence of nitrite on meat flavor. Prog. Meat Ind. Res. Conf., Chicago, Illinois, p. 29–45.

Bard, J. C., 1973. Effect of sodium nitrite and sodium nitrate on botulinal toxin production and nitrosamine formation in wieners. Proc. Meat Ind. Res. Conf., Chicago, Illinois, p. 61–68.

Christiansen, L. N., R. W. Johnston, D. A Kautter, J. W. Howard & W. J. Aunan, 1973. Effect of nitrite and nitrate on toxin production by *Clostridium botulinum* and on nitrosamine formation in perishable canned comminuted cured meat. Appl. Microbiol. 25: 357–362.

Duncan C. H. & E. M. Foster, 1968. Role of curing agents in the preservation of shelf stable canned meat products. Appl. Microbiol. 16: 401–405.

Fiddler, W., J. W. Pensabene, I. Kushnir & E. G. Piotrowski, 1973. Effect of frankfurter cure ingredients on N-nitrosodimethylamine formation in model system. J. Food Sci. 38: 714–715.

Friedman, L., 1973. Problem arising from the use of nitrite in food processing. Proc. Meat Ind. Res, Conf., Chicago, Illinois, p. 11–20.

Grecz, N., A. A. Walker, A. Anellis & D. Berkowitz, 1971. Effect of irradiation temperature in the range – 196 to 95°C on the resistance of spores of *Clostridium botulinum* 33A in cooked beef. Can. J. Microbiol. 17 (2) 135–142

Greenberg, R. A., 1972. Nitrite in the control of *Clostridium botulinum.* Proc. Meat Ind. Res. Conf., Chicago, Illinois, p. 25–34.

Herring, H. K., 1973. Effect of nitrite and other factors on the physico-chemical characteristics and nitrosamine formation in bacon. Proc. Meat Ind. Res. Conf., Chicago, Illinois, p. 47–60.

Josephson, E. S., A. Brynjolfsson & E. Wierbicki, 1968. Engineering and economics of food irradiation. Trans. New York Acad. Sci. Series II, 30 (4) 600–614.

Josephson, E. S. & E. Wierbicki, 1973. Radiation preservation of food: past, present and future. Activities Report 25 (1) 48–59 (U.S. Army Natick Labs., Natick Mass. 01760, U.S.A.).

Killoran, J. J., 1972. Chemical and physical changes in food packaging materials exposed to ionizing radiation. Radiation Res. Rev. 3: 369–388.

Lijinsky, W. & S. S. Epstein, 1970. Nitrosamines as environmental carcinogens. Nature 225: 21–23.

Olsman, W. J. & B. Krol, 1972. Depletion of nitrite in heated meat products during storage. 18th Meet Meat Res. Workers, Guelph, Canada, Vol. 2: 409–415.

Peryam, D. R. & F. J. Pilgrim, 1957. Hedonic scale method for measuring food preferences. Fd Technol. 11 (9), Supplement: 9–14.

Pivnick, H. & C. Thacker, 1970. Effect of sodium chloride and pH on initiation of growth by heat-damaged spores of *Clostridium botulinum.* Can. Inst. Fd Technol. 3: 70–75.

Rowley, D. B., H. M. El-Bisi, A. Anellis & O. P. Snyder, 1968. Resistance of *Clostridium botulinum* spores to ionizing radiation as related to radappertization of foods. Proc. First U.S. – Japan Conf. Toxic Microorganisms, Honolulu, Hawaii, October 7–10, p. 459–467.

Tannenbaum, S. R. & T. Y. Fan, 1973. Uncertainties about nitrosamine formation in and from foods. Proc. Meat Ind. Res. Conf., Chicago, Illinois, p. 1–10.

Tarladgis, B. G., B. M. Watts & M. T. Younathan, 1960. A distillation method for the quantitative determination of malonaldehyde in rancid foods. J. Am. Oil. Chem. Soc. 37: 44–48.

Thomas, M. H. & E. S. Josephson, 1970. Radiation preservation of foods and its effect on nutrients. The Science Teacher, 37 (3): (5 pages).

Thomas, M. H. & E. Wierbicki, 1971. Effect of irradiation dose and temperature on the thiamine content of ham. Technical Report 71–44–FL: U.S. Army Natick Laboratories, Natick, Mass., 01760, U.S.A.

Wasserman, A. E., W. Fiddler, E. C. Doerr, S. F. Osman & C. J. Dooley, 1972. Dimethylnitrosamines in frankfurters. Fd Cosmet. Toxicol. 10: 681–686.

Wasserman, A. E. & F. Talley, 1972. The effect of sodium nitrite on the flavor of frankfurters. J. Fd Sci. 37: 536–538.

Wierbicki, E., M. Simon & E. S. Josephson, 1965. Preservation of meats by sterilizing doses of ionizing radiation. Radiation Preservation of Foods, National Academy of Sciences – National Reserarch Council, Washington, D. C.; Publication No. 1273: 383–409.
Wolff, I. A. & A. E Wasserman, 1972. Nitrates, nitrites and nitrosamines. Science 177: 15–19.

Discussion

Influence of nitrite/nitrate on taste of ham

A taste panel of the Central Institute for Nutrition and Food Research TNO at Zeist gave the following marks for colour, odour and taste of irradiated ham with low and high nitrite levels (25 and 150 mg/kg): 7.8 and 7.9, 6.6 and 6.8, 6.1 and 6.1, respectively.

Nitrite seems to give a more firm product. Nitrate is necessary to prevent rancidity.

Influence of NH_4^-

It is known that nitrite as well as nitrate can be formed during irradiation of solutions of NH_4Cl with electrons.

However, Wierbicki did not determine the content of ammonia. Formation of nitrate and nitrite from ammonia in the irradiated ham samples is highly improbable since non-irradiated control samples contained comparable amounts of nitrite and nitrate.

Conclusions and recommendations of the chemical & technological session 13th September 1973

1. The general conclusion is that the problem of nitrite is so important that every effort must be made to fill gaps in our knowledge.

2. As to the loss of nitrite when added to meat only part of it could be accounted for by the proposed mechanisms. Much remains to be learned as to how nitrite, or related compounds, become bound to the meat fractions. In both of the above cases elucidation and quantification are needed.

3. It appears to be now possible to estimate the quantities of nitrite appreciably necessary to produce characteristic cured meat flavour. Disappointingly little progress has been made on the chemistry of cured meat flavour; the methodology is now at hand to make progress with this problem.

4. Bacon is currently the most critical product because, unlike other meat products, fried bacon has consistently been found to contain relatively large quantities of nitrosamine, particularly nitrosopyrrolidine.

5. Ascorbates at present are the best means to suppress or reduce the formation of nitrosamines in cured meat products. It is critically important to establish the optimal balance of nitrite and ascorbates in bacon.

6. Preliminary data indicate that the anti-botulinal effect of nitrite in cured meat products is not changed by the presence of at least 1000 mg per kg (ppm) of sodium ascorbate. Further work is required to establish this and to determine the general effect on the antimicrobial activity of nitrite.

7. Irradiation is a conceivable way of reducing the quantities of nitrite in processed meat products.

Toxicological session

Reporters: P. J. Groenen, V. J. Feron

Proc. int. Symp. Nitrite Meat Prod., Zeist, 1973. Pudoc, Wageningen.

Toxicity of nitrite and N-nitroso compounds

R. Preussmann

Institut für Toxicologie und Chemotherapie, Deutsches Krebsforschungszentrum, Postfach 449, D 6900, Heidelberg, W-Germany

Abstract

Toxic effects of chemical compounds may be roughly divided into acute, subchronic (or subacute) and chronic. There is general agreement among experts that the main toxicological problem of N-nitroso-compounds and nitrites is carcinogenesis, after chronic exposure to low quantities of N-nitroso-compounds, or the formation of such carcinogens from nitrite and nitrosatable amino compounds in the human gastrointestinal tract (see J. Sander p. 243–254). Therefore, in this short review, the *chronic* toxicity of the nitrite and N-nitroso compounds are stressed.

Nitrite, nitrates and nitrous gases

The main acute or subchronic toxic effect of ingested nitrites and nitrates (which are easily reduced to nitrites mainly by bacteria (Sander & Seif, 1969; Sander & Schweinsberg, 1972)) or are reduced in food samples to nitrite (Philips, 1971) is the formation of methemoglobin. Prominent symptoms are vomiting, cyanosis, shock and unconsciousness in man (Greenberg et al., 1945). Büsch (1952–1954), Pribilla (1965) and Bakashi et al. (1967) described nitrite poisoning in man due to illegal application of high concentrations of nitrite in meat products, mainly sausages. High nitrite reduction in infants under six months of age, due to low acid secretion in the stomach and the subsequent easy bacterial nitrate reduction, has led to serious intoxication and even death, especially after ingestion of nitrate-rich drinking water or spinach (Fassett, 1966; Phillips, 1971; Sander & Schweinsberg, 1972; all papers are summaries and contain further literature).

Chronic oral administration of nitrite has no carcinogenic or teratogenic effects: Druckrey et al. (1963) gave 100 mg/kg bodyweight daily to BD rats over the whole life-span of three generations. Except for a (probably not significant) shortening of the life-span and a reduction in hemoglobin concentration, no adverse toxic effects of the treatment were noted; reproduction of treated animals was not affected and no teratogenic effects were observed in the offspring. Sander (1971) added 1% $NaNO_2$ to the diet of rats for 117 days and observed the animals until natural death; he saw no tumors or other serious toxic effects. Lehman (1958) mentioned studies that showed sodium nitrate to be tolerated in the diet of rats for a lifetime at a level of 1% with no effects and only growth depression at 10%. Dogs tolerated

2% $NaNO_3$ for 3–4 months.

Chronic inhalation of NO_2 in rats did not lead to tumors (Ross & Henschler, 1963). The lung histology of rats inhaling air containing 2.9 mg NO_2/kg for 9 months was described by Arner & Rhoades (1973).

From such results a FAO/WHO expert committee (1962) considered a daily intake of 0.4 mg nitrite/kg bodyweight as acceptable; under special circumstances even doses as high as 0.8 mg/kg were considered tolerable. This evaluation was made at a time, when the possibility of a formation of carcinogenic N-nitroso-compounds in vivo had not been proven. In view of this, however, such a daily intake seems too high to be acceptable today and needs reconsideration.

N-nitroso compounds

According to their different chemical reactivities, organic N-nitroso compounds of the general formula

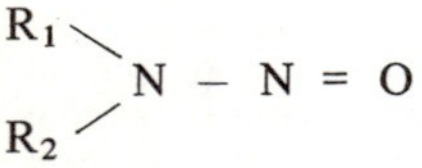

must be divided into two groups. The nitrosamines, in which R_1 and R_2 are alkyl, aralkyl and/or aryl residues, are chemically stable compounds, which are converted to reactive metabolites, very probably alkylating agents, by enzymatic processes in the mammalian organism (for review see Magee & Barnes, 1967). The nitrosamides usually have one alkyl residue R_1 and an acyl residue as R_2 and are chemically reactive compounds, which are split by hydrolysis more or less easily to form alkylating diazoalkanes.

Chemical characteristics

The chemical and biochemical characteristics of the two groups also govern their toxicity. Nitrosamines are mainly systemically acting agents, while nitrosamides react locally as well as systemically. Acute toxicities range from a LD_{50} of 18 mg/kg bodyweight for the highly toxic methylbenzylnitrosamine to a LD_{50} of more than 7500 mg/kg for the practically non-toxic diethanolnitrosamine. *N*-methyl-*N*-nitroso-urethane had a LD_{50} of 4 mg/kg, while butylnitrosourea had an LD_{50} of 1200 mg/kg (Druckrey et al., 1967; Magee & Barnes, 1972). The main organ to be affected by nitrosamines is the liver (centrilobular necrosis) and the lungs (hemorragic oedema), while with nitrosamides damage of the bone marrow and lymphatic tissue is predominant (Druckrey et al., 1967).

Biological effect

The main biological effect of N-nitroso compounds without any doubt is their potent carcinogenic activity. More than 100 N-nitroso compounds have been tested until now and more than 80 of them were more or less potent carcinogens in experimental animals. Some of the important data are presented in Tables 1–5. The

Table 1. N-nitroso compounds. Structural requirements for carcinogenicity.

Formula	Carcinogenicity
$\begin{matrix} R_1 \\ R_2 \end{matrix} \rangle N-NO$	+++
$\begin{matrix} R_1 \\ R_2 \end{matrix} \rangle N-NH_2$	(+)
$\begin{matrix} R_1 \\ R_2 \end{matrix} \rangle N-NO_2$	(+)
$\begin{matrix} R_1 \\ R_2 \end{matrix} \rangle N-CHO$	–
$\begin{matrix} R_1 \\ R_2 \end{matrix} \rangle CH-NO \rightleftharpoons \begin{matrix} R_1 \\ R_2 \end{matrix} \rangle C=N-OH$	?
$\begin{matrix} R_1 \\ R_2 \end{matrix} \rangle N-N=S$	–
$NO_2(-)$	–
$\begin{matrix} R_1 \\ R_2 \end{matrix} \rangle NH$	–

studies have shown that this group of carcinogens has very versatile carcinogenic activities, inducing malignant tumors in almost all important organs, such as liver, lungs, kidney, urinary bladder, oesophagus, stomach, small intestine, brain and nervous system. The organ specifity of the action depends mainly on the chemical structure of the compound and to a minor degree on the animal species, the route of application and the dose. Many nitroso compounds are active in single dose experiments (Druckrey et al., 1967; Magee & Schweinsberg, 1972).

The carcinogenic activity of dimethylnitrosamine, for example, has been studied in mouse, hamster, rat, guinea pig, rabbit and rainbow trout. In all these different species, the compound is a potent carcinogen (IARC, 1972).

The next higher homologue, diethylnitrosamine, has been shown to be a carcinogen in 12 animal species, including subhuman primates. The 12 species were mouse, hamster, rat, guinea pig, rabbit, dog, pig, rainbow trout, the aquarium fish Branchydanio rerio, grass parrakeet and monkey. So far no animal species tested has been found to be resistant to the carcinogenicity of this compound (IARC, 1972; Schmähl & Osswald, 1967).

Methylnitrosourea, as an example of a nitrosamide, has been tested in mouse, hamster, rat, guinea pig, rabbit and dog. It is powerful carcinogen in all of these animals (IARC, 1972).

Many of the other N-nitroso carcinogens have not been studied as extensively, but for almost all of them reliable and reproducible animal data are available.

There are further data, which are relevant to the problem of extrapolation of animal data to man. Since nitrosamines most likely require enzymatic activation to form the proximate and/or ultimate carcinogen, comparative metabolism in animal

Table 2. Carcinogenicity of symmetrical nitrosamines.

Nitrosamine	Formula	Carcinogenic action	Application	Main target organ
Dimethyl-	$O{=}N{-}N(CH_3)(CH_3)$	+++	p.o.	liver
Diethyl-	$O{=}N{-}N(CH_2{-}CH_3)(CH_2{-}CH_3)$	+++	p.o.	liver
Di-n-propyl-	$O{=}N{-}N((CH_2)_2CH_3)((CH_2)_2CH_3)$	+++	p.o.	liver
Di-isopropyl-	$O{=}N{-}N(CH(CH_3)_2)(CH(CH_3)_2)$	+	p.o.	liver
Di-n-butyl-	$O{=}N{-}N((CH_2)_3{-}CH_3)((CH_2)_3{-}CH_3)$	+++	p.o.	liver (urinary bladder)
Butyl-butanol-n-	$O{=}N{-}N((CH_2)_3{-}CH_3)((CH_2)_3{-}CH_2{-}OH)$	+++	sc. sc.	liver (urinary bladder) urinary bladder
Di-pentyl-	$O{=}N{-}N((CH_2)_4CH_3)((CH_2)_4CH_3)$	++	p.o. sc.	liver lung
Dibenzyl-	$O{=}N{-}N(CH_2{-}C_6H_5)(CH_2{-}C_6{-}H_5)$	–	p.o.	– – –
Diphenyl-	$O{=}N{-}N(C_6H_5)(C_6H_5)$	–	p.o.	– – –

p.o.: by mouth
sc.: subcutaneous
iv.: intravenous

Table 3. Carcinogenicity of unsymmetrical nitrosamines.

Nitrosamine	Formula	Carcinogenic action	Application[1]	Main target organ
Methyl-vinyl-	$O{=}N{-}N(CH_3)(CH{=}CH_2)$	+++	p.o.	oesophagus
Methyl-pentyl-	$O{=}N{-}N(CH_3)((CH_2)_4CH_3)$	+++	p.o. sc.	oesophagus
Methyl-benzyl-	$O{=}N{-}N(CH_3)(CH_2{-}C_6H_5)$	+++	p.o.	oesophagus
Ethyl-butyl-	$O{=}N{-}N(C_2H_5)((CH_2)_3CH_3)$	+++	p.o. iv.	oesophagus
NO-sarkosinester	$O{=}N{-}N(CH_3)(CH_2{-}C({=}O){-}OC_2H_5)$	+++	p.o.	oesophagus
Methyl-allyl-	$O{=}N{-}N(CH_3)(CH_2{-}CH{=}CH_2)$	++	iv.	kidney
Ethyl-tert.butyl-	$O{=}N{-}N(C_2H_5)(C(CH_3)_3)$	–	p.o.	– – –

1. Meaning of abbreviations: see Table 2.

and man can give further evidence to facilitate the discussed extrapolation of animal data to man. Montesano & Magee (1970) in a comparative study in vitro with dimethylnitrosamine showed that this compound is metabolized in a qualitatively similar manner in both human and rat liver. Quantitatively, it was shown that the rate of metabolism in human liver slices was comparable to that in rat liver and that similar levels of nucleic acid methylation occur in both species.

Last, but not least, there is one direct observation in man, although not related to carcinogenicity. Dimethylnitrosamine has induced acute toxic liver damage in workers in the chemical industry who were exposed to this compound (Barnes & Magee, 1954). Centrilobular necrosis is similarly produced in most animal species treated with high doses of nitrosamines in acute toxicity experiments.

From the present evidence, it is clear that N-nitroso compounds act as carcinogens in many animal species, including subhuman primates. Until now no animal species has been found to be resistant to the carcinogenic effect. Metabolism as well as the acute toxic effects are similar or identical in man and experimental animals. Taking all facts into consideration, one can conclude that man will probably react in a manner similar to that of experimental animals, which means that N-nitroso compounds are almost certainly carcinogenic in man. The opinion of Lijinsky & Epstein (1970) that nitrosamines and nitrosamides 'seem to be a major class of carcinogens that are likely to be causally related to human cancer' is now shared by many. The widespread concern about carcinogenic nitrosamines in the human environment is, therefore, justified.

A more difficult problem is, what are the levels of nitrosamines that should be considered dangerous to man. In general, there is agreement that for carcinogens only a 'zero tolerance' can be accepted in principle. However, it is well-known that this concept is not always feasible in general practice. Moreover, from the analytical point of view, 'zero' is always a matter of analytical methodology, which is not a constant. Therefore, a rough estimation of 'no-effect' doses from animal experiments is necessary.

There is one dose-response study available for dimethylnitrosamine in the rat

Table 4. Carcinogenicity of cyclic nitrosamines.

Nitrosamine N-nitroso-	Formula	Carcinogenic action	Application[1]	Main target organ
-morpholine	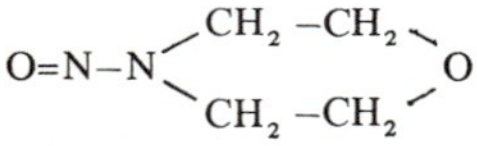 $O=N-N\begin{matrix} CH_2-CH_2 \\ CH_2-CH_2 \end{matrix}O$	+++	p.o. iv.	liver
-pyrrolidine	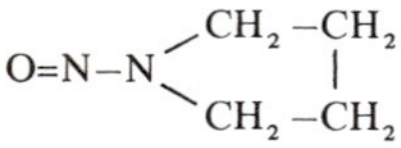 $O=N-N\begin{matrix} CH_2-CH_2 \\ \| \\ CH_2-CH_2 \end{matrix}$	+	p.o.	liver
-piperidine	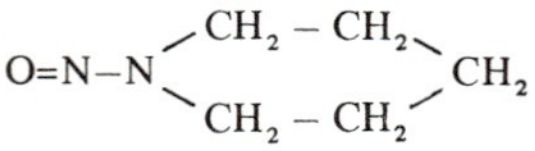 $O=N-N\begin{matrix} CH_2-CH_2 \\ CH_2-CH_2 \end{matrix}CH_2$	+++	p.o. sc. iv. sc. iv.	oesophagus nasal cavity
Di-nitroso-piperazine	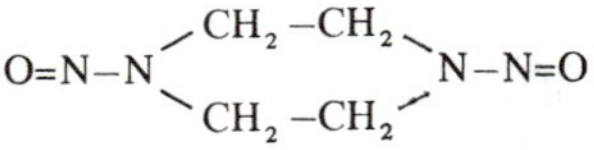 $O=N-N\begin{matrix} CH_2-CH_2 \\ CH_2-CH_2 \end{matrix}N-N=O$	+++	p.o. sc.	oesophagus
-N-carbethoxy-piperazine	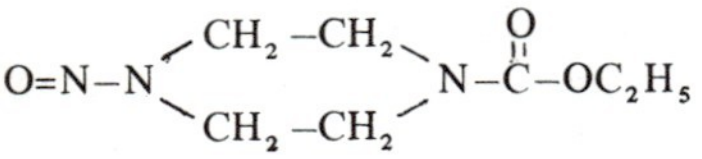 $O=N-N\begin{matrix} CH_2-CH_2 \\ CH_2-CH_2 \end{matrix}N-\overset{O}{\underset{\|}{C}}-OC_2H_5$	++	sc.	liver
-N-methyl-piperazine	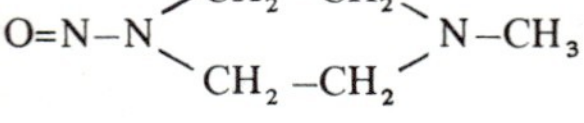 $O=N-N\begin{matrix} CH_2-CH_2 \\ CH_2-CH_2 \end{matrix}N-CH_3$	−	p.o.	− − −
-proline-ethylester	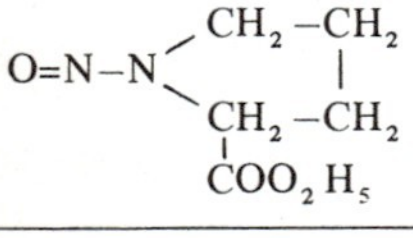 $O=N-N\begin{matrix} CH_2-CH_2 \\ \| \\ CH_2-CH_2 \\ \| \\ COO_2H_5 \end{matrix}$	−	p.o.	− − −

1. Meaning of abbreviations: see Table 2.

Table 5. Carcinogenicity of nitrosamides.

Nitrosamide N-nitroso-	Formula	Carcinogenic action	Application[1]	Main target organ
-methylacetamide	$O{=}N{-}N\langle^{CH_3}_{CO-CH_3}$	++	p.o.	fore-stomach
-N-methylurethane	$O{=}N{-}N\langle^{CH_3}_{CO-OC_2H_5}$	+++	p.o. iv.	fore-stomach lung
-methylurea	$O{=}N{-}N\langle^{CH_3}_{CO-NH_2}$	+++	p.o. iv.	fore-stomach brain
-dimethylurea	$O{=}N{-}N\langle^{CH_3}_{CO-NHCH_3}$	+++	p.o.	brain, nervous system, spinal cord
-trimethylurea	$O{=}N{-}N\langle^{CH_3}_{CO-N(CH_3)_2}$	+++	p.o. iv.	periph. nerves spinal cord
-methyl-N'-acetyl-urea	$O{=}N{-}N\langle^{CH_3}_{CO-NH-COCH_3}$	+++	p.o.	glandular stomach

1. Meaning of abbreviations: see Table 2.

(Terracini et al., 1967). Dietary concentration ranged between 2 and 50 mg/kg. At 2 and 5 mg/kg, incidences of liver tumors among survivors at 60 weeks were 1/26 and 8/74, respectively. With higher concentrations more than 70% of the rats had liver tumors. Therefore, for continuous feeding studies, a concentration of 1 mg/kg in the diet could be considered a 'threshold dose'. Single doses of 20 mg/kg body wt. are carcinogenic in the rat (Magee & Barnes, 1959).

A dose-response study, involving oral administration of diethylnitrosamine in the rat, was done by Druckrey et al. (1963). Daily doses ranged between 14.2 and 0.075 mg/kg in nine dosage groups. Total doses administered until death were between 14 and 0.075 mg/kg. All daily doses higher than 0.15 mg/kg gave a tumor yield of 100%. Doses of 0.15 mg/kg gave a tumor incidence of 27/30 liver carcinomas. At 0.075 mg/kg 20 rats lived longer than 600 days; 11/20 had benign or malignant tumors of the liver and the oesophagus. All four of the animals living longer than 940 days at this dose level had tumors. Therefore, 0.075 mg/kg per day diethylnitrosamine, which corresponds approximately to 0.5 – 0.75 mg/kg in the diet, is clearly carcinogenic and above the 'threshold concentration'. The marginal effect dose could be estimated to be at 0.5 mg/kg.

An extremely low single dose of 1.25 mg/kg diethylnitrosamine leads to kidney tumors in the rat (Mohr & Hilfrich, 1972).

It is generally accepted international practice in establishing tolerated doses, to observe a safety margin of 100 when extrapolating animal data to man. Therefore, a level of 5–10 μg/kg should be considered as a 'tolerable' dose of nitrosamines of low molecular weight. Hence this is also the concentration that must be detected and determined in a chemical analysis of such N-nitroso compounds in environmental media. This detection limit is generally accepted now.

It must be stressed, however, that such 'calculations' are rather unsatisfactory and can by no means lead to 'safe' levels of nitrosamines and nitrosamides. For one, it is well known that single-dose experiments with *N*-nitroso compounds have led to malignant tumors in animals. Unfortunately, no dose-response studies for such experiments are available. On the other hand, many experiments, especially those by Schmähl and his colleagues (1970) and Montesano & Saffiotti (1968), have shown that subthreshold doses of diethylnitrosamine still give rise to tumors when administered together with other carcinogens. This synergistic, additive effect of different groups of chemicals with the same organotropic carcinogenic effect (syncarcinogenesis) clearly is similar to the human situation, where a population almost certainly is never exposed to only one single carcinogen, but to minute quantities of many different carcinogenic compounds.

Extremely low doses of certain *N*-nitroso compounds have a transplacental carcinogenic effect, producing tumors in the offspring of mothers treated during pregnancy (Ivankovic & Druckrey, 1968). Many compounds of this group have also shown mutagenic and teratogenic effects.

References

Arner, E. C. & R. A. Rhoades, 1973. Long-term nitrogen dioxide exposure. Archs envir. Hlth 26: 154–160.

Bakashi, S. P., J. L. Fahey & L. E. Pierce, 1967. Sausage Cyanosis. Acquired methemoglobin nitrite poisoning. New Engl. J. Med. 277: 1072.

Büsch, O., Sammlung von Vergiftungsfällen 1952–1954, 14: 53.

Druckrey, H., R. Preussmann, S. Ivankovic & D. Schmähl, 1967. Organotrope carcinogene Wirkung bei 65 verschiedenen N-nitroso-Verbindungen an BD-Ratten. Z. Krebsforsch. 69: 103–200.

Druckrey, H., A. Schildbach, D. Schmähl, R. Preussmann und S. Ivankovic, 1963. Quantitative Analyse der carcinogenen Wirkung von Diäthylnitrosamin. Arzneimittelforschung 14: 841–851.

Druckrey, H., D. Steinhoff, H. Beuthner, H. Schneider & P. Klärner, 1963. Prüfung von Nitrit auf chronisch-toxische Wirkung an Ratten. Arzneimittelforschung 13: 320–323.

FAO/WHO 1962. Technical Report WHO.

Fassett, D. W., 1966. Nitrates and nitrites. In: Toxicants occuring naturally in foods. Nat. Res. Council, Fd Prot. Comm., Public. No. 1354 Nat. Acad. Sci., Washington 250–256.

Greenberg, M., W. B. Birnkraut & J. J. Schiftner, 1945. Outbreak of sodium nitrite poisoning. Am. J. Publ. Health 35: 1217–1220.

International Agency for Research on Cancer, 1972. Monographes on the evaluation of carcinogenic risk of chemicals to man. Vol. 1.

Ivankovic, S. & H. Druckrey, 1968. Transplacentare Erzeugung maligner Tumoren des Nervensystems. I. Äthylnitrosoharnstoff. Z. Krebsforsch. 71: 320–360.

Lehmann, A. J., 1958. Nitrates and nitrites in meat products. Drug Officials U.S. Q. Bull. 22, 136–151.

Lijinsky, W. and S. S. Epstein, 1970. Nitrosamines as environmental carcinogens. Nature 225: 21–23.

Magee, P. N., 1971. Toxicity of nitrosamines: Their possible human health hazards. Fd Cosmet. Toxicol. 9: 207–218.

Magee, P. N. and J. M. Barnes, 1967. Carcinogenic nitroso compounds. Adv. Cancer Res. 10: 163–246.
Magee, P. N. and J. M. Barnes, 1959. The experimental production of tumors in the rat by dimethylnitrosamine. Acta Un. int. Cancr. 15: 187.
Mohr, U. & J. Hilfrich, 1972. Effect of a single dose of N-diethylnitrosamine on the rat kidney. J. Nat. Cancer Inst. 49: 1729–1731.
Montesano, R. & P. N. Magee, 1970. Metabolism of dimethylnitrosamine by human liver slices. Nature 288: 173.
Montesano, R. & W. Saffiotti, 1968. Carcinogenic response of the respiratory tract of Syrian Golden Hamsters to different doses of diethylnitrosamine. Cancer Res. 28: 2197–2210.
Phillips, U. E. J., 1971. Naturally occurring nitrate and nitrite in foods in relation to infant methaemoglobinaemia. Fd Cosmet. Toxicol. 9: 219–228.
Pribilla, O., 1965. Gerichtsmedizinische Bemerkungen zum Nitrit-Gesetz. Beitr. gerichtl. Med. 23: 207–225.
Ross, W. & D. Henschler, 1963. Wirkung langdauernder Einatmung niedriger Konzentrationen von Stickstoffdioxyd auf die Mäuselunge, Naturwissenschaften 50: 503–504.
Sander. J., 1971. Untersuchungen über die Entstehung cancerogener Nitrosoverbindungen im Magen von Versuchstieren und ihre Bedeutung für die Menschen. Arzneimittelforschung 21: 1572–1576; 1707–1713; 2034–2039.
Sander, J. & F. Schweinsberg, 1972. Wechselbeziehungen zwischen Nitrat, Nitrit un kanzerogenen N-Nitroso-Verbindungen. Zentbl. Bakt. ParasitKde A-bt. I. Orig. 156: 299–340.
Sander, J. & F. Seif, 1969. Bakterielle Reduktion von Nitrat im Magen des Menschen als Ursache einer Nitrosamin-Bildung. Arzneimittelforschung 19: 1091–1093.
Schmähl, D., 1970. Experimentelle Untersuchungen zur Syncarcinogenese. Z. Krebsforsch. 74: 457–466.
Schmähl, D. & H. Osswald, 1967. Carcinogenesis in different animal species by diethylnitrosamine. Experientia 23: 497–498.
Shearer, M. P. H., J. R. Goldsmith, C. Young, O. A. Kearns & B. R. Tamplin, 1972. Methemoglobin levels in infants in an area with high nitrate water supply. Am. J. publ. Hlth 62: 1174–1180.
Terracini, B., P. N. Magee & J. M. Barnes, 1967. Hepatic pathology in rats on low dietary levels of dimethylnitrosamine. Brit. J. Cancer 21: 559–565.

Discussion

Sources of nitrate and nitrite ingestion

From preliminary estimates it appears that in the United Kingdom half of the quantity of nitrate/nitrite ingested by humans comes from saliva, one quarter from the drinking water and one quarter from cured meat products and other foods.

Possible role of the diet

Tumours or other serious ill-effects have never been observed in animals after long-term oral administration of nitrite alone. Apparently no effective amounts of carcinogenic nitrosamines were formed in the stomach. This might be due to the absence of sufficient quantities of amines that can be nitrosated in the diets used.

More knowlegde about N-nitrosopyrrolidine

At present N-nitrosopyrrolidine seems to be a weaker carcinogen than many other volatile nitrosamines. It has, however, not yet been tested thoroughly, because the particular importance of this compound was not known until recently. Further experiments with N-nitrosopyrrolidine are planned in the Deutsches Krebsforschungs Zentrum.

Carcinogenicity of nitrosoproline

Whereas the ethyl ester of nitrosoproline had been found to be non-carcinogenic, the free acid was unstable and difficult to handle, and had therefore not been tested for carcinogenicity. It should be, however.

Realistic nitrite levels in experiments

The relevance of results obtained from studies, in which unrealistically high nitrite and amine concentration have been used, was questioned. The purpose of such model studies was, however, to find out whether the formation of nitrosamines from nitrite and amines took place at all in the stomach. This appeared to be the case. Further experiments must now be conducted under more realistic conditions.

Inhibiting effect of ascorbic acid on nitrosation

Recent experiments by Ivankovic, Zeller, Schmähl and Preussmann (Naturwissenschaften, in press) have shown that the carcinogenic effect observed in animals following the feeding of ethyl-urea and nitrite was prevented by the simultaneous feeding of ascorbic acid. Since the simultaneous feeding of ascorbic acid and ethylnitroso urea did not inhibit tumor formation, it is clear that ascorbic acid prevented the formation of nitroso compounds, but had no effect on preformed *N*-nitroso compounds.

Need for more research

Because most environmental carcinogens are present in low to very low concentrations, the possible synergistic effects of such carcinogens must be investigated.

Since, however, very little data are available on the carcinogenicity of non-volatile nitroso-compounds, which might be formed in food products, carcinogenicity studies with these compounds, also have high priority.

A tolerance level for nitrosamines?

Though the question of tolerance levels for nitrosamines is of current interest in many advisory committees, doubt was expressed whether in Germany a recommendation as to a tolerance level for nitrosamines will be given.

Proc. int. Symp. Nitrite Meat Prod., Zeist, 1973. Pudoc, Wageningen.

Philosophy of 'no effect level' for chemical carcinogens

R. Kroes, G. J. Van Esch and J. W. Weiss

National Institute of Public Health, P.O. Box 1, Bilthoven, the Netherlands

Abstract

Several factors may have an effect on the setting of tolerance levels of carcinogens. Such factors are:
- modifying factors i.e. factors influencing the action of carcinogens,
- the existence of powerful and weak or possible carcinogens,
- the existence of a dose-response relationship of carcinogens,
- the question whether carcinogen-induced changes are always irreversible,
- epidemiological evidence especially in man for the carcinogenity of the compounds and
- the question of residues, especially with respect to the detection levels of carcinogens.

According to present knowledge the majority of human cancers are caused by chemicals. Therefore it is necessary to consider every step leading to the establishment of tolerance levels very carefully. However it seems illogical and inappropriate to ban every compound found to be a carcinogen in animal.

When evaluating tolerance levels we have to consider the possibility of the existence of many known and unknown factors which may augment or inhibit the carcinogenic action.

The fixing of tolerance levels for certain carcinogens, using a safety factor which is 10 times that used for toxic agents, is discussed for certain conditions.

Introduction

In the last decade numerous papers have been published about the threshold in carcinogenesis (Mantel, 1962; Weisburger & Weisburger, 1968; Roe, 1968; Bryan & Lower, 1970; Hatch, 1970; Goldberg, 1971; Hueper, 1971; Ledbetter, 1971; Shabad, 1971; Stokinger, 1971; Neiman, 1972; Shubert, 1972; Weil, 1972; Oser, 1973 and Shabad, 1973).

Although individual differences occur, most countries in the world have legislation prohibiting the use of carcinogenic compounds as food additives. These regulatory decisions were originally based on the Delaney Amendment, part of the Federal Food, Drug and Cosmetic Act (Public law 85–929).

The amendment reads as follows:

'No such regulation shall issue if a fair evaluation of the data before the Secretary– ...... fails to establish that the proposed use of the food additive, under the conditions of use to be specified in the regulation, will be safe: provided, that no additive shall be deemed to be safe if it is found to induce cancer when ingested by man or animal or if it is found, after tests which are appropriate for the evaluation of the safety of food additives, to induce cancer in man or animal, exept that this proviso shall not apply with respect to the use of a substance as an ingredient of feed for animals which are raised for food production, if the Secretary finds (i) that, under the conditions of use and feeding specified in proposed labeling and reasonably certain to be followed in practice, such additive will not adversely affect the animals for which such feed is intended, and (ii) that no residue of the additive will be found .... in any edible portion of such animal after slaughter or in any food yielded by or derived from the living animal; ....'

It becomes immediately clear that this clause raises many problems especially if one realizes the diversity of opinions which exist in the world on the meaning of concepts such as 'induction of cancer', 'appropriate tests' and 'no residues'. Furthermore the clause does not indicate any dose, so that compounds fed at unrealistically high levels and found to be carcinogenic at that level, are banned from use as an additive in food. The different opinions about this clause and the concept of a 'safe' level for carcinogens can not be better illustrated than by quoting Hueper (1971): 'The concept of a 'safe' dose for carcinogens is without scientific basis, is deceptive and, if legally adopted, represents a potential highly dangerous public health policy' and Roe (1973): 'This regulation would make good sense (1) if it were possible to distinguish absolutely between carcinogens and non-carcinogens by means of animal tests (2) if it were possible to devise a diet that was free of naturally occurring carcinogens, and (3) if the effects of carcinogens were really irreversible' and 'The present preoccupation with new chemical additives to the exclusion of natural food constituents and natural contaminants is illogical if not absurd'.

When considering the threshold problem one must realize that there are many factors which may have an effect on the 'safe' level of a carcinogen.

1. Factors influencing the action of the carcinogen(s).
2. The existence of powerful or real carcinogens, and weak or possible carcinogens, as well as differences in action mechanism among the various carcinogens.
3. The dose-response relationship of carcinogens.
4. The existence of reversible effects of carcinogens.
5. Epidemiological studies in man, especially on the possible correlation of potential carcinogens already present in the environment and the cancer incidence in man.
6. The residue problem: what amount is detectable?

Factors influencing the action of carcinogen(s)

These factors, also known as modifying factors, may cause inhibition, augmentation or a qualitative change in a particular biological process (Berenblum, 1969).

Inhibition

Inhibition can manifest itself in a reduction of the tumor yield and/or a longer latency period. Suppression of carcinogenic action may be caused by nutritional factors, inhibition or stimulation of enzyme systems and hormonal and immunological status of the host.

With regards to nutritional factors, it is known that caloric restriction inhibits tumor development, as do dietary fat, protein and histamin deficiency. However excess of some constituent also seems to reduce tumor growth (Tannenbaum, 1958; White, 1961; Kalbe et al., 1968; Madhaven & Gopalan, 1968; Magee, 1969; Warwick 1971).

Inhibitory effects of the stimulation or suppression of enzyme systems are known for several chemical carcinogens. Inhibition may occur through increased metabolism to inactive derivatives (Miller & Miller, 1971). There are numerous examples of increased metabolism to inactive derivatives, sometimes via a stage in which the ultimate carcinogen is formed, but quickly metabolized (Wattenberg, 1966; Hadjiolov, 1971; Conney et al., 1971; McLean & Marshall, 1971; Yamamoto et al., 1971; Conney & Burns, 1972; Den Tonkelaar et al., 1972). Decreased metabolism to ultimate carcinogens was shown by Weisburger et al. (1972).

Hormonal status may affect the tumor induction by carcinogens. Although several reports are available demonstrating that estrogens are carcinogenic by themselves (Herbst et al., 1971), a few papers have been published on the inhibitory effect of anti-ovulatory drugs on tumor induction (Stern & Mickey, 1969; Kistner, 1969 and Thomas et al., 1972). Epidemiological studies in humans revealed a decrease of neoplasms in women treated with estrogen or anti-ovulatory drugs (Wilson, 1962 and Doll, 1972).

The importance of the immunological status of the host on inhibition of tumor development or tumor growth was shown by the work of, among others, Mathé (1973).

An interesting findings (Lacassagne et al., 1945) was that one carcinogen may inhibit the effect of another. Hill et al. (1951), Richardson et al. (1952) Miller et al. (1958) and Likhachev (1968) found similar effects.

Augmentation or co-carcinogenesis

Augmentation is an increase in tumor incidence or a shortening of the latent period or both. Augmentation can happen in many ways (Berenblum, 1969):
– by addition; i.e. cumulation of the effects of two or more carcinogens;
– by synergistic action, when the combined effect exceeds the sum of their separate effects;
– by incomplete carcinogenic action, either as initiator or as promotor;
– and by a change in specific properties such as sensitivity of organs, or systems (e.g. endocrines, transplacental exposure), the absorption rate of the carcinogenic compound, or by activation of certain viruses.

Augmentation may be induced by nutritional factors, changed metabolism of the carcinogen, immunological status of the host, certain compounds which are not carcinogenic by themselves, and many other factors. The age of the host at the time of exposure is very important. Transplacental exposure to some carcinogens yields very high tumor incidences (Napaikov, 1971 and Tomatis et al., 1971) while exposure to carcinogens at a young age also seems to enhance tumor incidence (Terracini & Testa, 1970, and Vesselinovitch et al., 1972). Augmentation is without doubt one of the most important problems that have to be considered when the possibility of threshold dosages or 'safe' levels is evaluated. Especially the promoting action of certain compounds and the synergism among carcinogens are of importance. Bingham & Falk (1969) showed a 1000 fold enhancement of the potency of benzo(a)pyrene and benz(a)anthracene when the non-carcinogen-dodecane was used as diluent. The promoting effects of croton oil (Berenblum, 1969; Hecker, 1971, 1972) and tobacco smoke (Van Duuren, 1968 and Wynder & Hoffmann, 1969) are well documented. Some vegetables fats and oils are potent co-carcinogens (Sinnhuber et al., 1968). Interesting are the findings of Gaudin et al. (1972), who showed that co-carcinogens such as croton oil inhibit DNA repair, which might be the possible mechanism of their co-carcinogenic action.

Synergism has been observed in experimental animals as well as in humans. Schmähl et al. (1963) and Schmähl & Thomas (1965) indicated that carcinogens with the same target organ may act synergistically whereas carcinogens with a different target do not. When four different hepatocarcinogens were given to rats in daily doses which could be 'subthreshold' if given alone to rats, liver cancer occurred in 34% of the animals (Schmähl et al., 1970). Likhachev (1968), however, also noted a synergistic effect of carcinogens with a different 'target' organ. Synergistic action in skin carcinogenesis was extensively studied by Neiman (1967 and 1968), who found that painting of skin at one side with carcinogen sensitizes the whole skin for a second group of applications. Montesano (1970) reviewed the synergistic and additive effects in respiratory carcinogenesis. Of interest is the study of Deichmann et al. (1967) who gave four (weak) hepatocarcinogens to rats simultaneously. These compounds were Aramite, DDT, methoxychlor and thiourea or aldrin. Surprisingly no additive effect was noted. On the contrary an antagonistic type of effect was found because the survival time of these rats was longer and their liver tumor incidence lower.

In epidemiological studies in humans, synergistic or additive effects were found for uranium and smoking (Kuschner and Laskin, 1971), asbestos, tar and chromium (Bittersochl, 1971), asbestos and smoking (Gilson, 1973), and betelleaf chewing and smoking (Jussawalla, 1973) while the synergistic or additive effect in animals usually was no more than several fold, in human this increased tumor risk varied from twofold to thirty-fold.

Another form of augmentation of carcinogenic action can be induced by the immunological status (Roe & Rowson, 1968). Evidence is available that a suppressed immune system leads to increased tumor incidence in humans (Balner, 1970; Reis, 1971; Fudenberg et al., 1971 and Walder et al., 1971). In experimental animals treatment with immunosuppressive agents such as antilymphocyte serum or azathioprine was followed by an increase in tumor incidence and a shorter latency period of chemically induced tumors (Balner & Dersjant, 1969; Woods, 1969; Haran-Chera & Lurie, 1971; Sheehan et al., 1971 and Baroni et al., 1972). Thymectomy also proved to increase the tumor rate of chemical induced tumors (Sheehan & Shklar, 1971). Others, however, found no effect of immunosuppressants on tumor yield in chemical carcinogenesis (Frankel et al., 1970; Schmähl et al., 1971 and Wagner, 1971). There is evidence that certain carcinogens are immunosuppressive themselves, thus enhancing (or causing) their own action (Sternswärd, 1969; Ball, 1970; Barconi et al., 1970; Bluestein & Green, 1970). In a review Mäkelä (1972) states that immunosuppressive agents may provoke a higher tumor incidence but usually do not. There is however good evidence for a decrease in latency period and the number of tumors per animal may be increased. In a recent review Gleichman & Gleichman (1973) state that the immunosurveillance theory (i.e. lack of cellular immune response against tumor-specific antigens) might not fit as a mechanism and that perhaps two other mechanisms: an alteration of tissue proliferation and a lowered resistance against viruses, may be involved.

Powerful or real carcinogens and weak or possible carcinogens

Although there is no doubt about the existence of powerful carcinogens, such as aflatoxin, nitrosamines, polynuclear aromatic compounds, aromatic amines and azo-dyes, there are numerous examples of compounds which have only a weak carcinogenic action. Weak carcinogenic action means that usually high doses of the compound are necessary over a long time to provoke a carcinogenic effect.

Cyclamate, carbon tetrachloride, metals like Pb and Cd, DDT, Aldrin, Dieldrin and several more of the pesticides and industrial chemicals investigated by Innes et al. (1969) can be considered as (weak) carcinogens. It may be asked whether these weak carcinogens are carcinogens at all, or just compounds which augment 'spontaneous' tumor incidence by indirect action. Calcium cyclamate or ethylene glycol (Golberg, cited by Roe, 1968) cause bladderstones which through irritation give rise to bladder papillomas and bladder carcinomas. DDT might be an immunosuppressant as are some of the dithiocarbamates tested by Innes et al. (unpublished results, Van Logten & Kroes, 1970). Recent findings of Hicks et al. (1973) indicate that saccharine most probably is a promoting agent. If we assume this is so, it might be suggested that saccharine fed at a rate of 7.5% in the diet which has induced some bladder tumors (cited by Oser, 1973) promotes irritation of the bladder epithelium due to concrements.

Special attention must be given to hormones and antihormonal agents as carcinogens. Do they act directly? Diethylstilbestrol (Herbst et al., 1971) seems to be a transplacental carcinogen in humans, although only at high levels. Some goitrogenic compounds have been named carcinogens because they induce nodular hyperplasia in thyroids subsequently giving rise to tumors. The same hyperplastic lesions and tumors are produced when rats are given a low iodine diet (unpublished results, Kroes, 1969). Is such a compound a carcinogen? Or an 'indirect' carcinogen?

There is a need for a classification on properties and 'direct' or 'indirect' action. This classification will not be easy at all, but that is no reason not to undertake it.

Such a classification has nothing to do with the classification of carcinogens in a distinction between primary or proximate carcinogens and secondary or procarcinogens, which refers to properties of action at the point of application or elsewhere (Weisburger & Weisburger, 1968).

The dose-response relationship of carcinogens

Especially the brilliant work of Druckrey (1967b) threw a light on the existence of a dose-response relationship of carcinogens. He noted that dependency of the induction time t on the dose d resulted in a linear function according to the formula dt^n= constant. The values of the exponent n were usually 2–4.

From these and other studies (Bock, 1968; Weisburger & Weisburger, 1968;

Shabad, 1971) it can be concluded that very low doses might not induce tumor in animals because the induction time exceeds the average lifespan of the animals. Weil (1972) quoted the work done on the dose-response relationship for radiation-induced cancer.

The dose-response curve however may vary in slope and intercept depending on several factors, such as type of animal, sort of neoplasm, distribution of the compound in space and time etc. That a dose-response relationship also exists in humans has been found in several epidemiological studies; prolonged exposure to a certain hazard increased the risk of tumor development (Case, 1966; Boyland, 1969; Gilson, 1973).

Another example is the existence or a correlation between the increased risk of lung cancer and the higher number of cigarettes smoked (Anon., 1970 and Wynder & Mabuchi, 1972).

In the evaluation of the carcinogenic action of compounds, when inducing an increased tumor incidence, it is important to establish any dose-response relation.

The existence of reversible effects of carcinogens

It is generally believed that carcinogens induce irreversible alterations (Weisburger & Weisburger, 1968; Bryan & Lower, 1970). However it is also believed that the concept of irreversibility might not always be valid. Recently Roe et al. (1972) showed evidence of reversibility in the initiation of carcinogenesis by dimethylbenz(a)anthracene.

The concept of irreversibility is also made less likely by the occurrence of spontaneous tumor regression and by the fact that DNA repair exists and immunosurveillance is present. Everson (1964) evaluated many cases of spontaneous regression in human cancer and concluded that in 130 cases spontaneous regression was probable.

Epidemiological studies in man, especially on the possible correlation of potent carcinogens already present in the environment and the cancer incidence in man

From the many epidemiological studies of human cancer it can be concluded that these studies are very important but also very difficult. So many factors may contribute to the observed results that is usually very difficult to establish evidence for a single factor influencing the tumor incidence. Nevertheless many relationships could be found from this type of study such as the relation between lung cancer and smoking, bladder cancer and the rubber industry, lung cancer and uranium, mesothelioma and asbestos, lung cancer and mustard gas, and liver cancer and aflatoxin (Epstein, 1972).

It is also clear that weak carcinogens will not be identified as easily as strong carcinogens by epidemiological studies. Special attention must be directed to epidemiological studies on people or groups of people who were involved in production of certain carcinogens or exposed in another way to carcinogens (i.e. on drugs). For example it would be of great importance if a good study on exposure to nitrosamines (time of exposure, direct or indirect contact) showed that nitrosamines are also carcinogenic in man.

The residue problem, what amount is detectable?

It is only in recent years that nitrosamines can be detected in smaller amounts than a mg/kg. Very delicate techniques to detect chemicals have been developed so that levels can now be measured in the range of 10^{-3} mg/kg, depending on the compound.

Nitrosamines can be detected now in amounts equal to 10^{-2} mg/kg and more. Aflatoxins at one time detected only when present at a level of a mg/kg can now be determined in amounts well below 10^{-3} mg/kg. The detection level of DDT has dropped from the range of a mg/kg to less than 10^{-3} mg/kg thanks to a delicate electron capture method (Oser, 1973).

It should be emphasized that according to present knowledge about 80–90% of human cancers are caused by chemicals (Clayson, 1967; Boyland, 1969; Bryan & Lower, 1970). Therefore it is necessary to consider every step leading to the establishment of safe levels very seriously and carefully. However we are faced with the fact that a number of known powerful carcinogens (such as benz(a)pyrene, nitrosamines) are already present in the environment and that their presence in common foods does not seem etiologically correlated with the incidence of human cancer (Oser, 1973). This might indicate that trace amounts can be tolerated. In fact some official statements are already available with regard to the setting of levels for carcinogens. In the WHO publication on International standards for drinking-water (1971), the concentration of polynuclear aromatic hydrocarbons (including benz(a)-pyrene) in drinking-water may not exceed 0.2 μg/1. Furthermore, in WHO report no. 406 (1968) the following is stated: 'On the one hand the establishment of safe levels is difficult in the absence of general principles for extrapolation. On the other hand the approach based on the assumption of no threshold is possibly too conservative'. In a postscript to the UICC[1] Technical Report on Carcinogenicity Testing (1969) some change of attitude can also be perceived towards the concept of 'safe dose levels' mainly because of the importance of weighing benefits of a certain product against its potential carcinogenic hazard and the fact that rigid legislation for exclusion of all carcinogenic products for use by man might prove selfdefeating. In a recent WHO report (Food additives series Vol. 4, 1972) on the evaluation of, among others, lead and diethylpyrocarbonate tolerable intakes and maximum concentrations are set, of both compounds, although lead and urethane (one of the products of diethylpyrocarbonate hydrolysis) have been proved to be carcinogenic under certain conditions.

In the evaluation of safe levels we have to consider the possibility of the existence of modifying factors which inhibit or augment the carcinogenic action. As mentioned earlier these factors can change the carcinogenic action several fold. In a few exceptions of which we can doubt the feasibility in practice, a 1000 fold increase was found.

Secondly, it is important to know if a compound is a carcinogen or not. How to distinguish them from cocarcinogens? Roe (1968) already pointed out that the definition of a carcinogen as used by legislation does not concern the mechanism

1. Union Internationale Contre le Cancer.

but solely the relationship between cause and effect. That cocarcinogens might be mistaken for carcinogens if this stand is taken is obvious. The Food and Drug Administration Advisory Committee says in a Panel on carcinogenesis (1971): 'What can be the significance of the incidence of 'spontaneous' tumors of susceptible strains when one is not certain about the presence of carcinogenic contaminants in the diet of which the animals have been maintained.' and 'are tumors ever induced that do not occur on control animals? '.

Hence, it is necessary to have better definitions of chemical carcinogens than those given by WHO (1972): 'a substance that is known conclusively to induce or enhance neoplasia'. We have to be aware that carcinogenicity of a compound is established only if a compound induces tumors in more than one species or, if only one species is observed, the compound induces a specific type of tumor, which is not common in the test species and a relation between dose and effect exists. Of course it must be realized that in determining carcinogenic activity the route of administration is important. It has been known for some time that subcutaneous testing of compounds may result in interpretational difficulties unless care is taken to define the goal of study (Roe, 1968; Goldberg, 1971). Next to the question whether a compound is a carcinogen we are faced with its activity. As mentioned earlier many weak carcinogens are known. For most of them epidemiological studies in humans provide little indication that they could be a hazard. For cyclamate and saccharin for example, no correlation has been found between their use and the occurrence of bladder cancer in man (Inhorn & Meisner, 1969 and Price et al., 1970).

Isonicotinic hydrazide has incuded lung tumors in mice (Toth & Shubik, 1966). It has, however, already been used for decades in the treatment for tuberculosis in man as a very valuable drug. The therapeutic doses are high, nevertheless no increased occurrence of neoplasms in treated patients has been observed as yet (Hammond et al., 1967; Boyland, 1969; Ferebee, 1970).

Of course we have to realize that latency periods are usually long and may vary. Wada et al. (1968) found an average latency period of 24.4 years for mustard gas induced lung cancer. Boyland (1968) and Bryan & Lower (1970) mention latency periods for naphthylamine- induced and benzidine-induced bladder cancer ranging from 15–40 years. Herbst et al. (1971) found latency periods for vaginal adenocarcinoma transplacentally induced by diethylstilbestrol from 14–22 years. Asbestos-induced mesothelioma has a latency period of 20–30 years (Selikoff, 1973).

It has been argued that a dose-response relationship exists for most, if not all carcinogens. Furthermore, there is evidence of a dose-response relationship of carcinogenic hazards in man. Thus, theoretically if dosages could be found low enough so that the latency period exceeds the lifespan of human beings, it should be safe to use them when transplacental carcinogenic action can be eliminated. Some compounds are tested in animals at unrealistic high levels. For comparison: saccharine was found to induce bladder tumors at a concentration of 7.5% in the diet. This concentration when extrapolated to man without any safety factor, would mean a total uptake of 1368 kg saccharine (eating 1000 grams a day and 50 years on the 'diet'). Lead acetate induces renal tumors in rats and mice (van Esch et al., 1962 and van Esch & Kroes, 1969) at levels of 1 and 0.1% in the diet. For humans this equals without a safety factor, 182.5 kg or 18.25 kg lead acetate total intake over 50 years. It can never be said that man behaves in the same way as the tested spieces, moreover the sensitivity for the compound may be completely different. On the other hand almost all compounds which were found to be carcinogenic in man, were also found carcinogenic to animals. As however experiments with humans are immoral and impractical we are forced to extrapolate from animals to man.

Thus Weil's (1972) suggestion to introduce a safety factor for carcinogens does not seem illogical. He proposed a factor of 5000 using a 'minimum measured cancer producing level' (MIE) and not the 'no effect level'. This factor is 10 times higher than the usual safety factor on toxic compounds of 500 based on a MIE, or a 50-fold when based on a no effect level, which is 100 times a MIE. This additional factor is introduced especially on the assumption that carcinogenic action may be less reversible. Terracini et al. (1967) found that 2 mg/kg dimethylnitrosamine administered for lifetime to rats, still induced a liver tumor in one out of 26 rats. Druckrey (1967a) found that diethylnitrosamine (DENA) fed to rats at 0.075 mg kg^{-1} day^{-1} still induces tumors. Other experiments (Kroes et al., 1973) revealed no tumors after application during lifetime of DENA, 5 μg rat^{-1} day^{-1}, five times a week, in 58 surviving males and 40 surviving females. This represents about 0.175 mg/kg in the food which calculated without the safety factor would be an intake of 3193.6 mg of the compound in 50 years for man. Of course these extrapolations are very inaccurate because a tumor incidence as low as 1% may be missed even if only 108 rats are involved, and such a tumor incidence in human means 1000 cases per 100,000 individuals.

When using the information mentioned above for calculation of human risk for exposure to nitrosamines present in cured meat (taking DENA as representative for all nitrosamines) we can calculate that 18.9 mg of DENA should be the maximum total intake for humans. (Table 1).

In Table 1 some calculations are made for the total intake of 'nitrosamines' via cured meat when this meat contains 40, 20, 10 or 1 μg/kg of 'nitrosamines'. It is obvious that an average daily intake of 50 gram cured meat the total intake under certain conditions is already above the calculated maximum for humans. Moreover exposure to nitrosamines from other sources than cured meat is ignored. For this reason the levels of nitrosamines in cured meat should be kept as low as possible.

Table 1. Some calculations[1] for the total intake over 50 years (mg) of 'nitrosamines' via cures meat containing 40, 20, 10 or 1 μg/kg 'nitrosamine'.

Average daily intake cured meat (gr)	Nitrosamine level (μg/kg) 40	20	10	5
25	18.250	9.125	4.562	2.281
50	36.500	18.250	9.125	4.562

1. MIE DENA in rat is 0.075 mg kg^{-7} body weight day^{-1} (Druckrey, 1967). Calculated maximum tolerated intake 75/5000 μg/kg body weight day^{-1}, i.e. 18.9 mg/50 years.

Conclusion

It seems illogical and inappropriate to ban *every* compound, found to be a carcinogen in animals. Serious and careful judgement is required for each carcinogenic compound. It has to be established whether the compound is a real carcinogen, and what type of mechanism is involved by the induction of tumors. The dose levels have to be compared in relation with the levels of human exposure. It will be important to know whether the compound is stable or not. The compound has to be irreplaceable and essential for use. If alternative compounds are found which can replace the compound the use of the compound should be prohibited. Thus a 'safe level' should never be claimed for any possible carcinogen. In setting a safe level not only a safety factor and a MIE should be considered but available epidemiological information should equally be included. To enable calculations of tolerable exposure levels for potential carcinogenic stimuli epidemiological studies should be undertaken as well as correlation studies between cancer incidence and suspected stimuli. Such studies are especially indicated at the points of maximum possible human exposure such as production or use in concentrated form. More experience in 'low level' carcinogenesis is necessary. Also desirable are studies on interaction of a given test chemical with other materials likely to be used at the same time, or likely to be present in the environment.

It is a good thing to know that many laboratories are currently carrying out carcinogenicity studies with low levels of powerful carcinogens, thus providing more information about the possible hazard of these compounds.

References

Anonymus, 1970. Progress against Cancer. A report by the National Advisory Cancer Council. U.S. Dept. of Health. Education and Welfare 1970.

Ball, I. K., 1970. Immunosuppression and carcinogenesis: Contrasting effects with 7,12 dimethylbenz(a)anthracene, benz(a)pyrene and 3-methylcholantrene. J. natn. Cancer Inst. 44: 1–10.

Balner, H., 1970. Immunosuppression and neoplasia. Europ. J. clin. biol. Res. 15: 599–604.

Balner, H. and H. Dersjant, 1969. Increased oncogenic effect of methylcholanthrene after treatment with anti-lymphocyte serum. Nature 224: 376–378.

Baroni, D. C., P. Mingazzini, P. Pesando, A. Cavallero, S. Uccini & R. Scelsi, 1972. Effects of a single dose of rabbit anti mouse lymphocytic serum on tumors induced in mice by 7; 12-dimethylbenz(a)anthracene given at birth. Tumori 58: 397–408.

Baroni, C. P., R. Scelsi, P. C. Pesando & P. Mingazzini, 1970. Risposta immunitaria secondaria anti-eritrocity di montone in topi swiss iniettati alla nascita con un'unica dose di 7,12 – dimethylbenz(a) anthracene. Tumori 56: 269–278 (1970).

Berenblum, I., 1969. A re-evaluation on the concept of co-carcinogenesis. Prog. exp. Tumor Res. 11: 21–30.

Bingham, E. & M. L. Falk, 1969. Environmental carcinogens. The modifying effect of carcinogens on the threshold response. Archs. envir. Hlth 19: 779–783.

Bittersohl, G., 1971. Epidemiologische Untersuchungen über Krebserkrankungen in der chemische Industrie. Arch. Geschwulstforsch. 38: 198–209.

Bluestein, H. G. & I. Green, 1970. Croton oil induced suppression of the immune response of guinea-pigs. Nature 228: 871–872.

Bock, F. G., 1968. Dose response: experimental carcinogenesis. Natn. Cancer Inst. Monogr. 28: 47–63.

Boyland, E., 1969. The correlation of experimental carcinogenesis and cancer in man. Prog. exp. Tumor Res. 11: 222–234.

Bryan, G. T. & G. M. Lower, 1970. Diverse origins of ubiquitous environmental carcinogenic hazards and the importance of safety testing. J. Milk Fd Technol. 33: 506–515.
Case, R. A. M., 1966. Tumors of the urinary tract as an occupational disease in several industries. Ann. R. Coll. Surg. Eng. 39: 213–235.
Clayson D. B., 1967. Chemicals and environmental carcinogenesis in man. Eur. J. Cancer 3: 405–415.
Conney, A. H. & J. J. Burns, 1972. Metabolic interactions among environmental chemicals and drugs. Science 178: 576–586.
Conney, A. H., R. Welch, R. Kuntzman, R. Chang, M. Jacobson, A. D. Munro-Faure, A. W. Peck, A. Bye, A. Poland, P. J. Poppers, M. Finster & J. A. Wolff, 1971. Effects ofenvironmental chemicals on the metabolism of drugs, carcinogens and normal body constituents in man. Ann. N.Y. Acad.Sci. 179: 155–172.
Deichmann, W. B., M. Keplinjer, F. Sala & E. Glass, 1967. Synergism among oral carcinogens IV. Toxic. appl. Pharmac. II: 88–103.
Doll, Sir. R., 1972. Communication at the Conference on host-environment interaction in the etiology of cancer in man – Implementation in Research, 1972. Proceeding in press.
Druckrey, H., R. Preussmann, S. Ivankovic & D. Schmähl, 1967 a, Organotrope karzinogene Wirkungen bei 65 verschiedenen N-Nitroso-Verbindungen and BD-Ratten. Z. Krebsforsch. 69: 103–210.
Druckrey, M., 1967 b. Quantitative aspects in chemical carcinogenesis potential carcinogenesis hazards from drugs. Evaluation of risks. R. Truhaut (ed.) UICC Monograph Series 7. Springer-Verlag, Berlin, p. 60–78.
Van Duuren, B. L., A. S. I. Vak, L. Langseth, B. M. Goldschmidt & A. Segal, 1968. Initiators and promotors in tobacco carcinogenesis. Natn. Cancer Inst. Monogr. 28: 173–180.
Epstein, G. S., 1972. Environmental pathology, a review. Am. J. Path. 66: 352–373.
Esch, G. J. van, H. van Genderen & H. H. Vink, 1962. The induction of renal tumors by feeding basic lead acetate to rats. Br. J. Cancer 16: 289–297.
Esch, G. J. van & R. Kroes, 1969. The induction of renal tumors by feeding basic lead acetate to mice and hamsters. Br. J. Cancer 23: 761–771.
Everson, T. C., 1964. Spontaneous regression of cancer. Ann. N.Y. Acad. Sci. 114: 721–735.
Ferebee, S. M., 1970. Controlled chemoprophylaxis trials in tuberculosis. A general review. Adv. Tuberc. Res. 17: 28–106.
Food and Drug Administration Advisory Committee on protocols for safety evaluation: Panel on carcinogenesis report on cancer testing in the safety evaluation of food additives and pesticides, 1971. Toxic. appl. Pharmac. 20: 419–438.
Frankel H. H., R. S. Yamamoto, E. K. Weisburger & John. M. Weisburger, 1970. Chronic toxicity of azathioprine and the effect of this immunosuppressant on livertumor induction by the carcinogen *N*-hydroxy-*N*-2-fluorenyl acetamide. Toxic. appl. Pharmac. 17: 462–480.
Fudenberg. H. H., R. A. Good, H. C. Goodman, D. Hitzig, H. G. Kunkel, I. M. Roitt, F. S. Rosen, D. S. Rose, M. Seligman & J. R. Soothill, 1971. Primary immunodeficiences. Rep. of a WHO Committee. Pediatrics 47: 927–946.
Gaudin, D., R. S. Gregg & K. L. Yielding, 1971. DNA repair inhibition in a possible mechanism of action of co-carcinogens. Biochem. biophys. Res. Commun. 45: 630–636.
Gilson, J. C., 1973. Asbestos Cancer. Past and future hazards Proc. R. Soc. Med. 66: 395–403.
Gleichmann, H. & E. Gleichmann, 1973. Immunosuppression and Neoplasia, I and II. Klin. Wschr. 51: 255–265.
Golberg, L., 1971. Trace chemical contaminants in food: Potential for harm. Fd Cosmet. Toxic. 9: 65–80.
Hadjiolov, D., 1971. The inhibition of dimethylnitrosamine carcinogenesis in rat liver by aminoacetonitrile. Z. Krebsforsch. 76: 91–92.
Hammond E. C., I. J. Selikoff & E. H. Robitzek, 1967. Isoniazid theory in relation to later occurrence of cancer in adults and in infants. Br. med. J. 2: 792–795.
Haran -Ghera & M. Lurie, 1971. Effect of heterologous antithymocyte serum on mouse skin tumorigenesis. J. natn. Cancer Inst. 46: 103–112.
Hatch, T. F., 1971. Thresholds: Do they exist? Archs. envir. Hlth 22: 687–689.
Hecker, E., 1971. Isolation and characterization of the co-carcinogen principles from crotonoil. in: Methods in Cancer Research, H. Busch (ed.), Vol IV. Acad. Press New York – London, p. 439–484.

Hecker, E., 1972. Aktuelle Probleme der Krebsentstehung. Z. Krebsforsch. 78: 99–122.
Herbst, A. L., H. Ulfelder & O. C. Paskanzer, 1971. Adenocarcinoma of the vagina: Association of maternal stilbestrol therapy with tumor appearance in young women. New Engl. J. Med. 284: 878–881.
Hicks, R. M., J. St. J. Wakerfield & J. Chowanrei, 1973. Co-carcinogenic action of saccharin in the chemical induction of bladder cancer. Nature 243: 347–349.
Hill, W. T., D. W. Stranger, A. Pizzo, B. Riegel, P. Shubik & W. B. Wartman, 1951. Inhibition of 9, 10–dimethyl–12–benzanthracene skin carcinogenesis in mice by polycyclic hydrocarbons. Cancer Res. 11: 892–897.
Hueper, W. C., 1971. Public health hazards from environmental chemical carcinogens, mutagens and teratogens. Hlth Phys.: 21: 689–707.
Inhorn, S. L. & L. Meisner, 1969. Cyclamate ban. Science 166: 685.
Innes, I. R. M., B. M. Ulland, M. G. Valesco, L. Petrucelli, L. Fishbein, E. R. Hart, A. J. Palotta, R. R. Bates, H. L. Falk, J. J. Gart, M. Klein, I. Mitchell & J. Peters, 1969. Bio assay of pesticides and industrial chemicals for tumorogenicity in mice: A preliminary note. J. natn. Cancer Inst. 42: 1101–1114.
Jussawalla D. J., 1973. Cancer incidence patterns in the subcontinent of India. Proc. R. Soc. Med. 66: 308–312.
Kalbe, I., G. Brüschke, D. Kranz, U. Schmidt & O, Keim, 1968. Tierexperimentelle Studien mit Impftumoren bei Eisenmangel sowie bei Eisenüberladung. Dte Gesundh Wes. 23: 156–157.
Kistner R. W., 1969. The pill. Delacorte, New York.
Kroes, R., M. J. van Logten, J. M. Berkvens & G. J. van Esch, 1973. Studies on the effect of DENA on the possible carcinogenicity of lead arsenate and sodium arsenate. To be published.
Kuschner, M. & S. Laskin, 1971. Interaction of atmospheric agents with carcinogens from other sources. Oncologia, Proc. 10th Int. Cancer Congr. Vol 5 p 37–46.
Lacassagne, A., Bun-Hoi & Y. Rudali, 1945. Inhibition of the carcinogen action produced by a weakly active carcinogenic hydrocarbon on a highly active carninogenic hydrocarbon. Br. J. exp. Path. 26: 5–12.
McLean, A. E. M. & A. Marshall, 1971 Reduced carcinogenic effects of aflatoxin in rats given phenobarbitone. Br. J. exp. Path. 52: 322–329.
Ledbetter, J. O., 1971. Are our attempts at zero health risks futile? Archs. envir. Hlth 22: 513.
Likhachev, A. Ya., 1968. The combined effect of carcinogens (In Russian). Vop. Onkol. 14 (10) 114–124.
Madhaven, T. V. & C. Gopalan, 1968. The effect of dietary protein on carcinogenesis of aflatoxin. Archs. Path. 85: 133–137.
Magee, P. N., 1969. Growth and trophic factors in carcinogenesis. Envir. Res. 2: 380–396.
Mantel, N., 1962. The concept of threshold in carcinogenesis. Clin. Pharmac. Ther. 4: 104–109.
Mathé, G., 1973. La machinerie immunitaire, pour ou contre le cancéreux? Nouv. Presse med. 2: 553–555.
Mäkelä, O., 1972. Influence of immunological reactions on carcinogenesis. An overview. Conf. on host-environment interactions in the etiology of cancer in man – implementation in research. Proceeding in press.
Miller, E., J. Miller, R. Brown & J. McDonald, 1958. On the protective action of certain polycyclic aromatic hydrocarbons against carcinogenesis by aminoazodyes and 2-acetylaminofluorene. Cancer Res. 18: 469–477.
Miller, J. A. & E. C. Miller, 1971. Chemical carcinogenesis: Mechanism and Approaches to its Control J. natn. Cancer Inst. 47 (5) 5–13.
Montesano, R., 1970. Systemic carcinogens (N-nitroso compounds) and synergistic or additive effects in respiratory carcinogenesis. Tumori 56: 335–344.
Napalkov, N. P., 1971. Some experimental approaches to the problem of medical drugs in relation to environmental cancer in man. Oncologia, Proc. 10th Int. Cancer Congr. Vol 5, p. 230–238.
Neiman, J. M., 1967. On the problem of acceptable dosages of carcinogens. Vop. Pitan 26 (3) 3–7.
Neiman. J. M., 1968. The sensitizing carcinogenic effect of small doses of carcinogen. Europ. J. Cancer 4: 537–545.

Neiman, J. M., 1972. Maximum tolerable dosages of carcinogens (In Russian). Gig. Sanit. 37 (5) 90–93.
Oser, B. L., 1973. An assessment of the Delaney Clause after 15 years. Proc. Soc. Toxic., in press.
Price, J. M., G. G. Biava, B. L. Oser, E. E. Vogin, J. Steinfeld & H. L. Ley, 1970. Bladder tumors in rats fed cyclohexylamine or high doses of a mixture of cyclamate and saccharin. Science 167:1131–1132.
Reis, H. E., 1971. Die Bedeutung der Immunsuppression für die Entstehung und das Wachtsterm maligner Geschwülste. Dte med. Wschr. 96: 170–174.
Richardson, H. L., A. R. Stier & E. Borson-Nacht-Nebel, 1952. Livertumor inhibition and adrenal histologic responses in rats to which 3-methyl-4-dimethylaminoazobenzene and 20-methylcholantrene were simultaneously administered. Cancer Res. 12: 356–361.
Roe, F. J. C., 1973. Carcinogens and food: A broad assessment. Proc. roy. Soc. Med. 66: 23–26.
Roe, F. J. C., 1968. Carcinogenesis and sanity. Fd Cosmet. Toxic. 6: 485–498.
Roe, F. J. C. & K. E. K. Rowson, 1968. The induction of cancer by combinations of virusses and other agents. Int. Rev. exp. Pathol. 6: 181–227.
Roe, F. J. C., R. L. Carter, B. C. V. Mitchley, R. Peto & E. Hecker, 1972. On the persistence of tumorinitiation and the acceleration of tumorprogression in mouse skin tumorigenesis. Int. J. Cancer 9: 264–273.
Schmähl, D. & C. Thomas, 1965. Experimentelle Untersuchungen zur 'Syncarcinogenese' IV. Z. Krebsforsch. 67: 135–140.
Schmähl, D., 1970. Experimentelle Untersuchungen zur 'Syncarcinogenese' VI Z. Krebsforsch. 74: 457–466.
Schmähl, D., C. Thomas & K. König, 1963. Experimentelle Untersuchungen zur 'Syncarcinogenese' I. Z. Krebsforsch. 65: 342–350.
Schmähl, D., R. Wagner & H. R. Scherf, 1971. Der Einfluss von Immunosuppressiva auf die Cancerisierung der Rattenleber durch Diäthylnitrosamin und die Fibrosarkomentstehung durch 3, 4 – Benzpyren. ArzneimittelForsch. 21: 403–404.
Schubert, J., 1972. A program to abolish harmful chemicals. Ambio 1: 79–89.
Selikoff, I. J., 1973. Interaction of smoking and occupational hazards in the etiology of lung cancer. In: Second International Symposium on Cancer detection and prevention 1973, Italy. Excerpta Med. Int. Congr. Series No 275: p.10.
Shabad, L. M., 1971. The problem of maximum tolerable dosages (MTD) and maximum tolerable concentration (MTC) of carcinogens (In Russian). Gig. Sanit. 36 No 10: 92–95.
Shabad, L. M., 1973. Möglichkeiten, maximal zulässige Dösen (MZD) und maximal zulässige Konzentrationen (MZK) für karzinogene Stoffe festzulegen. Arch. Geschwulstforsch. 41: 217–220.
Sheehan, R. & G. Shklar, 1971. Thymectomy and its effect on chemically induced submandibular neoplasms. J. dent. Res. 50: 981.
Sheehan, R., G. Shklar & R. Tannenbaum, 1971. Azathioprine Effects on the development of hamster pouch carcinomas. Archs Path. 91: 264–270.
Sinnhuber, R. O., J. H. Wales, J. L. Ayres, R. H. Engelbrecht & D. L. Amend, 1968. Dietary factors and hepatoma in rainbow trout (Salmo gairdneri) I. Aflatoxins in vegetable protein feedstuffs. J. natn. Cancer Inst. 41: 711–718.
Stern, E. & M. R. Mickey, 1969. Effects of a cyclic steriod contraceptive regimen on mammary gland tumor induction in rats. Br. J. Canc. 23: 391–400.
Stjernswärd, J., 1969. Immunosuppression by carcinogens. Antibiotica Chemother. 15: 213–233.
Stokinger, H. E., 1971. Sanity in research and evaluation of environmental health. Science 174: 662–665.
Tannenbaum, A., 1958. Nutrition and cancer. In: Physiopathology of Cancer. F. Homburger (ed.), Cassell, London. p. 517–562.
Terracini, B. & M. C. Testa, 1970. Carcinogenicity of a single administration of N-nitrosomethylurea. A comparison between newborn and 5-week-old mice and rats. Br. J. Cancer 24: 588–598.

Terracini, B., P. N. Magee & J. M. Barnes, 1967. Hepatic pathology in rats on low dietary levels of dimethylnitrosamine. Br. J. Cancer 21: 559–565.
Thomas, C., H. Rogg & J. Bücheler, 1972. Die Krebserzeugenden Wirkung der N-Nitroso-methyl Harnstoffes nach Verabreichung hormonaler Kontrazeptica. Beitr. Path. 146: 332–338.
Tomatis, L., V. Turusov, D. Guibbart, B. Duperray, C. Malaveille & H. Pacheco, 1971. Transplacentral carcinogenic effect of 3-methylcholantrene in mice and its quantitation in fetal tissues. J. natn. Cancer Inst. 47: 645–651.
Tonkelaar, L. den, G. J. van Esch & R. Kroes, 1972. Influence of the microsomal liver enzyme induction by tetrasul on the carcinogenicity of dimethyl aminostilbene. Abstr. Toxicol. appl. Pharmacol. 23: 793.
Toth, B. & P. Shubik, 1966. Mammary tumors inhibition and lungadenoma induction by isonicotinic acid hydrazide. Science 152: 1376–1377.
Union Internat. Contre le Cancer, 1969. Carcinogenicity testing 1969. U.I.C.C. Technical Report Series 2: 55–56.
Vesselinovitch, S. D., N. Mihaslovich, G. N. Wogan, L. S. Lombard & K. V. N. Rao, 1972. Aflatoxin B, a hepatocarcinogen in the infant mouse. Cancer Res. 32: 2289–2291.
Wada, S., M. Miyanishi, Y. Nishimoto, S. Kambe & R. W. Miller, 1968. Mustard gas as a cause of respiratory neoplasia in man. Lancet 1: 1161–1163.
Wagner, J. L. & G. Haughton, 1971. Immunosuppression by antilymphocyte serum and its effect on tumors induced by 3-methylcholanthrene in mice. J. natn. Cancer, Inst. 46: 1–10.
Walder, B. K., M. R. Robertson & D. Jeremy, 1971. Skin cancer and immunosuppression. Lancet 2: 1282–1289.
Warwick, G. P., 1971. Metabolism of livercarcinogens and other factors influencing livercancer induction. In: Liver Cancer – International Agency for Research on Cancer, p. 121–157.
Wattenberg, L. W., 1966. Chemoprophylaxis of carcinogenesis: A review. Cancer Res. 26: 1520–1526.
Weil, C. S., 1972. Statistics, safety factors and scientific judgment in the evaluation of safety for man. Toxicol. appl. Pharmac. 21: 454–463.
Weisburger, J. H. & E. K. Weisburger, 1968. Food additives and chemical carcinogens: On the concept of zero tolerance. Fd Cosmet. Toxic. 6: 235–242.
Weisburger, J. H., R. S. Yamamoto, G. M. Williams, O. M. Gramtham, T. Matsushima & E. K. Weisburger, 1972. On the sulfate ester of *N*-hydroxy-*N*-2-fluorenylacetamide as a key ultimate hepatocarcinogen in the rat. Cancer Res. 32: 491–500.
White, F. R., 1961. The relationship between underfeeding and tumor formation, transplantation and growth in rats and mice. Cancer Res. 21: 281–290.
WHO, 1968. Research into environmental pollution (W.H.O.). Technical report no. 406, WHO, Geneva, p. 66–83.
WHO, 1971, International Standards on drinking-water. 3rd edn. WHO, Geneva, p. 31–44.
WHO, 1972. Health hazards of the human environment. WHO, Geneva, p. 219.
WHO, 1972. Evaluation on mercury, lead, cadmium and the food additives amaranth, diethylpyrocarbonate and octylgallate. WHO Food Additives Series no 4, p34–50 and 67–74.
Wilson, R. A., 1962 The roles of estrogen and progresterone in breast and genital cancer. J. Am. Med. Ass. 182: 327–331.
Woods, D. A., 1969. Influence of anti-lymphocyte serum on DMBA induction of oral carcinomas. Nature 224: 276–277.
Wynder, E. L. & D. Hoffmann, 1969. A study of tobacco carcinogenesis. Cancer 24: 289–301.
Wynder, E. L. & K. Mabuchi, 1972. Etiological and Preventive Aspect of Human Cancer. Prev. Med. 1: 300–334.
Yamamoto, R. S., J. H. Weisburger & E. K. Weisburger, 1971. Controlling factors in urethan carcinogenesis in mice. Effect of enzyme inducers and metabolism inhibitors. Cancer Res. 31: 483–486.

Discussion

The problem of setting human safety levels

It was stressed that the use of safety factors in calculating human risks for exposure to toxic agents is a necessity, because experiments with humans are not permissible. The application of higher safety factors for carcinogens (1 000 or 5 000) than for non-carcinogenic agents (100 or 500) is based on an arbitrary assessment of a greater hazard with carcinogens, mainly arising from the possibility of the irreversibility of carcinogenic action.

If it is not possible to distinguish between carcinogen or co-carcinogen from the results of a carcinogenicity study, the substance tested has to be regarded as a carcinogen.

Proc. int. Symp. Nitrite Meat Prod., Zeist, 1973. Pudoc, Wageningen

Formation of N-nitroso compounds in laboratory animals. A short review

J. Sander

Hygiene-Institut der Universität, D–74 Tübingen, W-Germany

Abstract

For secondary, tertiary and quarternary amines, amidopyrine and alkylamides a survey is given of reactivity towards nitrite, in vitro and in vivo. Secondary amines of low basicity are particularly nitrosated in the stomach if nitrite is present. Feeding with slightly basic amines combined with nitrite induced specific carcinomata. The author estimated the nitrosation reaction equation of tertiary amines in vitro. Carcinogenicity of their nitroso compounds has not yet been reported. Quarternary ammonium compounds are nitrosated in vitro. They do not seem to hold a health risk. Reactivity of alkylamides in vitro varies. Various alkylureas combined with nitrite caused specific carcinomata. So far alkylurethanes and other alkylamides combined with nitrite have shown no carcinogenicity. Halide ions, thiocyanate and some aldehydes catalyse nitrosation and cause a shift in optimum pH for nitrosation. Thiocyanate acts as a catalyst in the rat's stomach. Ascorbic acid in vitro and in vivo effectively inhibits nitrosamine formation from several precursors. Quantitative data are so far inadequate.

Introduction

In principle, carcinogenic nitroso compounds can be formed from secondary or tertiary amines, from enamines, from quarternary ammonium bases and from a variety of alkylamides. The amino-compounds react with nitrogen oxides which are usually derived from nitrite.

Secondary amines

$(R - \overset{\displaystyle H}{\overset{|}{N}} - R')$

secondary amine

$CH_3 - \underset{\displaystyle H}{\underset{|}{N}} - CH_3$

dimethylamine

high basicity, pK_A : 10.7

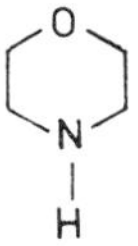

morpholine

moderate basicity, pK_A : 8.3

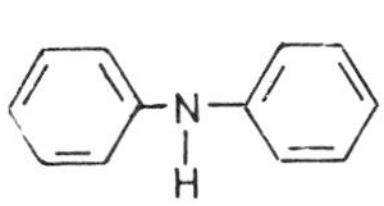

diphenylamine

low basicity, pK_A : 0.8

Reactivity in vitro ·

Secondary amines can be nitrosated in organic solvents or in water. In an organic solvent a high basicity will increase the reactivity of the amines towards the electrophilic nitrosating agent. In water the nitrosating mechanism is similar but the nitrosamine yield is strongly influenced by side reactions. It was shown that formation of a salt, or the addition of protons to the free electron pair of the amines, considerably inhibits the nitrosamine formation (Sander et al., 1968). In vivo a nitrosamine formation will mainly take place in an aqueous environment, but the organic phase must not be neglected.

While salt formation of amines is increased by lowering the pH of the solution, thus inhibiting the nitrosation, a low pH value allows formation of a greater proportion of free nitrous acid and its equilibrium products which are necessary for the nitrosamine formation. Maximum yields per time unit are therefore neither obtained at very high nor at very low pH values, but at a certain optimum pH. This was found for most secondary amines to be near pH 3.3. Some amino acids with a secondary amine group showed a maximum reaction near pH 2.25– 2.5 (Mirvish, 1972).

With respect to the optimum pH, good conditions for the formation of nitrosamines could be expected in the vertebrate stomach, although of course a synthesis in vitro cannot be directly compared with that in vivo where several other factors have to be taken into account.

Animal experiments

Animal experiments showed, although with very great differences in yield, that all secondary amines can be nitrosated in the stomach when nitrite is present. Secondary amines of a high basicity were, corresponding to the in vitro results, very poorly nitrosated, but the nitroso derivatives of amines of low basicity were easily formed.

Methods used to demonstrate the nitrosamine formation were the analysis of the stomach contents, the determination of nitrosamines excreted with the urine, the production of toxic effects typical for nitrosamines and, of course, the induction of tumors.

Despite the fact that the nitroso derivatives of highly basic amines could be detected in the stomach in very low yield, cancer was not induced by such amines and nitrite (Druckrey et al., 1963; Sander, 1971; N.P. Sen, 1973, personal communication). Feeding of less basic amines and nitrite, however, resulted in the induction of tumors typical for the corresponding nitrosamines. For example methylbenzylamine and nitrite caused tumors in the oesophagus of rats. Rats receiving morpholine and nitrite developed liver tumors (Sander & Bürkle, 1969). Methylaniline and nitrite caused a marked increase in the number of lung adenomas in mice (Greenblatt et al., 1971). Concurrent feeding of the amino acids proline and hydroxyproline and nitrite did not produce tumors in rats. This is probably because of a lack of in vivo decarboxylation of the obviously non carcinogenic nitroso-amino-acids (Greenblatt et al., 1973).

Tertiary amines

With tertiary amines experiments have not yet been carried out to the same extent as with secondary amines. They should be distinguished from enamines, because of different chemical behaviour.

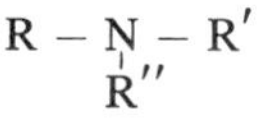
tertiary amine

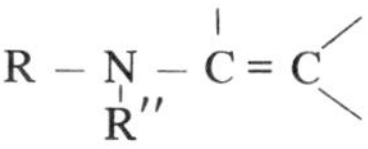
enamine

Reactivity in vitro

Our experiments (Schweinsberg & Sander, 1972) with tertiary amines of low molecular weight demonstrated that the nitrosation in dilute aqueous solution at 100°C obeys the following equation:

$$\text{rate} = k\,[\text{amine}] \times [\text{nitrite}]^3$$

The first step in the nitrosamine formation from tertiary amines is an oxidative desalkylation, the nitrosation of the resulting secondary amine being the second step. In our experiments tertiary amines formed nitrosamines much slower than secondary amines. This finding is partly opposite to results of Lijinsky et al. (1972), who described a quick nitrosamine formation out of a variety of tertiary amines. Lijinsky and his co-workers used much higher concentrations of the amines and nitrite and a much longer reaction time (16 h 30 min).

Feeding trials

Induction of tumors by concurrent feeding of tertiary amines and nitrite has not been reported so far. In our experiments, feeding of a diet containing 0.5% triethylamine and 0.5% $NaNO_2$ did not produce tumors within a year, nor did it cause a toxic reaction exceeding that of nitrite alone.

It should be kept in mind that the nitrosation of tertiary amines needs further investigation. The problem of nitrosamine formation from the corresponding amine oxides has also not received sufficient attention.

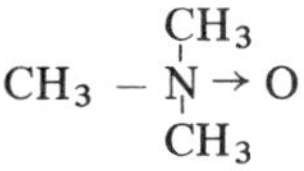
trimethylamine oxide

Enamines

Among the enamines, amidopyrin, a drug which has been widely used, easily reacts with nitrite in vitro and in vivo, leading to the formation of dimethylnitrosamine (Lijinsky & Greenblatt, 1972; Lijinsky et al., 1973).

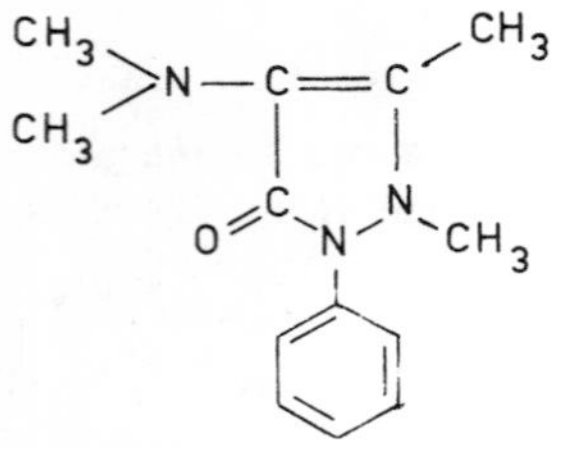

amidopyrin

Quarternary ammonium compounds

Only a few experiments on the nitrosamine formation from quarternary ammonium compounds have been done. Fiddler et al. (1972) demonstrated the formation of dimethylnitrosamine from a variety of naturally occurring substances. Tetramethylammonium chloride reacted less readily than trimethylamine hydrochloride which in turn yielded less than 10% of the dimethylnitrosoamine produced from dimethylamine hydrochloride. Fairly high amounts of dimethylnitrosamines were produced from 2-dimethylaminoacetate, 2-dimethylaminoethanol, *N,N*-dimethylglycine and the methylester of the latter.

With respect to the feeding experiments done with diethylamine and triethylamine and nitrite, it seems unlikely that feeding of realistic amounts of the quarternary ammonium compounds tested and nitrite will induce cancer in test animals.

Alkylamides

Reactivity in vitro

There is a great number of alkylamides which also can react with nitrite to produce carcinogenic nitroso compounds. The reactivity of the amides towards nitrite varies as much as with the amines. Alkylureas were nitrosated very easily, while alkylurethanes showed a slow reaction. The same applies to methylguanidine and citrullin (Mirvish, 1971). Several other amides like methylacetamide did not show any measurable reaction in dilute aqueous solutions (Sander et al., 1971).

As the alkylureas do not readily form salts they are nitrosated best at higher acid concentrations. The main nitrosating agent probably is the nitrous acidium ion $(H_2NO_2)^+$. The reaction rate proved to be proportional to the concentration of the alkylurea, the nitrous acid and the protons.

$$\text{rate} = k\,[\text{alkylurea}] \times [HNO_2] \times [H_3O^+]$$

The k values of the N-alkylurethanes were about one-thirtieth of those for the corresponding alkylureas, the k values of the ethyl compounds being about one-fourth of the methyl derivatives.

Experiments in vivo

Tumors have been induced in animals by feeding various alkylureas and nitrite (Sander, 1970, 1971; Ivankovic & Preussmann, 1970; Mirvish et al., 1972). These tumors were similar to those which had been previously induced by N-nitroso ureas.

Alkylurethanes and other alkylamides have not been reported to induce tumors when concurrently administered with nitrite, although it seems very likely that at least with methyl and ethyl urethan a positive result might be obtained. A diet containing 5 000 mg/kg citrullin given concurrently with drinking water containing 5 000 mg/kg $NaNO_2$ for about half a year did not induce tumors after about two years of observation (unpublished data).

Catalysis and inhibition of the formation of N-Nitroso compounds in vivo

The nitrosation reaction may be catalysed as well as it may be inhibited by several agents. Catalysing agents are known to be halide ions and the pseudo halide thiocyanate (SCN^-). The latter arises from several goitrogenic substances which are contained in certain vegetables, and also are formed in vivo from cyanide which is inhaled with cigarette smoke. Some aldehydes may also be important as catalysts, e.g. formaldehyde (Roller, personal communication). The catalysts change the reaction rates of the nitrosation reaction. In the uncatalysed reaction the nitrosation of secondary amines obeys the following equation (Mirvish, 1970):

$$\text{rate} = k\,[R_2NH] \times [HNO_2]^2$$

But the catalysed reaction is expressed by another equation (Fan & Tannenbaum, 1973):

$$\text{rate} = k\,[HNO_2] \times [H_3O] \times [Hal^-] \times [R_2NH]$$

The catalysed reaction is linearly dependent upon the nitrite concentration while the uncatalysed reaction is dependent upon the square of the nitrite concentration. Catalysis also causes a shift of the pH-optimum. Boyland et al. (1971) described the pH-optimum of the catalysed nitrosation of methylaniline to be near pH 1; For morpholine Fan & Tannenbaum (1973) found pH 2.3, for diethylamine Schweinsberg (1973) found 2.5.

By kinetic studies it was recently demonstrated that in the rat stomach the nitrosamine formation normally occurs as a catalysed reaction due probably to the presence of thiocyanate. Thus additional catalyst does not significantly change the reaction (Schweinsberg, 1973).

The chemical analysis is in good accordance with feeding experiments in rats. Feeding a diet containing amines and nitrite and thiocyanate did not produce a higher toxicity than the corresponding mixture without the catalyst (Sander et al., 1972).

Several experiments have been concentrated on the problem of the inhibition of the nitrosamine formation by ascorbic acid which was first described by Mirvish et

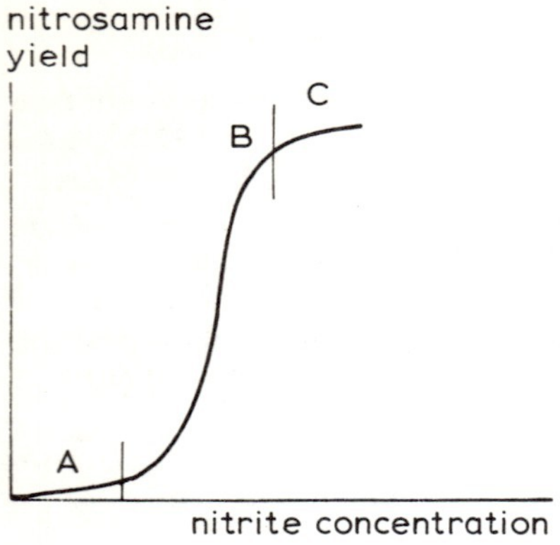

Fig. 1. Schematic description of the nitrosamine formation in the vertebrate stomach from a given amount of a secondary amine and different amounts of nitrite.

al., (1972). In vitro and in vivo, addition of ascorbic acid inhibited nitrosamine formation from dimethylamine, morpholine, piperazine, N-methylaniline and amidopyrin (Kamm et al., 1973; Greenblatt, 1973). With high doses of ascorbic acid the inhibition was complete. Feeding of piperazine or morofpholine with nitrite induced lung adenomas, but was ineffective when administered concurrently with ascorbic acid (Mirvish et al., 1973).

The development of hydrocephali in the offspring of pregnant rats which can be induced by administration of ethylurea (100 mg/kg) and nitrite (50 mg/kg) to the mothers, was completely prevented by simultaneous application of a high dose of ascorbic acid (250 mg/kg) (Ivankovic et al., 1973).

Kinetic studies (unpublished data) demonstrate that the nitrosation of amines in the vertebrate stomach does not obey a simple equation, because inhibition by nitrite destroying agents and catalysis are active at the same time. In principle, the nitrosamine formation from a given amount of an amine can be described by a curve as shown in Fig. 1.

In Section A the nitrosamine formation is very low because a significant part of the nitrite is destroyed by ascorbic acid or related compounds. Destruction of nitrite becomes less important with greater amounts of nitrous acid. The steepness of the curve in Section B is highly influenced by the concentration of the catalyzing ions. The curve levels off (Section C) with amounts of nitrite much higher than equimolar to the amine.

To summarize the data reviewed here it can be said that experiments in vitro and in animals have contributed much during the last few years to the understanding of the nitrosamine formation in the stomach. We have good qualitative information on the nitrosation of amines and amides in vivo, but we need more quantitative measurements.

Some experiments in this direction are now underway in our laboratory.

References

Boyland, E., E. Nice & K. Williams, 1971. The catalysis of nitrosation by thiocyanate from saliva. Fd Cosmet. Toxicol. 9: 639–643.

Druckrey, H., D. Steinhoff, H. Beuthner, H. Schneider & P. Klärner, 1963. Prüfung von nitrit auf chronisch toxische Wirkung an Ratten. Arzneimittel Forsch. (Drug Res.) 13: 320–323.

Fan, T. Y. & S. R. Tannenbaum, 1973. Factors influencing the rate of formation of nitrosomorpholine from morpholine and nitrite: Acceleration by thiocyanate and other anions. J. Agr. Fd 21: 237–240.

Fiddler, W., J. W. Pensabene, R. C. Doerr & A. E. Wasserman, 1972. Formation of N-nitrosodimethylamine from naturally occurring quarternary ammonium compounds and tertiary amines. Nature 236: 307.

Greenblatt, M., V. R. C. Kommineni & W. Lijinsky, 1973. Brief Communication: Null effect of concurrent feeding of sodium nitrite and amino acids to MRC rats. J. Nat. Cancer Inst. 50: 799–802.

Greenblatt, M., 1973. Ascorbic-acid of aminopyrine nitrosation in NZO/BL mice. J. Nat. Cancer Inst. 50: 1055–1056.

Greenblatt, M., S. S. Mirvish & B. T. So, 1971. Nitrosamine studies: Induction of lung adenomas by concurrent administration of sodium nitrite and secondary amines in Swiss mice. J. Nat. Cancer Inst. 46: 1029–1034.

Ivankovic, S., R. Preussmann, D. Schmähl & J. Zeller, 1973. Verhüting von nitrosamidbedingtem Hydrocephalus durch Ascorbinsäure nach parentaler Gabe von Äthylharnstoff und Nitrit an Ratten. Z. Krebsforsch. 79: 145–147.

Ivankovic, S. & R. Preussmann, 1970. Transplacentare Erzeugung maligner Tumoren nach oraler Gabe von Äthylharnstoff und Nitrit an Ratten. Naturwissenschaften 9: 460–461.

Kamm, J. J., T. Dashman, A. H. Conney & J. J. Burns, 1973. Protective effect of ascorbic acid on hepatotoxicity caused by sodium nitrite and aminopyrin. Proc. Nat. Acad. Sci. 70: 747–749.

Lijinsky, W., H. W. Taylor, C. Snyder & P. Nettesheim, 1973. Malignant tumors of liver and lung in rats fed aminopyrin or heptamethyleneinmine together with nitrite. Nature 244: 176–178.

Lijinsky, W., L. Keefer, E. Conrad & R. van de Bogart, 1972. Nitrosation of tertiary amines and some biological implications. J. Natl. Cancer Inst. 49: 1239–1249.

Lijinsky, W., M. Greenblatt, 1972. Carcinogen dimethylnitrosamine produced in vivo from nitrite and aminopyrin. Nature (New Biol.) 236: 177–178.

Mirvish, S. S., L. Wallcave, M. Eagen & P. Shubik, 1973. Ascorbate-nitrite reaction: Possible means of blocking the formation of carcinogenic N-nitroso compounds. Science 177: 65–68.

Mirvish, S. S., A. Cardesa, L. Wallcave & P. Shubik, 1973. Effect of sodium ascorbate on lung adenoma induction by amines plus nitrite. Proc. Amer. Assoc. Cancer Res., Abstract paper.

Mirvish, S. S., 1972. Kinetics of N-nitrosation reactions in relation to tumorigenesis experiments with nitrite plus amines and ureas. in: Bogovski, P., R. Preussmann & E. A. Walker (eds.): N-Nitroso Compounds, analysis and formation, IARC Scientific publications 2: 104–108, Lyon 1972.

Mirvish, S. S., M. Greenblatt & V. R. C. Kommineni, 1972. Nitrosamide formation in vivo: Induction of lung adenomas in Swiss mice by concurrent feeding of nitrite and methylurea or ethylurea. J. Nat. Cancer Inst. 48: 1311–1315.

Mirvish, S. S., 1971. Kinetics of nitrosamide formation from alkylureas, N-alkylurethans and alkylguanidines: Possible implications for the etiology of human gastric cancer. J. Nat. Cancer Inst. 46: 1183–1193.

Mirvish, S. S., 1970. Kinetics of dimethylamine nitrosation in relation to nitrosamine carcinogenesis. J. Nat. Cancer Inst. 44: 633–639.

Sander, J., G. Bürkle & F. Schweinsberg, 1972. Induction of tumors by nitrite and secondary amines or amides in: Nakahara, W., S. Takayama, T. Sugimura, & S. Odashima: Topics in chemical carcinogenesis. Proceedings of the 2nd Int. Symp. Princess Takamatsu Cancer Res. Fund, Tokyo, 297–310.

Sander, J., G. Bürkle, L. Flohe & B. Aeikens, 1971. Untersuchungen in vitro über die Möglichkeit einer Bildung cancerogener Nitrosamide im Magen. Arzneimittel Forsch. (Drug Res.) 21: 411–414.
Sander, J., 1971. Untersuchungen über die Entstehung cancerogener Nitrosoverbindungen im Magen von Versuchstieren und ihre Bedeutung für den Menschen. Arzneimittel Forsch. (Drug Res.) 21: 1572–1580, 1703–1707, 2034–2039.
Sander, J., 1970. Induktion maligner Tumoren bei Ratten durch orale Gabe von *N,N'*-Dimethylharnstoff und Nitrit. Arzneimittel Forsch. (Drug Res.) 20: 418–419.
Sander, J. & G. Bürkle, 1969. Induktion maligner Tumoren bei Ratten durch gleichzeitige Verfütterung von Nitrit und sekundären Aminen. Z. Krebsforsch. 73: 54–66.
Sander, J., F. Schweinsberg & H.-P. Menz, 1968. Untersuchungen über die Entstehung cancerogener Nitrosamine im Magen. Hoppe-Seyler's Z. physiol. Chem. 349: 1691–1697.
Schweinsberg, F., 1973. Catalysis of nitrosamine synthesis. Report, to be given at the IARC-meeting on nitroso Compounds, Lyon, Oct. 1973.
Schweinsberg, F. & J. Sander, 1972. Kanzerogene Nitrosamine aus einfachen tertiären Aminen und Nitrit. Hoppe-Seyler's Z. Physiol. Chem. 353: 1671–1676.

Proc. int. Symp. Nitrite Meat Prod., Zeist, 1973. Pudoc, Wageningen.

Formation of nitroso compounds in man: evaluation of the problem

J. Sander

Hygiene-Institut der Universität, D–74 Tübingen, W-Germany

Abstract

There are differences between the test animal and human stomach, implying a different suitability for formation of carcinogenic *N*-nitroso compounds. The possibility of nitrosamine formation in the human stomach is pointed out. Precursors, catalysts, but also inhibitors of the nitrosation reaction, like ascorbic acid, may easily be ingested in the daily diet.

To measure *N*-nitrosation in man, epidemiological investigations as well as with patients taking medical products, smokers and volunteers are feasible. In tests on his own body, the author did not find any dinitroso piperazine after ingestion of piperazine hexahydrate.

However, as long as the harmlessness of nitrite has not been proved, ingesting nitrite and amine should be held as risky. Ascorbic acid may play a protective role.

Introduction

There is little doubt that *N*-nitroso compounds are carcinogenic for man as they are for animals. It is also quite certain that, under suitable conditions nitrosamines or nitrosamides will be formed in the human stomach. But it is still unclear to what extent the ingestion of nitrosamine precursors may contribute to the induction of cancer in man.

There are several differences between the anatomy and biology of the stomach of man and experimental animals, therefore making it impossible to calculate the nitrosation of amines or amides in the human stomach just on the basis of animal experiments. Direct investigations in man will be necessary.

Differences in stomach anatomy and biology

Experiments on the nitrosation of amines or amides have to take into account that the vertebrate stomach is divided into functional sections. This division may or may not be strongly marked. The stomach of rodents for example is divided into a forestomach which serves mainly as a food store, and a glandular stomach where acid and gastric juice are produced. The human stomach does not show this distinct anatomic separation.

Experimental animals are normally healthy and usually young. In man, however, all ages and several diseases have to be considered. The gastric physiology in man is known to change with age. For example, in babies acid secretion is very low. It is high in young adults but with old age the production of acid and other components of gastric juice is reduced. This has great influence on the nitrosamine formation for two reasons. One is the dependence of the nitrosation reaction on the pH of the medium. The second reason is the influence of the acid concentration on the metabolism of bacteria in the stomach. A neutral reaction allows bacteria to form nitrite from nitrate which is present in almost every meal. The amount of nitrite formed is dependent on the concentration of nitrate, the time available for the nitrate reduction and the number of bacteria present in the stomach contents. Food remains in the stomach only for a few hours. Therefore the multiplication of the bacteria in the stomach may be less important than the ingestion of nitrate reducing bacteria with food or saliva. Although complete anacidity yields maximum nitrite, the nitrosamine formation depending on bacterial nitrate reduction is favoured most by a moderate subacidity. Under these conditions at momentary optimum pH values nitrosation follows nitrite formation (Sander & Seif, 1969).

Precursors, catalysts and inhibitors

A great number of nitrosable amines and amides are known to occur in food. We do not yet have enough information, however, on the concentrations. Some nitrosable amino compounds are ingested in high doses as drugs (Sander, 1971; Lijinsky et al., 1972). Nitrite is found in the saliva and often also in gastric juice. It is formed, as mentioned, by bacteria in food or in the human organism from nitrate. Nitrate in turn is a component of drinking water and plant material. Besides this, nitrite and, to some extent nitrate, are used as food additives (Sander & Schweinsberg, 1972). It is quite likely that nitrite and nitrosable amino compounds do come together in the stomach in amounts sufficient to form a significant quantity of nitrosamines. The yield, however, will be strongly influenced by catalysts as well as by inhibitors.

As thiocyanate seems to be present in almost every human gastric juice (Schweinsberg, 1973), catalysis will always be an important factor in nitrosamine formation in the human stomach. Thiocyanate is not only formed in vivo in smokers from inhaled cyanide, but also independently from the smoking habit. Several precursors do occur in food, e.g., glucobrassicin which was found in cabbage (*Brassica* species). Savoy cabbage yields up to 300 mg/kg, cauliflower up to 100 mg/kg SCN^- (Lang, 1970).

Inhibition of nitrosamine formation is also very likely to be regularly an important factor in the nitrosation of amines or amides in the human stomach. The gastric juice contains very often substances which reduce 2,6-dichlorphenol-indophenol. This reduction may be due partly to the presence of ascorbic acid, partly to other lactones with similar structures, e.g. glucuronolactone and partly to other reducing agents (unpublised data). Ascorbic acid is also a constituent of many foodstuffs (Table 1). One mole of ascorbic acid destroys two moles of nitrite (Dahn et al., 1960).

Table 1. Amount of Vitamin C (mg/kg) in some selected food stuffs. (According to Souci-Fachmann-Kraut: Die Zusammensetzung der Lebensmittel, Stuttgart, 1962).

milk of cows	15
liver of cows	280
brain of pigs	180
fish	3–20
potatoes	120
kohlrabi *(Brassica oleracea,* var. gongyloides L.)	360
radish	180
asparagus	150
curly kale	540
Brussels sprouts	840
spinach	370
tomato	230
apple	110
grapefruit	110
bread	0
rice	0

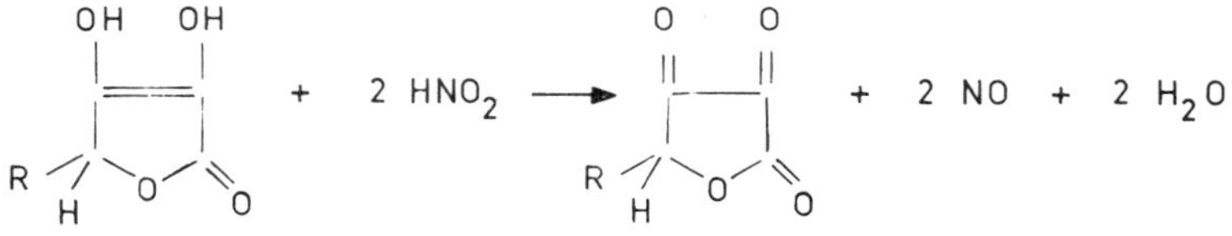

ascorbic acid dehydro ascorbic acid

The reaction between nitrite and ascorbic acid is fast enough to inhibit the nitrosation of most secondary amines in vitro. In animals the inhibition is almost complete if equimolar concentrations of the amine, of ascorbic acid and of nitrite are applied (unpublished data). Very often the amount of ascorbic acid in the food exceeds the nitrite content. In these cases the nitrosamine formation will be very low. A very strong inhibition of the nitrosation of piperazine was also found in rats receiving ascorbate in a concentration lower than equimolar to nitrite (unpublished data). Because of the high reactivity of ascorbic acid towards nitrite, the concentration of Vitamin C and related compounds has to be measured in all experiments on nitrosamine formation in biological materials.

Ascorbic acid is very reactive not only towards nitrite but also towards a variety of oxidizing agents which may occur in food. Some foodstuffs contain a significant amount of Fe (III), e.g. spinach which readily is reduced by ascorbic acid to Fe (II). Very low concentrations of copper ions catalyse the autoxidation of ascorbic acid (Lehmann, 1971). The concentration of ascorbic acid in food is further strongly influenced by cooking, canning or storing.

Experiments in man

In addition to animal experiments, some experiments in man will be necessary, and they can be performed under certain conditions. Human gastric juice, urine or blood can easily be obtained for analyses. But who should be examined?

1. There are many patients who, for medical reasons, have to use certain drugs which may be nitrosated in vivo when ingested with normal food. Piperazine, for example, which is prescribed all over the world and which is taken orally in high doses as a vermicide offers an extremely good opportunity for examinations. Piperazine has the further advantage that it yields a relatively high amount of dinitrosopiperazine in urine, as has been shown in experiments with dogs (Sander et al., 1973). There are more drugs which can also be included in such an analysis (Lijinsky et al., 1972). Certainly the measurement of the corresponding nitrosamines in the gastric contents or in the urine will be the best analytical procedure in some cases, in others it will be more useful to look for metabolites. Besides that a toxicologic analysis of side effects of nitrosable drugs may help to solve the problem. There is extensive literature on the toxicity of such drugs. It is very likely that patients will co-operate in such a programme, as no disturbance of health will result from the examinations as such.

2. Volunteers are easily found among smokers who permit an analysis of their gastric juice. Experiments with smoking students which will soon be published, are underway in our laboratory to find out what is the meaning of nitrosamine formation from amines in tobacco smoke (Schweinsberg, unpublished).

3. There will be a few volunteers who will permit experiments with concurrent ingestion of amines and nitrite. I myself tried to demonstrate a nitrosation of piperazine in my saliva and in my stomach. In the first two experiments I took 10 mg of piperazine hexahydrate and the same amount of sodium nitrate. After ten minutes the saliva which was then strongly positive for nitrite, was alkalized and extracted twice with dichloromethane. Dinitrosopiperazine was not found, neither by thin-layer nor by gas chromatography.

Four more experiments were done to examine whether a nitrosation of piperazine in the stomach may be demonstrated. I swallowed 100 mg piperazine hexahydrate (Eraverm^{R}) twice with and twice without boiled ham (so called 'gekochter Schinken') which was always found to be exceptionally rich in nitrite. The highest amount was 60 g boiled ham. The nitrite content was 120 mg/kg. Ham and the drug were chewed and swallowed together. The gastric content was collected for 15 minutes, 25 minutes after swallowing the material. In the experiments without ham the gastric juice was collected by a stomach tube, in the other cases by vomiting. The pH value was in all cases below 3, although there were always some parts of the material, containing mainly saliva which reacted more or less neutrally. No dinitrosopiperazine was found.

4. Epidemiologic investigations may be among the most convincing examinations which can be done in man. They are, however, not easy to perform because they have to be based on a sufficient understanding of the interrelationship of the relevant chemical and biological factors influencing the nitrosamine formation.

Conclusions

The question as to what a possible nitrosamine formation in the human stomach may mean for the induction of cancer in man is far from being solved. Therefore we do not yet know how great the risk of ingesting nitrite and/or amines will be. Nitrite as a food additive has to be considered hazardous as long as its innocuous use has not been proved. Ascorbic acid may serve to inhibit nitrosamine formation not only in foodstuffs, but it is especially suitable to prevent nitrosation in vivo.

References

Dahn, H., L. Loewe & C. A. Bunton, 1960. Über die Oxidation der Ascorbinsäure durch salpetrige Säure. Teil VI: Übersicht und Diskussion der Ergebnisse. Helv. chim. Acta 43: 320–333.

Lang, K., 1970. Biochemie der Ernährung. 2. Auflage, D. Steinkopf Verlag, Darmstadt.

Lehmann, G.. 1971. Lebensmittelchemische Betrachtungen über die Ascorbinsäure. Ernährungsumschau 18: 282–285.

Lijinsky, W., E. Conrad & R. Van de Bogart, 1972. Carcinogenic nitrosamines formed by drug / nitrite interactions. Nature 239: 165–167.

Sander, J. & F. Seif, 1969. Bakterielle Reduktion von Nitrat im Magen des Menschen als Ursache einer Nitrosaminbildung. Arzneimittel-Forschung (Drug Res.) 19: 1091–1093.

Sander, J. 1971. Untersuchungen über die Entstehung cancerogener Nitrosoverbindungen im Magen von Versuchstieren und ihre Bedeutung für den Menschen. ArzneimittelForsch. (Drug Res.) 21: 1572–1580, 1707–1713, 2034–2039.

Sander, J. & F. Schweinsberg, 1972. Wechselbeziehungen zwischen Nitrat, Nitrit und kanzerogenen Nitrosoverbindungen. Zbl. Bakt. Hyg., I. Abt. Orig. B 156: 299–340.

Sander, J., F. Schweinsberg, M. Ladenstein, H. Benzing & S. H. Wahl, 1973. Messung der renalen Nitrosaminausscheidung am Hund zum Nachweis einer Nitrosaminbildung in vivo. Hoppe-Seyler's Z. physiol. Chem. 354: 384–390.

Schweinsberg, F. (1973). Rhodanid als Katalysator der Nitrosaminsynthese in vivo (in press).

Discussion

Nitrite formation in human saliva

Nitrite in the saliva of man is formed by bacterial reduction of nitrate taken in with food and drink. Whether nitrate gets into the mouth from the salivary glands is not known. Nitrite has not been found in saliva direct from the salivary glands. In saliva of dogs Sander could not detect nitrite.

Role of acidity of the stomach

Nitrosamine formation in the stomach, depending on bacterial nitrate reduction, is favoured most by a moderate sub-acidity. There might be a relation between this experimental finding and the epidemiological evidence that patients suffering from sub-acidity or anacidity more often develop tumours of the stomach and other organs than other people.

Inhibition of nitrosation in vitro

The nitrosation of secondary amines was not prevented by the simultaneous presence of primary amines. The same applied to amino acids and sugars. Ascorbate inhibited nitrosation very effectively. Comment of Dr Walters: cysteine and reduced glutathione did decrease nitrosamine formation, though with less efficiency than ascorbate. Dr. Sander had not tested these sulphydryl compounds.

Discussion of a contribution from E.O. Haenni[1]

Division of Chemistry & Physics, Office of Sciences, Bureau of Foods, Food & Drug Administration, 299 'c' St., S.W., Washington, D.C. 20204, USA

Drawbacks of high-resolution mass spectrometry

High-resolution mass spectrometry (resolution up to 14 000) was considered to be the most unambiguous method for the identification of nitrosamines. Comments of Dr Wasserman: high-resolution mass spectrometry failed to discriminate dimethylnitrosamine from the trimethylsilyl ion (difference in $^m/_e$ only 3 parts in 14 000). From existing literature it is known, that with low-resolution mass spectrometry dimethylnitrosamine may be erroneously assumed to be present, where in fact hydroxyacetone is present. In both cases, careful consideration of the complete low-resolution mass spectrum could prevent misidentification.

Nitrosamines and nitrite in bacon

In Dr Haenni's laboratory bacon samples are routinely examined for about 12 volatile nitrosamines. However, only dimethylnitrosamine and nitrosopyrrolidine have been found. This applies to fried bacon. None of these 12 nitrosamines have been detected in raw bacon.

It was not known, whether there was any difference in residual nitrite levels in the commercial bacon samples that contained widely differing amounts of nitrosopyrrolidine after frying. Nitrite had not been determined in these samples. In the literature there are no data, that indicate a correlation between levels of residual nitrite and of nitrosamines. Participants in the discussion strongly disagreed about this subject. There was a casual remark, that even 1 mg/kg of residual nitrite might perhaps be sufficient to produce nitrosamines in amounts of a few μg/kg.

It might well be, that nitrosoproline is present in raw bacon and that it is the precursor of the nitrosopyrrolidine in fried bacon, but both propositions remain speculative as yet.

1. This discussion was based on an article by Fazio, T., R. H. White & L. R. Dusold, 1973. Nitrosopyrrolidine in cooked bacon. IAOAC 56: 919–921.

The role of fat

Dr Haenni thought that nitrosamines tended to concentrate in the fat. Perhaps re-use of rendered fat was not advisable; perhaps it had better be discarded. In Dr Schram's view nitrosamines are actually formed in the fat, and probably in the fat only. When fat was rendered out of bacon fatty tissue at low temperature and then heated to 170°C, N-nitrosopyrrolidine was formed.

Conclusions and recommendations of the toxicological session 14th september 1973

1. Many N-nitroso compounds are powerful and versatile chemical carcinogens.

2. There is evidence that trace quantities of such carcinogens can be present in food among others in some cured meat products.

3. Carcinogenic N-nitroso compounds may be formed from nirosating agents such as nitrite and nitrosatable amino compounds in the gastrointestinal tract.
Model experiments with high concentrations of nitrite and certain amino compounds have shown to induce tumours characteristic for the corresponding N-nitroso compound. However additional chemical and biological experiments with more realistic amounts of precursors are lacking and have to be carried out.

4. There is a need for epidemiological studies concerning the human health hazard arising from N-nitroso compounds.

5. Where the continuous improving of the sensivity of analytical methods is leading to the revealing of so far undetectable quantities of carcinogens it is suggested to induce a safety factor for these carcinogens.
This safety factor should be much higher than that used for non-carcinogenic agents, having regard to unknown modifying factors such as synergism and promotion. Results from transplacental and single dose experiments and the knowledge that irreversible effects often occur, further strengthen this proposition.

6. From the present status of scientific knowledge it appears desirable to reduce the amount of added and residual nitrite where possible, but not in defiance of the relevant conclusions of other sessions, especially those of the microbiological session (protection against botulism hazard).

Resolutions

The participants of the International Symposium on Nitrite in Meat Products, held at Zeist, the Netherlands, from 10th till 14th September 1973, have accepted the following resolutions:

1. The discussion of problems concerning nitrite in meat products with a restricted number of scientists and public health authorities, from various countries and disciplines was considered fruitful.

2. All relevant information indicates that nitrite is currently an indispensable inhibitor of pathogenic microorganisms *(Clostridium botulinum)* in many meat products.
In addition nitrite plays a keyrole in colour formation and flavour development.
No adequate substitute is yet known.

3. The presence of nitrosamines in some meat products has been established. Sampling procedures and some analytical methods have to be improved to make results more representative and accurate.

4. Epidemiological studies related to ingestion of nitrosamines are desirable.

5. An inventory is needed of all meat products according to their nitrite content and relevant aspects of general composition, of manufacturing and merchandizing and of use by the consumer.
This could facilitate division of foods into categories appropriate for separate consideration as regards the necessity to use nitrite.

6. There are preliminary indications that the use of ascorbates decreases the risk of nitrosamine formation without impairing antimicrobial function.

Appendix

Nitrate and nitrite allowances in meat products

Summary of some countries' regulations, made up September 1973

J. Meester

Central Institute for Nutrition and Food Research TNO,
Dept. Netherlands Centre for Meat Technology, Utrechtseweg 48, Zeist

Permitted	Products	Limits in mg/kg on addition = A in finished product = F
European Community (Directive Preservatives in Foods)		
KNO_3, $NaNO_3$, pure or mixed with NaCl $NaNO_2$, only mixed with NaCl	*no further specifications*	
Members European Community	*up to now no common regulation*	
Belgium		
KNO_3, $NaNO_3$	all	500, F, calc. as KNO_3
KNO_2, $NaNO_2$, only mixed (max. 0.6%) with NaCl	all	200, F, calc. as $NaNO_2$
Denmark		
KNO_3, $NaNO_3$	products not heated above 60°C	500, F, calc. as KNO_3
KNO_2, $NaNO_2$, only mixed (min. 0.4, max. 0.6%) with NaCl	all	100, F, calc. as $NaNO_2$
France		
KNO_3, $NaNO_3$, only mixed (max. 10%) with NaCl	all	no limit
$NaNO_2$, only mixed (0.6%) with NaCl	all	150, F, calc. as $NaNO_2$ (in freshly prepared brines mx. 210 g nitrite-salt per liter)

Permitted	Products	Limits in mg/kg on addition = A in finished product = F
West Germany		
KNO_3, $NaNO_3$	all, except raw unfermented comminuted products	600 KNO_3, A, or 500 $NaNO_3$, A, calc. on amount of meat and fat used
$NaNO_2$, only mixed (min. 0.5, max. 0.6%) with NaCl	all	no limit (combined use of nitrate and nitrate-salt only permitted in curing large pieces of meat, provided their ratio is max. 1 : 100)
Ireland		
KNO_3, $NaNO_3$	all	no limit
KNO_2, $NaNO_2$	all	200, F, calc. as $NaNO_2$, in cooked meat other than cured or pickled pork
Italy		
KNO_3, $NaNO_3$	all	250, F
$NaNO_2$, only mixed with NaCl	all	150, F
United Kingdom		
KNO_3, $NaNO_3$	bacon, ham pickled meat	500, F
KNO_2, $NaNO_2$	bacon, ham, pickled meat	200, F
Luxemburg		
KNO_3, $NaNO_3$	all	2000, F, calc. as KNO_3
$NaNO_2$, only mixed (max. 0.6%) with NaCl	all	no limit
the Netherlands		
KNO_3	all	2000, F
$NaNO_2$, only mixed (max. 0.6%) with NaCl (conditional exemptions for pure nitrite are granted)	all	500, F, calc. as $NaNO_2$
Some other European countries		
Austria		
KNO_3, $NaNO_3$	all	500 F
$NaNO_2$, only mixed (min. 0.5, max. 0.6%) with NaCl	all	200, F. calc. as $NaNO_2$ (combined use of nitrate and nitrite-salt only permitted in curing large pieces of meat, provided their ratio is max. 1 : 100)

Permitted	Products	Limits in mg/kg on addition = A in finished product = F
Finland		
KNO_3, $NaNO_3$	all	500, F
$NaNO_2$, only mixed (max. 0.6%) with NaCl (also $NaNO_2$ in max. 10% solution was allowed, but probably cancelled for 1973)	all	150, F
Norway		
KNO_3, $NaNO_3$	– some specified products (cured and dried meats e.g. bacon, dried ham; semi-preserves	500, A
	– some other specified products (e.g. meat-rolls, cooked ham, saveloy)	250, A
$NaNO_2$, only mixed (0.5–0.6%) with NaCl	– some specified products (cured and dried meats e.g. bacon, dried ham; semi-preserves)	200, A (provided no nitrate has been used)
	– some other specified products (e.g. meat-rolls, cooked ham, saveloy)	120, A (provided no nitrate has been used)
Sweden		
KNO_3, $NaNO_3$	all	500, F
$NaNO_2$, only mixed (max. 0.6%) with NaCl	all	200, F
Switzerland		
KNO_3, $NaNO_3$	all	60 g, A, per kg NaCl used in the product
$NaNO_2$, only mixed (max. 0.6%) with NaCl	all	200, F

Permitted	Products	Limits in mg/kg on addition = A in finished product = F
Some non-european countries		
Canada		
KNO_3, $NaNO_3$	all	no limit
KNO_2, $NaNO_2$	all	200, F, calc. as $NaNO_2$
Proposed revision (July 1973):		
KNO_3, $NaNO_3$	– dry and semi-dry sausage products	200, A, calc. as $NaNO_3$
	– slow-cure and specialty products (to be enumerated)	200, A, calc. as $NaNO_3$
KNO_2, $NaNO_2$	– cooked sausage products	200, A, calc. as $NaNO_2$
	– dry and semi-dry sausage products	200, A, calc. as $NaNO_2$
	– side bacon	150, A, calc. as $NaNO_2$
	– cured primal cuts, canned meats, slow-cure, and specialty products	200, A, calc. as $NaNO_2$
Japan		
KNO_3, $NaNO_3$	bacon, ham, sausage, corned beef	no limit
$NaNO_2$	bacon, ham, sausage, corned beef	70, F
United States		
KNO_3, $NaNO_3$	all	2188, A to meat (dry cure) 1719, A to chopped meat (in brines max. 8.4 g per liter, calc. at 10% pump level)
KNO_2, $NaNO_2$	all	625, A to meat (dry cure) 156, A to chopped meat (in brines max. 2.4 g per liter, calc. at 10% pump level) 200, F. calc. as $NaNO_2$